palgrave advances in
witchcraft historiography

Palgrave Advances

Titles include:

Jonathan Barry and Owen Davies (*editors*)
WITCHCRAFT HISTORIOGRAPHY

H.G. Cocks and Matt Houlbrook (*editors*)
THE MODERN HISTORY OF SEXUALITY

Saki R. Dockrill and Geraint Hughes (*editors*)
COLD WAR HISTORY

Patrick Finney (*editor*)
INTERNATIONAL HISTORY

Jonathan Harris (*editor*)
BYZANTINE HISTORY

Marnie Hughes-Warrington (*editor*)
WORLD HISTORIES

Helen J. Nicholson (*editor*)
THE CRUSADES

Alec Ryrie (*editor*)
EUROPEAN REFORMATIONS

Richard Whatmore and Brian Young (*editors*)
INTELLECTUAL HISTORY

Jonathan Woolfson (*editor*)
RENAISSANCE HISTORIOGRAPHY

Forthcoming:

Katherine O'Donnell, Leann Lane and Mary McAuliffe (*editors*)
IRISH HISTORY

Palgrave Advances
Series Standing Order ISBN 1–4039–3512–2 (Hardback) 1–4039–3513–0 (Paperback)
(*outside North America only*)

You can receive future titles in this series as they are published by placing a standing order.
Please contact your bookseller or, in the case of difficulty, write to us at the address below
with your name and address, the title of the series and the ISBN quoted above.

Customer Services Department, Macmillan Distribution Ltd, Houndmills, Basingstoke,
Hampshire RG21 6XS, England

---

palgrave advances in witchcraft historiography

edited by
jonathan barry and owen davies

First published 2007 by
PALGRAVE MACMILLAN
Houndmills, Basingstoke, Hampshire RG21 6XS and
175 Fifth Avenue, New York, N.Y. 10010
Companies and representatives throughout the world

PALGRAVE MACMILLAN is the global academic imprint of the Palgrave
Macmillan division of St Martin's Press LLC and of Palgrave Macmillan Ltd.
Macmillan® is a registered trademark in the United States,
United Kingdom and other countries. Palgrave is a registered
trademark in the European Union and other countries.

ISBN-13   978–1–4039–1175–9 hardback
ISBN-10   1–4039–1175–4 hardback
ISBN-13   978–1–4039–1176–6 paperback
ISBN-10   1–4039–1176–2 paperback

This book is printed on paper suitable for recycling and
made from fully managed and sustained forest sources.
Logging, pulping and manufacturing processes are expected to
conform to the environmental regulations of the country of origin.

A catalogue record for this book is available from the British Library.

A catalogue record for this book is available from the Library of Congress.

10 9   8   7   6   5   4   3   2   1
16 15 14 13 12 11 10 09 08 07

Printed and bound in Great Britain by
Antony Rowe Ltd, Chippenham and Eastbourne

contents

notes on contributors

Jonathan Barry is Senior Lecturer in History and Head of the School of Historical, Political and Sociological Studies at the University of Exeter. He has published widely on urban society and culture in early modern and eighteenth-century England. He is co-editor of *Witchcraft in Early Modern Europe* (1996), and is currently preparing volumes on *Witchcraft and Demonology in South-West England* (University of Exeter Press) and *Religion in Bristol c. 1640–1775* (Redcliffe Press).

Willem de Blécourt is Honorary Research Fellow at the Huizinga Institute of Cultural History, Amsterdam. He has written numerous articles on witchcraft, popular culture and irregular medicine, published in Dutch, German and English journals such as *Social History*, *Medical History* and *Gender & History*. His most recent book is *Het Amazonenleger* ('The Army of Amazons') (1999), which deals with irregular female healers in the Netherlands, 1850–1930. He is currently working on *The Cat and the Cauldron*, a history of witchcraft in the Northern and Southern Netherlands from the Middle Ages to the twentieth century.

Owen Davies is Reader in Social History at the University of Hertfordshire. He has published numerous articles on the history of witchcraft and magic in eighteenth- and nineteenth-century England and also France. He is also the author of *Witchcraft, Magic and Culture 1736–1951* (Manchester University Press, 1999), *A People Bewitched* (1999) and *Cunning-Folk: Popular Magic in English History* (2003). He has also co-edited, with Willem de Blécourt, *Beyond the Witch Trials* (2004) and *Witchcraft Continued* (2004). His latest book is *Murder, Magic, Madness: The Victorian Trials of Dove and the Wizard* (2005).

Peter Elmer is Senior Lecturer in the History of Science, Technology and Medicine at the Open University. His main field of interest lies in the early modern history of medicine, and its relation to the broader religious, political and cultural movements of the period. He is currently preparing a major study of the politicisation of medicine and healing in seventeenth-century England, with particular reference to the role of witchcraft in that process. As part of this study, he also offers the first major, in-depth study of the miracle healer, Valentine Greatrakes, as well as tracing the links between healing and witch hunting in early modern England.

Marion Gibson is Senior Lecturer in English Literature at the University of Exeter and is based at its new Tremough Campus in Penryn, Cornwall. She is the author of *Reading Witchcraft* (1999) and has edited a number of collections of early modern writings on witchcraft, including *Early Modern Witches* (2000) and *Witchcraft and Society in Early Modern England and America 1550–1750* (2003). Her new monograph is *Possession, Puritanism and Print: Darrell, Harsnett, Shakespeare and the Elizabethan Exorcism Controversy* (2006). She is finishing a monograph on representations of witchcraft in American culture from 1620 to the present for Routledge.

Katharine Hodgkin is Principal Lecturer in the School of Social Sciences, Media and Cultural Studies, University of East London. Recent publications include *Madness in Seventeenth-Century Autobiography* (2006), and 'The Witch, the Puritan and the Prophet: Historical Novels and Seventeenth-Century History', in Heilmann and Llewellyn (eds), *Metanarrative and Metahistory in Contemporary Women's Fiction* (2007). She is co-editor with Susannah Radstone of *Contested Pasts: the politics of Memory, and Regimes of Memory* (2003), and has published essays on various aspects of seventeenth-century English cultural history, including gender, religion, witchcraft, writing and madness. She is currently editing a seventeenth-century manuscript, *Women, Madness and Sin: the auto-biographical writings of Dionys Fitzherbert* (Ashgate, forthcoming 2007).

Richard Jenkins is Professor of Sociology at the University of Sheffield. He has carried out ethnographic field research in Northern Ireland, England, Wales and Denmark. Among his recent publications are *Pierre Bourdieu* (2nd edn, 2002), *Social Identity* (2nd edn, 2004), *Rethinking Ethnicity* (1997), *Questions of Competence* (1998), and *Foundations of Sociology* (2002). He is currently doing research on the meeting of 'traditional' and 'New Age' supernatural beliefs and practices in England.

Brian P. Levack is John Green Regents Professor in History at the University of Texas at Austin. His publications include *The Civil Lawyers in England, 1603–1641: A Political Study* (1973), *The Formation of the British State: England, Scotland and the Union, 1603–1707* (1987) and *The Witch-Hunt in Early Modern Europe* (3rd edn, 2006). He is the co-author of *Witchcraft and Magic in Europe: the Eighteenth and Nineteenth Centuries* (1999) and editor of *The Witchcraft*

Sourcebook (2004). His book on witch hunting in Scotland will be published in 2008.

Peter Maxwell-Stuart is a Lecturer in History in the School of History, University of St Andrews. His particular interests are the occult sciences in the Middle Ages and early modern Europe. Recent publications include an edited translation of Martin Del Rio's *Disquisitiones Magicae* (2000); *Satan's Conspiracy: Magic and Witchcraft in Sixteenth-Century Scotland* (2000); *Witch Hunters: The Occult in Mediaeval Europe* (2003); and *An Abundance of Witches: The Great Scottish Witch Hunt, 1658–1662* (2005). Forthcoming publications include a new edited translation of the *Malleus Maleficarum*.

Marko Nenonen is Assistant Professor of History at the University of Tampere. He has published widely on Finnish history. His work on witchcraft in English includes the entry for Finland in the *Encyclopedia of Witchcraft* (2006), and 'Envious Are All the People, Witches Watch at Every Gate' (*Scandinavian Journal of History*, 1993). He is currently working on a study of the historiography of the witch hunts. Some themes regarding this work are presented in his article 'Witch-Hunt Historiography from the 18th Century Encyclopaedias to the Present Research: A New Geography', in Aradas and Pappas (eds), *Themes in European History* (2005).

Jo Pearson studied History and Religious Studies at Lancaster University, and has previously held posts at the Open University, Cardiff University and Liverpool Hope. She is the author of *Wicca – the Christian Heritage: Ritual, Sex and Magic* (2007) and is editor of *Nature Religion Today* (1998) and *Belief Beyond Boundaries: Wicca, Celtic Spirituality and the New Age* (2002). She currently lives and writes in Salisbury.

Raisa Maria Toivo is currently researching witchcraft and women's power and editing a historical atlas of Finland at the University of Tampere. Her recent publications on the social history of witchcraft include: 'Marking (Dis)Order: Witchcraft and the Symbolics of Hierarchy in Late Seventeenth- and Early Eighteenth-Century Finland', in Davies and de Blécourt (eds), *Beyond the Witch Trials* (2004), and 'Naiset Roviolla' ('Women at Stake') in Sari Katajala-Peltomaa and Raisa Maria Toivo (eds), *Paholainen, noituus ja magia – kristinuskon kääntöpuoli* (2004). She has also published on the subject of the history of traffic and communications in Finland.

Christa Tuczay is a Lecturer in Medieval Literature at the University of Vienna and a researcher at the Austrian Academy of Science on the major project 'Motif-Index of the Secular German Narrative Literature from the Beginnings to 1400'. She has written a book on medieval magic, *Magie und Magier im Mittelalter* (1992), many articles on varied topics such as urban legends and revenants, and numerous entries for the *Encyclopedia of Witchcraft* (2006). She is currently researching trance and ecstasies in the Middle Ages and vampires and revenants in German literature.

Juliette Wood is Associate Lecturer in the School of Welsh at the University of Cardiff and Secretary of the Folklore Society at the Warburg Institute, London. She is the author of several books on Celtic mythology, including *The Celts: Life, Myth and Art* (1999), and has published widely on the subject of folk tales, medieval traditions, magic and New Age movements. She has also co-edited a collection and appraisal of the correspondence and research notes of Margaret Murray, *A Coven of Scholars* (1998).

1
introduction

owen davies and jonathan barry

Scholars of witchcraft have often been pioneers of new forms of historical study and interdisciplinary developments, as the subject touches upon many fundamental issues regarding the human experience both in the past and the present.[1] The witch *trials* cannot be understood properly without considering the development of science, medicine, religion and the political and economic apparatus of the modern European state. The analysis of witchcraft *accusations* demands an understanding of the processes of social negotiation, the structure of communities, and the nature of gender relations. Decoding the meaning of witchcraft *beliefs* requires grappling with the research and theories of anthropology, folkloristics and psychology, even if their application raises serious issues regarding the application of contemporary notions and behaviour to people in the past. Few topics in historical research invite so much interdisciplinary engagement, demand such a broad exploration of historical processes and yet give so much free reign to our historical imagination.

The rewards are obvious. Witchcraft is one of the few topics guaranteed to excite the interest of the public (and therefore publishers). It provides the historian with an ideal platform for emphasising the importance of engaging with the past. The subject serves to highlight that events and beliefs that may seem extraordinary, cruel, stupid, ignorant, the products of a distant credulity, have to be understood in the context of the perceived realities of past societies, and yet they still have ramifications for our own times. While this is generally the message that contemporary witchcraft historians wish to convey, as we shall see, in the past the history of witchcraft has been employed for less dispassionate purposes. It has been used as a tool of religious propaganda, a dubious marker of societal progress and a highly-charged, and at times inappropriate, political label. The history of witchcraft is clearly not just about the past but about how the past is constantly being reinterpreted. So for those already exploring the history of witchcraft, and those embarking on such an exciting venture, this book will provide a valuable guide to the many paths that others have taken over the centuries. It explores the successive reinterpretations of the historical pattern of belief in and persecution of witchcraft from medieval to contemporary Europe, which remains the focus of most research and teaching in the field.

In addition to synthesising historiographical traditions, the following chapters embody significant new research on the context and explanation for the widely varying traditions of historiography, relating them to the ideological and cultural background of those concerned and the evolving nature of the disciplines and research traditions involved. A particular focus is the shifting borders between historical scholarship and other approaches, both within and beyond academe. The impact of historical knowledge on the public's understanding of the subject and on the growth of modern movements (occult, pagan and 'Wicca') which identify themselves with aspects, at least, of witchcraft historically understood, is also a key theme.

The team of contributors from Europe and America has been brought together to reflect the diversity and depth of scholarly research in the field of witchcraft. Their distinctive approaches to their respective historical themes are shaped by their expertise in certain periods, regions and disciplines. The structure of the book is broadly chronological, tracing the emergence of a series of interpretations from the period of the witch trials to the present day. But most chapters also examine an approach or tradition which has persisted within the historiography and trace the earlier and later developments of that approach, as well as a central core of authors in one period. Each chapter combines an exposition of the main features of that interpretative tradition, with an explanation of how the approach had developed and also an evaluation of its contribution to our overall understanding of the subject, including how far its insights have stood up to criticism. In most cases, one or two key works or authors are explored in some depth in order to convey the material more concretely and to offer readers a critical introduction to the classic texts of the historiography.

There is certainly no shortage of accessible surveys of the history of witchcraft. Publishers see them as guaranteed money-makers in a crowded history market. Brian Levack set the standard with his successful, comprehensive survey, *The Witch-Hunt in Early Modern Europe*, which was first published in 1987 and is now in its third updated edition. Levack has also edited two major collections of articles on witchcraft, which reflect historiographical trends over the last few decades.[2] Darren Oldridge's *Witchcraft Reader* (2001) provides a far less comprehensive but more affordable collection of useful essays on the subject. The six volumes of the Athlone *History of Witchcraft and Magic in Europe*, under the general editorship of Bengt Ankarloo and Stuart Clark, which cover the subject from antiquity to the present through the lengthy essays of influential historians in the field, add up to an impressive contribution to our understanding. Much briefer syntheses of the subject also serve a valuable purpose in stimulating student and public audiences to delve further into the subject.[3] While these studies are generally sensitive to the historiography, their primary aim is to provide overviews of the rise and fall of the witch trials and the various interpretations that have emerged to explain them. The way in which this book is structured enables the reader to get a good sense of the general historical trends in witchcraft history, but it will be best appreciated as a companion to any of these broad surveys. The profusion of such overviews in recent years is an interesting historiographical trend in itself, and its contemporary significance needs to be understood in the context of the venerable history of witchcraft histories.

As the chapters in this book emphasise, there is a clear geography of historiography over the last 150 years, with research traditions shifting most obviously between France, the US, Germany and England up until the mid twentieth century. Even in the last couple of decades regional research clusters have come and gone. Why such patterns of intensive study occur is another significant historiographical issue, reflecting in part the nature of academic funding in different countries, the stimulating effect of ground-breaking publications in regional contexts, and the enthusiasm and influence of individual scholars. In the 1980s and early 1990s Dutch historians were particularly active, spurred on by the formation in 1982 of the interdisciplinary study group 'Witchcraft and Sorcery in the Netherlands'. In Hungary Gábor Klaniczay and Éva Pócs have, for over two decades, orchestrated and encouraged an impressive community of researchers, who have published a series of studies locating the history of witchcraft in central and south-eastern Europe within a wider European context.[4] Recently, Swedish scholars have been particularly active, in part due to a project run by the Dalarna Research Institute which funded research on the region's influential place in the country's experience of witchcraft.[5] German scholars have, for a number of years, been amongst the most prolific and organised researchers. The *Arbeitskreis interdisziplinäre Hexenforschung* (AKIH), or Workshop for Interdisciplinary Witchcraft Research, acts as an important forum for the study of witchcraft not only in Germany but internationally, putting on regular conferences and producing collaborative publications. It has consequently had a significant influence on current historiography.[6] While they have never adopted such a collective organisational approach to research, English historians of witchcraft have also been very productive following the renewed enthusiasm for the subject that emerged in the mid 1990s, which saw a timely revision of the influential studies of Keith Thomas and Alan Macfarlane.[7]

Elsewhere in Europe, in recent years the quantity but not the quality of historiography has been patchy. In Belgium Jos Monballyu continues to mine the trials of Dutch-speaking Flanders.[8] Witchcraft historians in Mediterranean countries, such as María Tausiet in Spain and Oscar di Simplicio in Italy, also tread a rather solitary path compared to their colleagues in the US and other parts of western and northern Europe.[9] There is plenty of interest in and research going on regarding the records of the Inquisitions, but as they rarely dealt with witchcraft accusations, there has been more focus on their preoccupation with 'superstitious' folk magical practices as distinct from the prosecution of *maleficium*. In eastern Europe work has also only appeared sporadically, with the language barrier further hindering most English scholars from incorporating their findings into their own research and teaching.[10] In this respect one looks forward to forthcoming publications issuing from recent doctoral studies on Ukraine and Poland, supplementing the important English language studies of Russell Zguta and W. F. Ryan on Russia and the Ukraine.[11] French historians, who in the past played an important role in shaping the historiography on the witch trials, have at present receded into the background of European scholarly endeavour. The most significant recent French-language publications have been from Swiss historians working on the early trials in the Alpine region.[12] English-language publications, by the likes of William Monter, Robin Briggs and Sarah Ferber, have made the most impact on our understanding of early modern French witchcraft over the last few years.[13]

Across the Atlantic, while there are relatively few judicial records and other sources to work on, historians of the North American witch trials have produced some important interdisciplinary and theoretical models over the years, which have had a significant impact on the way historians interpret the dynamics of witchcraft accusations and trials more generally.[14] Although recently there has been a lull in research on colonial American witchcraft, the fascination with the Salem witch trials continues to act as a magnet for populist and scholarly research. This is not surprising considering the symbolic status of Salem in the US's historical tradition. Unlike in Europe, early modern witchcraft, as represented by Salem, is an integral part of the school curriculum.[15] This undoubtedly generates a commendable popular awareness of the basic historical issues, though the downside is that the country's experience of witchcraft tends to get projected through one major event, overshadowing the wider pattern of colonial witch prosecutions. One of the distinctive dynamics in the history of American witchcraft is obviously the influence of colonial repression and intolerance of indigenous cultures in shaping the persecution and perception of witches. While such conflicts, an amalgam of political, religious and economic imperatives, have been identified as a significant factor in the generation of the tensions that led to the trials in Salem and Essex County,[16] they are more starkly evident in revealing new work on witchcraft in colonial Central and South America.[17] These analyses of witchcraft in colonial contexts also provide useful interpretive lessons for those studying the history of European witchcraft. The popular religious beliefs of subjects in European territories sometimes seemed as equally alien to 'civilised' Christian elites as those overseas.

One of the purposes of this book is to provide the student of witchcraft with a means of easily situating current developments and available resources within the historiography of the past. There has never been a better time for students and scholars to study witchcraft. A substantial range of primary sources have recently become readily accessible to those fortunate enough to belong to institutions with the income or foresight to invest in their purchase. The six-volume edited and annotated collection of facsimile reprints of early modern English works on witchcraft, published under the general editorship of James Sharpe, is one such valuable resource.[18] Key continental demonological texts have recently been reprinted, translated and edited, albeit some in abridged form.[19] Another recent major development is the online digitised collections of English printed texts, Early English Books Online (EEBO), Eighteenth Century Collections Online (ECCO) and Literature Online (LION), which open up even greater possibilities. Their search functions enable historians to identify and examine opinions and discourses on witchcraft in early modern texts that, from their titles, would not seem concerned with the subject and therefore remained unobserved by researchers.[20] While access to such sources is obviously restricted, over the next few years the insights they provide will filter through into more accessible publications.

It is important to note that there are also considerable resources available to scholars and the general public who do not have academic privileges. Some of Cornell University Library's invaluable collection of early modern books and pamphlets on witchcraft are available to all via the internet.[21] Various national libraries are also increasingly making primary and secondary sources available online for free. The French *Bibliothèque nationale*, in particular, has been at the forefront of the library digitisation revolution and offers a range of rare books and pamphlets on witches and witch trials, as well as later ethnographic texts.[22] The Survey of Scottish Witchcraft, a searchable database of all known witchcraft cases in the country between 1563 and 1736, is another valuable project that provides both scholars and the public with new research possibilities.[23] Moving away from the sources to the historiography of witchcraft, readers can get a good sense of its huge breadth and depth from consulting the Witchcraft Bibliography Project Online, which lists hundreds of books and articles by country.[24] Recent publications and historiographical developments can also be traced through reading the contributions from international scholars to the German academic email forum Hexenforschung ('Witchcraft research').[25] There are many more private or institutional internet sites providing relevant information and sources, though the quality varies enormously.[26] One of the pitfalls of the internet for the uninformed is that it has given a new medium of exposure to erroneous data propagated by early witchcraft historiography. Then again, it has, at the same time, helped bridge the information divide that existed between witchcraft historians and sections of the Neo-Pagan community.[27]

It will be interesting to see how the internet and digitisation projects shape the historiography of witchcraft in the future. For obvious financial reasons most major digitised collections are concerned with English-language texts. Yet, in terms of the number of witch trials, England and the American colonies were not of major significance in a European context. English demonologists also made only a minor contribution to the development of the theological demonological framework that helped generate the witch trials. The bulk of the trials, and surviving records, were in the numerous states and principalities of what is now Germany, while most influential tracts were written by continental Catholic clergy, magistrates and lawyers. There are still many court cases to be analysed in the German archives and naturally much of current German scholarship is focused on the analysis of trial records and patterns of prosecutions in the multifarious states and principalities that existed in the early modern period. In England, by contrast, there are few new insightful court records to be mined and not much more to be read into the statistics of the comparatively few cases. One suspects that English witchcraft historians, exploiting the potential of the print and digitised collections, will, in the next few years, focus on textual analyses, situate witchcraft in broader studies of magic, examine related beliefs within a more ethnographic framework, and increasingly consider the role of witchcraft beyond the early eighteenth century. A major project by the British Library to digitise a wide range of English local newspapers will, for example, open up new possibilities for researching witchcraft accusations in the eighteenth and nineteenth centuries. The publications mentioning witchcraft currently available via Google's digitisation initiative, which consist mostly of American editions dating to the first half of the nineteenth century, also highlight potentially new research terrain. They suggest there is more work to be done exploring the nineteenth-century representation of and public response to the seventeenth-century New England trials, and also on the continuance of witch beliefs in nineteenth-century American society.

While it is important to recognise the exciting possibilities web-based research presents, print continues to provide the main means of disseminating cutting-edge research on witchcraft in the present and the immediate future. During the writing and editing of this book the four-volume *Encyclopedia of Witchcraft: The Western Tradition*, edited by Richard M. Golden, has been published. This makes available to the scholarly community a huge and comprehensive survey of the history of European witchcraft, with entries from 172 scholars representing 28 countries.[28] In the last year or so, two books have appeared that build upon Charles Zika's innovative work by highlighting the significance of the artistic representation of witches in the early modern period.[29] As an international witchcraft conference held at the University of Essex in the spring of 2006 confirmed, the topic of masculinity and witchcraft, highlighted in Chapter 11 of this book, has also developed as a major new trend in witchcraft research.[30] Arguments for a rapprochement with anthropology and a more comparative global approach to the subject have re-emerged.[31] A new academic journal has also come into being. *Magic, Ritual, and Witchcraft*, under the editorship of Michael D. Bailey, is set to become an important forum for future research.

As these recent developments indicate, witchcraft continues to generate some of the most innovative and lively scholarly activity in the discipline of history. Witchcraft historians are ever willing to move forward in new directions, yet they display a commendable self-reflexivity and a cautious respect for the endeavours of their predecessors. This book provides an opportunity to pause and reflect on that historiography, though its authors are aware that it, too, will soon become part of the tradition. This is positive recognition that the history of witchcraft is far from exhausted, and we look forward to seeing where new sources, interpretations and interdisciplinary approaches lead the subject in the future.

notes

1. Our thanks to Willem de Blécourt for his informative comments and suggestions on an early draft of this introduction. Good recent overviews of witchcraft historiography are provided by Thomas A Fudge, 'Traditions and Trajectories in the Historiography of European Witch-Hunting', *History Compass* 4, 3 (2006) 488–527; Wolfgang Behringer, 'Historiography', in Richard M. Golden (ed.), *Encyclopedia of Witchcraft: The Western Tradition*, 4 vols (Santa Barbara and Oxford, 2006), Vol. 2, pp. 492–8.
2. Brian Levack (ed.), *Articles on Witchcraft, Magic and Demonology*, 12 vols (New York, 1992); *New Perspectives on Witchcraft, Magic and Demonology*, 6 vols (London, 2001).
3. Peter Maxwell-Stuart, *Witchcraft in Europe and the New World, 1400–1800* (Basingstoke, 2001); Maxwell-Stuart, *Witchcraft: A History* (Stroud, 2000); Geoffrey Scarre and John Callow, *Witchcraft and Magic in Sixteenth- and Seventeenth-Century Europe*, 2nd edn (Basingstoke, 2001); Robert W. Thurston, *Witch, Wicce, Mother Goose* (London, 2001).
4. See, for example, Éva Pócs (ed.), *Demonológia és boszorkányság Európában* (Budapest, 2001); Gábor Klaniczay and Éva Pócs (eds), *Communicating with the Spirits* (Budapest, 2005); Pócs, *Between the Living and the Dead* (Budapest, 1999).
5. See, for example, Marie Lennersand and Linda Oja (eds), *När oväsendet tystnat. Efterspelet till uppror och religiösa konflikter, 1670–1860* (Uppsala, 2004); Marie Lennersand, 'The Aftermath of the Witch-Hunt in Dalarna', in Owen Davies and Willem de Blécourt (eds), *Beyond the Witch Trials: Witchcraft and Magic in Enlightenment Europe* (Manchester, 2004), pp. 61–9; Linda Oja, 'The Superstitious Other', in Davies and Blécourt, *Beyond the Witch Trials*, pp. 69–81; Richard L. Sjöberg, 'The Outbreak of Mass Allegations of Satanist Child Abuse in the Parish of Rättvik, Sweden, 1670–71: Two Texts by Gustav J. Elvius', *History of Psychiatry* 15, 4 (2004) 477–87; Kristina Tegler Jerselius, *Den stora häxdansen: Vidskepelse, väckelse och vetande i Gagnef 1858* (Uppsala, 2003).

In neighbouring Norway Rune Blix Hagen runs an impressive campaign publicising and disseminating information about his country's witchcraft history. His latest publication is *Samer er trollmenn i norsk historie* (Kárášjohka, 2005). See also Gunnar W. Knutsen, 'Norwegian Witchcraft Trials: A Reassessment', *Continuity and Change* 18 (2003) 185–200.

6. Examples of recent publications on Germany are Rita Voltmer, *Hexenverfolgung und Herrschaftspraxis* (Trier, 2005); Kurt Rau, *Augsburger kinderhexenprozesse 1625–1730* (Wien, 2006).

7. For a contextual discussion of this revisionist development see Jonathan Barry, 'Introduction: Keith Thomas and the Problem of Witchcraft', in Jonathan Barry, Marianne Hester and Gareth Roberts (eds), *Witchcraft in Early Modern Europe: Studies in Culture and Belief* (Cambridge, 1996); James Sharpe, *Instruments of Darkness: Witchcraft in England 1550–1750* (London, 1696), 'Introduction'.

8. His most recent publications include *Heksen en hun buren in Frans Vlaanderen. 16de tot 18de eeuw* (Ieper, 2004) and 'Het proces van Leyn Weckx in Eksel – Hechtel in 1725. Het laatste heksenproces in België?', in *Hechtel-Eksem. Historische en naamkundidige bijdragen, Jubileumboek 3 Heemkundige Kring Hechtel-Eksel* (2005) 129–55. See also his website, which has some English content: <www.kuleuven-kortrijk.be/facult/rechten/Monballyu/Rechtlagelanden/Homepage.htm>.

9. Their most recent books are: María Tausiet, *Los posesos de Tosos (1812–1814). Brujería y justicia popular en tiempos de revolución* (Zaragoza, 2002); Oscar di Simplicio, *Autunno della stregoneria. Maleficio e magia nell'Italia moderna* (Il Mulino, 2005).

10. The problem of scholarly monoglottism regarding witchcraft is well made in Maxwell-Stuart, *Witchcraft in Europe*, p. viii.

11. On Ukraine see Kateryna Dysa, 'Attitudes towards Witches in the Multi-Confessional Regions of Germany and Ukraine', in Eszter Andor and István György Tóth (eds), *Frontiers of Faith. Religious Exchange and the Constitution of Religious Identities, 1400–1750* (Budapest, 2001), pp. 285–9. On Poland see, for example, Wanda Wyporska, 'Witchcraft, Arson and Murder: The Turek Trial of 1652', *Central Europe* 1 (2003) 41–54. Her book *Motive and Motif: Representations of the Witch in Early Modern Poland* is forthcoming in the Palgrave Macmillan 'Cultures of Magic' series.

12. See, for example, George Modestin, *Le diable chez L'évêque. Chasse aux sorciers dans le diocese de Lausanne (vers 1460)* (Lausanne, 1999); Martine Ostorero, *L'imaginaire du sabbat: Edition critique des texts les plus anciens (1430c–1440c)* (Lausanne, 1999). For the later period see Fabienne Taric Zumsteg, *Les sorciers à l'assaut du village Gollion (1615–1631)* (Lausanne, 2000).

13. The work of Monter and Briggs are now part of the canon of witchcraft studies and will be discussed in later chapters. Sarah Ferber's most recent publication on the topic is *Demonic Possession and Exorcism in Early Modern France* (London, 2004).

14. Most notably the respective sociological, psychological and gender approaches of Paul Boyer and Stephen Nissenbaum, *Salem Possessed: The Social Origins of Witchcraft* (Cambridge, Mass., 1974); John Putnam Demos, *Entertaining Satan: Witchcraft and the Culture of Early New England* (Oxford, 1982); Carol. F. Karlsen, *The Devil in the Shape of a Woman: Witchcraft in Colonial New England* (New York, 1987).

15. Books for children include Marilynne K. Roach, *In the Days of the Salem Witchcraft Trials* (Boston, 1996; reprinted 2003); Jane Yolen, *The Salem Witch Trials: An Unsolved Mystery from History* (New York, 2004); Tamra Orr, *The Salem Witch Trials* (San Diego, 2004).

16. Mary Beth Norton, *In the Devil's Snare: The Salem Witchcraft Crisis of 1692* (New York, 2002); Richard Godbeer, *The Devil's Dominion: Magic and Religion in Early New England* (Cambridge, 1992).

17. Laura A. Lewis, *Hall of Mirrors: Power, Witchcraft and Caste in Colonial Mexico* (Durham, 2003); Laura de Mello e Souza, *The Devil and the Land of the Holy Cross: Witches, Slaves and Religion in Colonial Brazil*, trans. Diane Grosklaus Whitty (Austin, 2004); Nicholas Griffiths and Fernando Cervantes (eds), *Spiritual Encounters: Interactions Between Christians and Native Religions in Colonial America* (Birmingham, 1999).

18. James Sharpe (ed.), *English Witchcraft, 1560–1736*, 6 vols (London, 2003). For a thoughtful review of this resource see William Monter, 'Re-contextualizing British Witchcraft', *Journal of Interdisciplinary History* 35, 1 (2004) 105–11. See also the annotated primary sources presented in Marion Gibson (ed.), *Witchcraft and Society in England and America, 1550–1750* (London and Ithaca, 2003); Marion Gibson, *Early Modern Witches: Witchcraft Cases in Contemporary Writing* (London, 2000).

19. For example, Johann Weyer, *On Witchcraft: De præstigiis dæmonum*, edited by Benjamin G. Kohl and H. C. Erik Midelfort, trans. John Shea (Asheville, 1998); P. G. Maxwell-Stuart, *Investigations into Magic: Martin Del Rio* (Manchester, 2000); Nicolas Rémy, *La démonolâtrie*, trans. Jean Boës (Nancy, 1998); Friedrich von Spee, *Cautio Criminalis, or a Book on Witch Trials*, trans. Marcus Hellyer (Charlottesville, 2003); Wolfgang Behringer, Günter Jerouschek and Werner Tschacher, *Der Hexenhammer. Malleus Maleficarum* (Munich, 2000). The latter is a more accurate translation than the popular version by Montague Summers. Useful translated extracts from primary sources have also been made available in Brian Levack (ed.), *The Witchcraft Sourcebook* (London, 2002) and Peter Maxwell-Stuart (ed.), *The Occult in Early Modern Europe* (Basingstoke, 1999).

20. EEBO and LION are produced by Chadwyck-Healey, and ECCO by Thomson Gale.

21. <http://historical.library.cornell.edu/witchcraft/witchcraft_D.html>.

22. <http://gallica.bnf.fr>.

23. Julian Goodare, Lauren Martin, Joyce Miller and Louise Yeoman, 'The Survey of Scottish Witchcraft', <www.arts.ed.ac.uk/witches>.

24. 'Witchcraft Bibliography Project Online', <www.witchcraftbib.co.uk>. It is currently maintained by Jonathan Durrant.

25. <www.uni-tuebingen.de/IfGL/akih/akih.htm>. The forum is managed by Klaus Graf.

26. A good guide to internet witchcraft resources is that compiled by Wanda Wyporska for the Humbul Humanities Hub: <www.humbul.ac.uk/topics/witchcraft.html>.

27. Although its coverage of the historiography is now a bit out of date, the Stella Australis resource site for Australian pagans is a commendable example of this process of engagement: <www.geocities.com/Athens/2962/>.

28. Golden (ed.), *Encylopedia of Witchcraft*.

29. Claudia Swan, *Art, Science, and Witchcraft in Early Modern Holland: Jacques de Gheyn II (1565–1629)* (Cambridge, 2005); Linda C. Hults, *The Witch as Muse: Art, Gender, and Power in Early Modern Europe* (Philadelphia, 2005); Charles Zika, *Exorcising our Demons: Magic, Witchcraft and Visual Culture in Early Modern Europe* (Leiden, 2003).

30. A recent contribution to the debate is J. Wijaczka, 'Men Standing Trial for Witchcraft at the Lobzenica Court in the Second Half of the Seventeenth Century', *Acta Poloniae Historica* 93 (2005) 69–86. Forthcoming publications in the Palgrave Macmillan 'Cultures of Magic' series include an English translation of Rolf Schulte's

pioneering *Hexenmeister. Die Verfolgung von Männern im Rahmen der Hexenverfolgung von 1530–1730 im Alten Reich* (Frankfurt, 2000) and Alison Rowlands (ed.), *Witchcraft and Masculinities in Early Modern Europe.*

31. See Wolfgang Behringer, *Witches and Witch-Hunts* (Cambridge, 2004); Ronald Hutton, 'Anthropological and Historical Approaches to Witchcraft: Potential for New Collaboration?', *Historical Journal* 47 (2004) 413–34; Ronald Hutton, 'The Global Context of the Scottish Witch-Hunt', in Julian Goodare (ed.), *The Scottish Witch-Hunt in Context* (Manchester, 2002), pp. 16–33.

2
the contemporary historical debate, 1400–1750

peter maxwell-stuart

One of the principal difficulties of combining magic and history lies in the complexity of definitions which need to be traversed in order to arrive at any kind of sensible conclusion. What does one mean by 'magic'? Natural, demonic, mathematical, ritual, folk, witchcraft? What does one mean by 'history'? Learned, exemplary, popular, oral, commemorative, narrative, cyclical, investigatory? The distinctions are important because they were made in the past, and because, although the boundaries between them were fluid, quite frequently people tried to blur them still further for specific ends: as, for example, when Jean Bodin, who was hostile to magic in any of its forms, argued for the reappraisal of certain magic practices as essentially demonic, thereby in effect transferring them from a category of licit magic to one which was illicit. My discussion of the relationship between magic and history is therefore going to be one closely involved in the business of definitions. I shall start with the concept of history as time, then with history as a process of recording and interpretation, and finally try to see how treatises on magic relate to these premises.

time

Michel de Certeau has offered everyone who deals with historiography a timely warning: 'Modern Western history essentially begins with differentiation between the present and the past ... Far from being self-evident, this construction is a uniquely Western trait.'[1] For the Middle Ages and the early modern period, however, the practice of constructing categories and then assuming that such categories have always existed would have been completely alien. They lived in a known and knowable world, only a very few thousand years old, which had been shaped by God and continued to be sustained by Him. Time was both sacred and limited. It had a beginning and an end recorded by Holy Scripture, and people might hear of the principal events from the Creation to the Last Judgement in sermons, read of them in their Bible, or see them illustrated in stained glass windows (such as those in York Minster), or in mosaics (such as those in the Baptistery in Florence). By the middle years of the seventeenth century, however, this comfortable history was beginning to creak under the weight of accumulating amounts of new factual material. Astronomy is perhaps the most obvious example of this, but we can see its effects reflected even in biblical commentary, where speculation based on the discovery of the inhabitants of the New World was set to undermine the structure imposed upon time by Judaeo-Christian theology.

Suppose we accept the opinion that there were people *before* Adam, wrote Isaac de la Peyrère in 1655, more than a century after Paracelsus had suggested something similar:

> the story [*historia*] of Genesis becomes much clearer. It is rendered consistent with itself. It is also rendered consistent in remarkable ways with all the secular records, whether they be ancient or more recent: for example, the Chaldaeans, the Egyptians, the Scythians, and the Chinese. The oldest creation of things, set out in the first chapter of Genesis, is rendered consistent with the Mexicans whom to Columbus came not so long ago. It is rendered consistent with those people from the South and North who have not yet been made known [to us] ... It is probable that they have been created, along with the land itself, in every country, and not propagated by Adam.[2]

By the middle of the eighteenth century, however, geological studies had begun to establish that the world was very much older than people had thought hitherto: 75,000 years, according to Georges de Buffon in his revised *Histoire Naturelle*, and about 2 billion, according to the flawed but interesting calculations of Benoit de Maillet in his *Telliamed* (1748). The Creation was rapidly vanishing into an almost unimaginable past; knowledge of the earth, not to mention the universe, was becoming unmanageably complex and subject to constant change, although the ancient notion of cycles of time can still be seen even in eighteenth-century cosmological speculation:

> Thus a perpetual succession of new created worlds is or may be continually flowing from the celestial regions to the Sun and as finally feeding its eternal fires ... Thus also by successive seats of new life in destined states past, present or to come and all proportioned to the sphere of glory in which they naturally exist ... This great period of nature cannot be imagined to take up less time than a million of millions of years.[3]

In spite of attempts such as Thomas Burnet's *Sacred History of the Earth* (1681–89) to defend biblical chronology while accepting at least some of these later speculations and discoveries, however, the enclosed world of demonology with its relative certainties was proving more and more friable with every passing decade. But did this affect the way in which treatises on magic and witchcraft were written? Can we see this changing concept of time reflected in disquisitions on the role of spirits and angels, and of the magical practitioners who invoked and made use of them? In fact, we do not. The spirit worlds of angels and demons remained constant *sub specie aeternitatis*, regardless of the century in which people wrote about them, and in consequence the relationship between the human and the non-human worlds was seen in just the same light. Hence, Simon Ravensberg felt able to say to a university audience in 1636:

> Can the first source of magic – i.e. where, when, etc. it first arose – be definitively demonstrated? No, it can't, any more than the origin of any idolatry, superstition, apostasy, or transition from the true worship of God to paganism ... Eusebius and Lactantius correctly say that evil demons invented this, along with other arts. They are quite clear that a first inventor among humans cannot be designated, and that we should not even look for one. This, however, can be affirmed – that it was born along with idolatry, as though from the same egg.

These basic sentiments were echoed by Thomas Vaughan, who wrote that magic was imparted to humanity as a kind of revelation or discovery, and therefore properly speaking had neither birth nor beginning.[4]

history

On history, two brief points may be useful. First, it was Herodotus who began the western notion that history deals with the truth of facts – as *historia*, 'asking questions' suggests – and thus stimulated the western preoccupation with coherence in history. To this, Thucydides added the supposition that human nature is constant and that events are cyclical (a notion overtly governing the end to many later Christian prayers, *per saecula saeculorum*, whose meaning is entirely lost in the conventional translation 'for ever and ever'), and one which can be glimpsed in its Christianised form in the assumption underlying witchcraft treatises, that humanity sees recurring assaults upon it by Satan and his followers (heretics, Jews, witches), in accordance with God's permission. To this we may add the Hebrew concept of history, as evidenced by the Old Testament, which saw it as providing an account of what had happened in the past serving as an illustration of what to expect from God in a variety of circumstances, and St Augustine's understanding that the Bible as a whole tells us about God's overall plan for the salvation of humanity.

Second, Aristotle drew a distinction between the historian whose job is to relate particular events and relate them to each other within the chronological period he is discussing, and the poet who writes about the kind of thing which *could* happen and describes the characteristics appropriate to certain kinds of people (*Poetics* 1451a–b, 1459a). What is more, history is remembered experience, a notion picked up by St Isidore of Seville who defined history as 'the narrative of something which has been done, and by whose means those things which have happened in the past are distinguished [from others] ... Among the ancients, no one used to write a history, except a person who had played a part in it and had been an eye-witness of what needed to be recorded' (*Etymologiae* 1.41.1). Later medieval historians agreed, but emphasised history as a written, rather than an oral narrative. History was largely commemorative and exemplary within a biblical framework, some writers extending the record of human deeds as far back as Adam and as far forward as the Last Judgement, thereby turning history into an exercise which not only sets down the past but seeks to predict the future. *Written* history thus becomes authoritative.

These commemorative writings, then, had to be interpreted, and it was the interpretation which lent the past significance, not the mere recording of fact or would-be fact, as implied in Francis Bacon's later use of the word *historiae* to mean 'narrations of phenomena'. Hence, accuracy of facts was important only in as far as it helped a 'better' interpretation; the past was a living legacy to be preserved for use in the present, and the past was neither remote nor alien – hence the willing acceptance of the fantastic genealogies of royal and noble houses. Interpretation of the past (and consequent selective use of its documentation) provided a bridge between the dead and the living which united them in one recognisable community. The early modern period inherited this historiographical cast of mind and continued to employ it, although not, of course, unchallenged. Petrarch, for example, saw the crucial turning point of history as the decline of the Roman Empire, and thus displaced the Incarnation as its pivotal event, while Machiavelli helped to remove historical discourse from considering 'the world' to concentrate upon a single city. Textual criticism begged both writer and reader to consider the reliability of historical sources, and thus became of increasing concern, to religious controversialists in particular, during the sixteenth century; and this can be seen in the care devoted to the etymologies of the various Hebrew, Greek and Latin words for different kinds of magical operator, which is exhibited by many writers of witchcraft treatises as they seek to establish a firm basis for their interpretation of what a witch is, or to controvert an opponent's view on the matter. But the influence of the developing sciences which were dominated by mathematics came to govern expectations of what the other disciplines might have to offer, and so helped to formulate the suspicions of those such as Descartes (1596–1650), who dismissed 'even the most faithful histories' as factually unreliable, and Thomas Hobbes (1588–1679), who rejected any notion of one's being able to have any absolute knowledge of fact, past or to

come, interpreting the study of facts as engagement with consequences. 'If th.
be, that is; if this has been, that has been; if this shall be, that shall be: which
is to know conditionally', a deductive method he called *science*. 'Science is the
knowledge of consequences, and dependence of one fact upon another.'[5]

If only the eighteenth century could have built upon this ideal of careful
establishment and assessment of fact from earlier record. Unfortunately, it did
not. It fell, instead, victim to the allurement illustrated by Voltaire's advice:

> Inculcate in the young a greater liking for the history of recent times, which
> is essential for us to know, than for ancient history, which serves only to
> satisfy our curiosity. Let them reflect that modern history has the advantage
> of being more certain, from the very fact that it is modern.[6]

Consequently, from conceits such as these it is all too easy to derive the
comforting assurance that 'modern' is synonymous with 'correct, advanced,
progressive, reliable', and so import into the past modern notions and pre-
conceptions which have no business being there – a flawed approach which
may be particularly and most damagingly present in discussions of witchcraft
and its prosecution.

witchcraft treatises

How far can these changing concepts of time and history be seen in writings
on witchcraft? Discussion of the development of magic itself plays, at best,
a very small role in these treatises. Institoris, Binsfeld and Grillando, for
example, more or less ignore it, while Del Rio, Weyer and Bodin keep it to a
minimum. Gabriel Naudé goes back as far as Adam in his historical study of
the occult sciences (*Apologie*, 1625), but Del Rio and the others are content to
start with the Persians or Zoroaster (thus reflecting the notion that wisdom
began in the East and thence moved to the West), and let their surveys peter
out somewhere in the High Middle Ages after they have demonstrated that
the Greeks and Romans practised the art and the Church Fathers were united
in condemning it. The Jesuit Martín Del Rio also distinguished between
natural (licit) magic and other varieties, and lists magicians who were inspired
by the Devil; while the Protestant Reginald Scot is keen to emphasise (for
controversial purposes) that the age of miracles and prophecy and oracles has
ceased. History enters these tractates, not as a chronological survey so much
as a series of *exempla* drawn from the past, intended to show that Satan has
been allowed to exercise his power in every age so far, including this one; and
to that extent the history of witchcraft is a frozen history, based upon the
notion that human nature is constant, and that Satan operates according to
recurrent cycles informed by the constant tension which exists between the
civitas Dei and the *civitas terrena*.

Writings on witchcraft take all this for granted. Their main purposes are directed elsewhere, and we can divide such books into three broad categories.

1. *Practical manuals.* These tended to be written for the clergy and the lawyers, as in the case of Johann Nider's *Formicarius* (c.1438, printed 1475) whose fifth book contains a large number of exempla relating to witches which priests could use in their sermons, showing people how to live well by providing them with illustrations of wickedness and the divine retribution which inevitably follows; Heinrich Institoris' *Malleus Maleficarum* (1486) which came out of his own failure to prosecute witches in Bern with any success; Jean Bodin's *Démonomanie des sorciers* (1580), the fourth book of which in particular addresses the question of how to establish judicial proof of the offence; and Del Rio's *Disquisitiones Magicae* (1599–1600) was aimed at 'theologians, legal experts, physicians, and scholars', according to the title page, and sought to give them authoritative guidance on all aspects of the occult sciences and their suppression. It was very successful, and after his immense and comprehensive work there was no real need for any further manual.

2. *Polemical treatises.* The largest number of writings on witchcraft actually belongs to this category, the intellectual warfare engaging different groups at different times and in different places. Thus, one can see Catholics controverting Protestants during the sixteenth and seventeenth centuries, the former arguing that magic followed heresy and that the outbreak of various Protestant ideologies therefore explained the apparent increase in the number of witches; the latter suggesting that Catholicism was riddled with magic and that, with the imminent arrival of the Last Judgement, societies needed to be purged of the Devil's servants who were seeking to aid him in the ultimate downfall of humanity. The late seventeenth and eighteenth centuries, by contrast, often made belief in the validity of witchcraft a test for people's belief in Christianity, the so-called enlightened few representing themselves as the vanguard of a more liberal and 'scientific' notion of how society should function. In consequence, there tended to be a strong political and anti-Catholic strain in their views and arguments. Narratives about magic can thus reveal the ways in which a community defined itself in relation to another. Support for and opposition to prosecution of witches, however, did tend to cross confessional lines. So we find that Del Rio, Peter Binsfeld and Heinrich Schulteis, Luther, Thomas Erastus and Joseph Glanvill were in favour of prosecuting witches, even though a number of them entertained doubts about some of the methods used to obtain proof of guilt, while Cornelius Loos, Friedrich Spee and Adam Tanner, Johann Weyer, Johann Meyfahrt, Johann Goedelmann and Christian Thomasius expressed doubts about some of the fundamental reasoning behind demonology itself, although

again, some of them tended to pay a greater or lesser lip-service to the theoretical possibility of Satan's active suspensions of the common laws of nature.

3. *Personal accounts*. Most witchcraft treatises were written by men who had at least some personal connection with witchcraft prosecution, but a few are based either entirely or in appreciable measure upon that personal involvement. Obvious examples are the *Tableau de l'inconstance des mauvais anges* by Pierre de Lancre who was ordered by Henri IV to deal with a major and highly disturbing outbreak of witchcraft in the Basque-speaking region of south-west France, Pays de Labourd, between 1609 and 1611; and the *Visitation Reports* of the Spanish Inquisitor, Alonso de Salazar Frías, who was investigating an equally alarming manifestation just over the border in Spain. It is interesting that their accounts came to more or less opposite conclusions: de Lancre having few doubts about the reality of what he was facing; Salazar Frías, on the other hand, having all too many. Nicolas Remy's *Daemonolatria* (1595), reviewing an intense series of witch prosecutions in the French Duchy of Lorraine, and Henri Boguet's *Discours des sorciers* (1602), recounting his activities in the Franche-Comté, clearly depend on much detailed information they both gathered during their times as *procureur général* and *grand juge* respectively.

What all the writers belonging to any of these categories have in common is, first, a personal agendum in reaction to immediate or looming events, and, second, a sense of anxiety caused by the proximity, real or apparent, of those same events. Thus, Nider and Institoris were both intensely aware of the danger that heresies were breeding a new and expanding cult which threatened the Church and manifested itself in magical practitioners, and their anxieties are also to be seen in the titles of other late fifteenth-century works: Nicolas Jacquier's *Scourge of Heretics* (1458), Alfonso da Spina's *Fortress of Faith against Enemies of the Christian Religion Everywhere* (1459) and Petrus Marmor's *Scourge of Workers of Harmful Magic* (1462). Institoris' violent title, *Hammer of Witches*, is thus part of a well-established tradition of fearfulness. Learned responses to this fear were both practical and charged with emotion. Nider, for example, wrote his *Formicarius* as a handbook to aid clergy in effecting a moral reform throughout Christian society, while Institoris composed *Malleus* in a paroxysm of personal anger that he had been thwarted in his attempt to prosecute a number of witches in Innsbruck, and out of a deep anxiety that Christendom was being subverted by men and women who had laid themselves open to corruption by Satan. Del Rio, too, wrote his *Disquisitiones* for more or less the same reason, although he lacked Institoris' rage and employed for his purpose the techniques of legal writing that heaped up disparate authorities and evidence to support his arguments – the distinctive feature of demonologists' writing as a whole. Bodin's *Démonomanie* was composed at a time when there was a rising fear of impiety in France, when the

Wars of Religion exacerbated everyone's anxiety, and 'atheists' were especially feared and condemned.

Writings on witchcraft are dramas, frequently composed in the manner of legal submission to the court of learned or public opinion, and derive their emotional power partly from the personality of the author, partly from the seismic event he sees threatening society – and one may note that the more effective doubters among them were those who argued from a legal rather than a medical basis. Friedrich Spee, for example, does a better job than Johann Weyer. That this genre did not survive the later eighteenth century may be attributed at least in part to significant changes, encouraged by the closer attention paid to textual evidence since the sixteenth century, which many legal systems introduced into their administration of the law, but more importantly, perhaps, to a shift in the intellectual preoccupations of opinion-makers. Thus, the learned tacitly agreed to change their perception of history from the contained Hebrew version tempered by Classical and New Testament literature to one informed by the burgeoning mathematical sciences which effectively isolated human beings in vast stretches of undetermined time and unlimited physical spaces. Magic, in all its variations, however, appeared to make sense only in the former universe and in consequence lost the interest of the *philosophes* and *bien pensants*. Only when twentieth-century astrophysics caught up with Giordano Bruno's proposition of multiple universes would magic be rethought and reconceptualised in a manner reminiscent of the earlier perception of interpenetrating and inter-reactive worlds.

It will now be helpful to look at one or two examples from each of these proposed categories of writings on witchcraft. The *Malleus Maleficarum* circulated in manuscript in 1485–86 and was then printed in the autumn of 1486. Although often attributed to two authors, Heinrich Institoris (Kramer) and Jakob Sprenger, it is now generally agreed to be the work, at least in the main, of Institoris.[7] Both men were Dominican Inquisitors, but Institoris' career was much more problematic than that of Sprenger, and his animus against female witches in particular seems to have sprung partly from a concern that Satan was using the alleged weaknesses of women to subvert them into his service and thus taint others with what was, at basis, an impetus to heresy, apostasy and idolatry, and partly from personal frustration attendant upon his failing, in the teeth of the local bishop's hostility to his methods, to achieve the successful extirpation of certain witches he was prosecuting. The *Malleus* is thus a tacit apologia for those methods, and a warning to all officials, both ecclesiastical and secular, of the present dangers embodied in a growing number of Satan's servants. It is perhaps worth bearing in mind that two years after its publication, Institoris produced another, shorter treatise (never published) directed at the legal concerns expressed by magistrates regarding the prosecution of witches in Nuremberg, a treatise whose rigorous recommendations the magistrates then ignored completely. His advice on how to increase the number of successful prosecutions, which is what the

Nuremberg treatise in particular was written to urge, was thus rejected by representatives of both Church and state.

The *Malleus* is divided into three parts: the first devoted to theological questions relating to witchcraft, the second to a description of witches' modus operandi and ways to counter them, and the third to the appropriate legal procedures to be adopted in prosecuting witches successfully. Although new, in the sense that it was the first printed manual of its kind, the *Malleus* was not without precedent, and one can see that inquisitorial and demonological handbooks, and anti-heretical texts such as Nicolas Eymeric's *Practica Inquisitionis* (1376), Johann Nider's *Formicarius* (1435), Nicolas Jacquier's *Flagellum Haereticorum* (1458), and Alfonso da Spina's *Fortalitium Fidei* (1459) will have influenced the climate in which the *Malleus* was produced, and afforded models, at least in part, for its composition. Institoris had a number of important points to make. First, witchcraft is a real, not an imaginary crime and therefore deserves to be treated like any other capital offence. Second, what makes it particularly evil is its combination of heresy and apostasy with murder and carnal lust. Third, most witches are women. Fourth, witchcraft is a matter for both ecclesiastical and secular courts because witches commit both spiritual and temporal crimes.[8] The *Malleus* is perhaps best known for its alleged misogyny, and certainly is the most quoted of witchcraft treatises in that connection. It can, however, be misunderstood. For example, it is perfectly true that the *Maleficarum* of the title is specifically feminine, but in his text Institoris recognises the contemporary situation (which remained true for later periods), that workers of harmful magic might be men as well as women, for he continually uses *malefici* to describe them, the grammatically masculine plural being gender-inclusive. To translate this as 'witches', as Montague Summers regularly does, is therefore to give the reader a false impression, because the English word 'witch' immediately suggests a woman. Moreover, Institoris' misogyny has been overemphasised by the habit of selective quotation and misleading translation, and Spierenburg has made a relevant and useful point in connection with this:

> A general increase in misogyny around 1500 has not been demonstrated so far ... The antifeminine cultural tradition is at least two thousand years older than witchcraft doctrine, and it is difficult to accept that this tradition would suddenly have led to witch persecutions between 1480 and 1680. Significantly, when Sprenger and Kramer discussed the inferiority of women, they were not looking for support from earlier authorities, as they had with other subjects. For them the point was simply self-evident.[9]

That said, however, it must be noted that sexual questions of one kind and another are given especial prominence in the book. Can children be born from intercourse between women and evil spirits? Are magical operators able to generate emotions of love or hatred in human beings? Can midwives produce an abortion? Can witches really and actually remove a man's penis, or is this merely an illusion? These are queries raised in the first part of the book and, in one form or another, discussed yet again in the second. Is this a result of some personal obsession on Institoris' part? It seems unlikely. For all his animus against witches, Institoris does not exhibit the same violence as St Bernardino of Siena preaching on the same subject and, as Hartmann Ammann pointed out long ago, sexual jealousy or marital infidelity played a notable part in the accusations of witchcraft levelled against the defendants in Brixen in 1485.[10] So Institoris had, from his experience as an Inquisitor, practical examples of the importance of this particular motive in hostile magic. An increasing emphasis on clerical celibacy also meant that women were in danger of being seen as both tempters and pollutants, in spite of the positive images of the Virgin and other female saints, and the rehabilitation of St Jeanne d'Arc which resounded throughout Europe in 1456, only three decades before the publication of the *Malleus*. Moreover, Stuart Clark has made the important observation that, from ancient times, Creation had been described in terms of opposites, and in consequence women could be regarded as the opposite of men. Hence, if men are strong and intelligent and noble, women are weak, feeble-minded and disreputable, and this is just the line which Institoris takes in explaining why women rather than men are likely to succumb to the Devil's blandishments and become his servants in evil.

From this widespread contemporary outlook, then, allied to deep concern about the spread of heresy, came Institoris' principal themes – that Satan was taking advantage of women's weakness to corrupt them and so extend his demonic kingdom, that maleficent magic provided the outward sign of a coherent organisation devoted to the worship of Satan, which thus constituted the mean whereby the Faith and the Church were subject to attack, and that it was imperative this attack be repelled by extirpating the human agents through whom the Devil was doing his work. The *Malleus* was therefore intended as a call to arms in the face of a deeply embedded conspiracy against the salvation of humankind, and thus against the benign intentions of God Himself. Heresy and apostasy were the book's principal targets, concerns which had been at the heart of the Council of Basel (1421–37) where Johann Nider and Martin le Franc were active in promoting several of the notions – transvection through the air to a Sabbat, for example – which later became standard features of many official European accounts of witches' behaviour. But in fact, Institoris shows no interest in the Sabbat, in witches' marks or in the blasphemous celebration of the Mass in a ritual of inversion. His aim, which on the whole he accomplishes effectively in spite of the repetitiousness of his writing, is to explain why Church and state should prosecute witches and how they can do it legally with best chance of success.

The *Malleus*, however, had little immediate effect – the German authorities simply ignored it – and it provoked no fresh prosecutions, contrary to what Institoris had intended. What it did was to provide a treasure-house of explanations for any authorities who wished to account for their having initiated prosecutions already, and a useful manual for judges and advocates who wanted to know the proper procedure they ought to follow. This proved to be Institoris' most durable legacy, and if we look at the *Disquisitiones Magicae* of Del Rio, first published in 1599–1600, we can see how frequently he refers to his Dominican predecessor and how much he owes to Institoris' formulation of desirable approaches to the problem of witchcraft. If the *Malleus* was the first coherent witchcraft manual, the *Disquisitiones* was the last. It is a large work in six volumes, covering the various types of magic ranging from 'artificial', which we would call 'conjuring tricks' or 'illusion'; 'natural', referring to magic using only lawful means found within nature to produce its effects; and 'demonic' which sought the assistance of non-human entities for that purpose. Divination receives an entire book to itself, and it is here that Del Rio discusses the swimming test used to help identify someone as a witch – a method he associates with German states in particular, and which he rejects as unreliable and unwarranted. In Book 5, he gives advice to judges and advocates on the correct legal procedures they should adopt in arresting, trying, sentencing and executing witches; and in Book 6 he offers guidance to confessors who have to deal with persons accused of and condemned for witchcraft. Witchcraft itself, therefore, forms only a part of his wide-ranging discussion of the occult sciences as a whole, and is firmly included in those sections of the work dealing with demonic and harmful magic.

Both Institoris and Del Rio have methodologies in common. They make extensive use of appeals to authority in the form of Greek and Roman writers, and the Church Fathers. Institoris, however, relies constantly upon St Augustine and St Thomas Aquinas; Del Rio, while using both, makes more frequent use of Classical authors and more recent commentators on magic from the fifteenth and sixteenth centuries. To some extent this is a difference one might expect. Institoris was writing in the Scholastic tradition, Del Rio in one more recent and perhaps more combative, so each mirrors his own age, with Del Rio having the advantage of a much greater body of work on the occult sciences on which to draw for illustration and support. Both men touch upon their personal experience of magical operators – Institoris mentions eye-witness testimonies he has seen at Innsbruck and Enningen, Del Rio his meetings with a young Spanish diviner – but neither relies on this to forward his argument or even provide it with added authenticity. Indeed, Institoris is noteworthy for his almost complete authorial absence from his text, thereby rendering attribution of the *Malleus* to him alone that little bit more difficult.

Appeal to tradition and authority rather than to personal narrative Institoris and Del Rio clearly regard as both sufficient and effective, Del Rio being the better (and more accurate) scholar. His aim, however, is similar to that of Institoris, since he sees his society subject to attacks by the same kind of combination of heresy and illicit magic as troubled Institoris. The prologue to the *Disquisitiones* – 'explaining why this treatise has been difficult to write, but why it was necessary to do so' – makes this point clear. The pride and malice of God's enemies is increasing. They have a thousand ways of doing harm and use innumerable weapons against humanity, of which magic is the most deadly. Evil spirits are on the loose, seeking to take possession of foolish and deluded souls. Never have there been as many witches as there are today, and the main reason for this is the faintness of and contempt for the Catholic faith. The heretics are strongly opposed by the Jesuits. The *Disquisitiones* is a weapon in that war. Change 'Jesuits' to 'Dominicans', and the sentiments could equally well have come from the pen of Institoris. So it is perhaps a mark of how self-contained was the universe of demonologists that Del Rio could still be used as a prime source of information and guidance on witchcraft as late as 1697 in Scotland, and still be quoted, if reservedly, in mid-eighteenth-century Germany. Del Rio had, in effect, said the last word on the subject, and after him any attempt to produce a handbook or even a large-scale disquisition on magic would risk being mere repetition.

Nevertheless, others did have things to say, and if some of their points and authorities reiterated his, their intended narrative, personal preoccupations and potential readership were really not the same. Del Rio, as Institoris, accepted the reality of witchcraft as a phenomenon, and therefore the reality of the threat posed by witches, although unlike Institoris he had his reservations about some aspects of both. Protestant writers on the subject had their own battles to fight. This can be seen in the disagreement between Johann Weyer (1515–1588) and Thomas Erastus (1524–1583).[11] Weyer was not a lawyer but a physician, and came to the question of witches from a quite different angle from that of the Dominican and the Jesuit, although in a letter to Duke Wilhelm V of Cleve, Jülich and Berg, his employer, he claimed to review its theological, philosophical and legal aspects, too. He made three basic assumptions: (1) witches tend to be women who are mad or senile or ill, so their witchcraft is the product of insanity or illness, and they should not have the death penalty imposed on them; (2) their illness is either rooted in natural causes, or comes from evil spirits who use natural causes as the medium to make women sick; and (3) the notion of a pact with the Devil is mistaken, silly, and carries no weight or importance. Weyer thus subscribed to the very much earlier notion that women are naturally weak and in consequence prone to succumb to and take seriously illusions created by Satan – a conclusion reflected in the title of his best-known treatise, *De Praestigiis Daemonum et Incantationibus et Veneficiis* (1563), 'The Deceitful Tricks of Evil Spirits, Incantations, and Poisons/Works of Poisonous Magic'.[12] The effect of Weyer's work was to make two important points: first, that witches' confessions sprang from an externally induced delusion, and, second, that the Devil was sufficiently powerful in his own right not to require the assistance of mentally unbalanced women. It is worth noting this second point, as it is sometimes lost in the rush to see Weyer as some kind of prototype of a modern sceptic.

Erastus, who was also a physician, took a different point of view, although his controversy with Weyer was conducted more or less courteously, especially by the somewhat scabrous conventions of the time. His arguments are fairly straightforward and are to be found in his *Repetitio Disputationis de Lamiis* (1578). Anyone who produces a preternatural effect is a magician. 'Magician' includes witch, diviner, and so forth. Magicians achieve their effects knowingly or unknowingly with demonic help. God has condemned all magical practices, therefore there is no such thing as licit magic. The demonic pact is real. Demons can play tricks so skilfully that their effects may seem to be genuine and even tangible. The witch is worse than any other type of magician because, while the latter may not realise she or he is being manipulated by demons who can hide their participation in the magical working, the demonic elements in witches' practices is overt and therefore she or he knows about it. The magician may thus be redeemable: the witch is not, since the demons work their harmful magic specifically at the request of a witch. Erastus also emphasises the perverse sexual relationship between witches and demons (thereby clearly showing he thought of witches as women: male witches were sometimes accused of incest or adultery during the Sabbat, but never sodomy in its modern sense), and added this to his reasons for demanding the death penalty for witches.

Both Weyer and Erastus, then, had in common a view that witches' powers were entirely subject to the laws of nature, but they differed in their interpretation of the biblical terms for 'magical operator' – Weyer arguing that witches were unknown in biblical times and therefore that modern witches could not be subject to Mosaic law – in the importance of Satan's role in the witchcraft phenomenon, and in the validity of the argument that *all* witches were mentally sick or enfeebled by illness. In short, Weyer tended to argue that witchcraft was an impossible crime; Erastus that, in spite of many exceptions to the rule, witches could be real enough and what they were accused of doing might also be real if the demons they used decided to make it so. The scepticism which can be seen in texts as early as the tenth-century *Canon Episcopi* was thus extended by Weyer into a basic principle of his approach to the whole question of human involvement in demonic magic, and therefore, by implication, to demonic magic itself. Erastus, on the other hand, stopped short, as most writers on witchcraft did, from denying the ability or even the wish of non-human entities to use human agents to effect and further their aims.

But if we regard these polemical treatises as arguments directed mainly against or in favour of contemporary legal practice, since those written by doctors for a medical audience are much fewer in number, one of the most potent must be the *Cautio Criminalis* of the Jesuit Friedrich Spee, published anonymously in 1631. It was heavily influenced by the work of another Jesuit, Adam Tanner, who broadly accepted the current theological position on witches and their powers, but questioned, for example, the validity of confessions obtained under torture, and went as far as to compare the sufferings of accused witches to those of the early Christian martyrs. Maintaining the uselessness of forced confessions naturally led him to abdicate a reform of the legal code, and it is these two points which were taken up in particular by Spee and heavily stressed in his book, which was addressed directly to the sovereign princes of Germany and illustrated by constant, detailed references drawn from his own experiences as a confessor in several German towns, although the worst of the German prosecutions happened before his time or in places in which he was not actually present. His book is vivid and, as one might expect from a Jesuit, argued with subtlety; but it is also repetitious, partly because he was eager to hammer home to his noble and regal audience the simple facts that abuses of the law were taking place. True, Spee censures almost everyone else – bishops, priors, confessors, judges, scholars, fellow Jesuits (especially Del Rio and Peter Binsfeld, one of the most active prosecutors of witches in late sixteenth-century Germany), and the gossiping, slanderous, superstitious mob – for their involvement in making a dreadful situation worse and perpetuating this type of injustice. But his principal targets were the princes, since they bore ultimate responsibility for godly rule within their states, and they more than anyone could remedy these brutal wrongs.[13]

So individual is Spee's book that it might almost be included in the third category of witchcraft treatises, the personal accounts, except that it is well-rooted in legal argumentation and is thus an essay in a certain type of persuasion. The personal accounts certainly argue a point of view, but they are principally the revelations of an individual's experience of the witchcraft phenomenon, and so differ fundamentally from the lucubrations of theologians and doctors, however forcefully and sometimes passionately those learned works might be expressed. Henri Boguet was a *grand juge* in the Franche-Comté, a region that enjoyed its own *parlement* and therefore a degree of independence from the Holy Roman Emperor's jurisdiction.[14] Institoris, Del Rio, Weyer, Erastus and Spee all had experience, whether at first or second hand or both, of witches in the German-speaking states and (in Del Rio's case) the Spanish Netherlands; but their experiences for the most part involved adults, mainly women. With Boguet, it was somewhat different. His *Discours execrable des sorciers* (1602) begins with the very first witch trial over which he presided in 1598. It had been provoked by the testimony of an eight-year-old child, and the form the witchcraft took was demonic possession.

Louise Maillat, suddenly paralysed without apparent cause, was diagnosed as *une possédée* ('one possessed'), and when she was asked who had bewitched her, she pointed to a woman in the crowd, whom she recognised, Françoise Secrétain. Françoise was questioned and at first seemed to be an unsuitable target. But the cross on her rosary was broken and she shed no tears during her interrogation, so they shaved her head, and stripped her, looking for the tell-tale mark of the Devil's grip; and in spite of the absence of any such sign, Françoise began to confess that long ago she had given herself to the Devil, attended many a Sabbat, and so on and so forth. Boguet was uneasy about the whole business. Was there, he asked himself (and his readers whom he was addressing in French, not Latin), sufficient evidence to arrest Françoise and put her in prison, considering that the case against her rested on the testimony of an eight-year-old child, and that therefore proof of Françoise's crime should be 'clearer than daylight'? We may be able to sympathise with his predicament. Police investigating child abuse may have similar difficulties with an underage witness. The rest of the *Discours* is devoted to justifying the action Boguet took in this particular case. He ranges over every aspect of traditional and contemporary witchcraft and, in the usual manner of lawyers at the time, piles example upon example, many from cases he himself had judged, to prove the validity of each tiny point. The more examples, the more overwhelming the proof. Describing what witches do and what happens at a Sabbat takes Boguet many chapters; but he always returns to Françoise, as when he devotes chapter 45 to explaining the broken cross on her rosary, chapter 46 to her failure to shed any tears, chapter 49 to the reasons for having a suspect's head, and chapter 50 to a discussion of marks on a suspect's body; and finally, in chapter 73, he explains why atrocious crimes committed in secret require a less rigorous standard of proof than other offences.

The picture that emerges from his account is the one implicit in *Malleus Maleficarum*, a scene of battle between the armies of God and the forces of evil, with the Church and the law ranged with the former against witches and other practitioners of magic on the side of the latter. So, just as Institoris not only gave expression to his own anxieties but also to the contemporary fear of heresy, so Boguet clearly wrestles with the nagging personal doubts raised by the Maillat–Secretain case, and mirrors the post-Tridentine Catholic Church's determination to face down and eliminate the apostasy and schism which were linking hands with Satan all over Europe. This last he does by dramatising the conflict, almost as though the world were a stage and its inhabitants both actors and spectators of the immense conflict between Heaven and Hell, which was embodied in the struggle between the Church and God's enemies.[15]

A similar theatrical note, but one in a different context, can be seen in Pierre de Lancre's *Tableau de l'inconstance des mauvais anges et démons* (1612).[16] In 1609, he was instructed by Henri IV to go to the Pays de Labourd, a Basque-speaking region in the south-west of France, and deal with what appeared to be an appalling outbreak of witchcraft there. The subtext of his commission was that he should bring French law and French civilisation to an area every bit as foreign and exotic as the Indies or Japan. Since the Labourdin men spent months at sea, de Lancre found on arrival that he was dealing mainly with children and women. Most of the men were either old or priests, and few if any of the inhabitants spoke French, so he was obliged to rely upon interpreters, especially a young, ambitious priest called Lorenzo de Hualde who not only wished to collaborate with visiting French officials in order to further his career, but was also a firm believer in the reality of witches and every aspect of witchcraft. De Lancre was no fool, and he worried about his reliance on intermediaries; but he had no choice, and pressed rapidly ahead with the task he had been set. The *Tableau* is, in effect, his report to the French government, and is a highly personal account of a puzzled and disapproving stranger in an exotic land and society. Because of the lack of men, he was obliged to take evidence from girls and boys and women. These painted for him an extraordinary world in which witchcraft was the norm and everybody, sooner or later, attended a Sabbat and worshipped the Devil. The apparent ubiquity of Satan and the ready acceptance that witches were everywhere produced in de Lancre a fascinated horror which springs vividly from the pages of his report and makes it comparable with the accounts of early travellers to the Americas or the Far East where everything was different and needed interpretation to make it comprehensible. But while those early travellers tended to turn to writers such as Herodotus or Pliny to make sense of the societies they were in the process of discovering, de Lancre explained to a French government what was happening in the Basque country by referring to the experience of witches in France itself and in other European countries. Finding himself in the hands of an interpreter who accepted the reality of witchcraft, a king who was determined to have the rule of French law imposed, by force if necessary, on an un-French and unruly portion of his realm, and teenage witnesses whose testimonies, regardless of whether they were to be trusted or not, were virtually all he had to work with, de Lancre did what was expected of him and set about extirpating the diabolic menace as rigorously as he could. So relentless was he, in fact, that the local archbishop wrote to the king, asking that de Lancre be recalled to Paris – an interesting parallel with the situation between Institoris and the Bishop of Brixen.

One feature of the *Tableau* stands out: the alleged involvement of priests in the Sabbats. Many were accused of attending them; several were said to have celebrated parodies of the Mass there. De Lancre was shocked and scandalised. If the Church itself could not be trusted to preserve its courage and integrity in this war against the Devil, what hope was there for humanity? From this sense of bewilderment and confusion, therefore, comes the title of de Lancre's book, 'A Picture of the Instability of Wicked Angels and Demons'. The first half of the seventeenth century was marked by a threefold crisis. Nascent political absolutism was faced by the ideal of power based on natural law and a contract between ruler and ruled; confessional ruptions were encouraging the impulse to transcend dogmatic difference and concentrate religious experience on personal interiority; and investigation of the natural world was divided between supporters of an enclosed, mechanistic universe and those who argued that the structure of the world argued reason, plan and purpose, and therefore an ultimate Planner. 'Instability' was thus a keynote of the age. De Lancre was able to recognise few landmarks in the bizarre and disturbing society into which he had been sent, and even the Church appeared to have lost its constancy. So rigorous imposition of French law – and de Lancre emphasises this point – seemed to be the only way to make sense of the barbarism and bring order out of chaos, an essentially political answer to what had usually been seen as a theological problem.

By the beginning of the seventeenth century, witchcraft treatises had become more or less self-referential. Each succeeding work quoted extensively from its predecessors, and a hundred years of theological and legal effort had built up a body of literature which very few would be happy to challenge. Moreover, its premises and conclusions were self-evidently supported by the acknowledged consistency of witches' confessions from all over Europe, and therefore the fact that witches were real, worked real harmful magic and belonged to a real diabolical conspiracy directed towards human ruin, formed a consensus which was scarcely to be questioned. Contemporary *exempla* and appeals to *experientia* in the works themselves reinforced this notion of a given truth which transcended geography and time, and linked the present with past experience, thereby creating a seamless continuity. This is not to say, of course, that doubts over a multitude of individual points were not expressed. Detailed reservations, indeed, were an integral part of the whole discussion of witchcraft from at least the tenth century onwards, and there is scarcely an author who does not strike some kind of negative note here and there in his argument, or raise a fundamental problem – how witches can be present at a Sabbat while they seem to be asleep at home, for example – which involves investigation of the relative powers of God and Satan, the relationship between physical and non-physical states of being, and the nature of reality and illusion. It is because of the intellectual uncertainties inherent in what one may call the demonological universe that Stuart Clark is right to say that these disquisitions on witchcraft and other forms of magic stood at the very frontiers of research, pushing the boundaries of knowledge to what seemed to be their limits.[17]

Indeed, in many seventeenth- and eighteenth-century eyes, those limits were often passed. The Dutch Protestant minister, Balthasar Bekker, published *De Betoverde Weereld*, a large work in four volumes, between 1691 and 1693. He was a follower of Descartes and therefore dismissed any notion that there could exist any powers which might limit or counter the power of God himself. Consequently, he concluded, if Satan actually exists, he is more of a thought or impulse, and can have no influence, of himself, upon humanity or the created world; and this being so, the notion of such things as the demonic pact, copulation with witches and all the rest of the phenomena associated with and described by those accused of witchcraft must necessarily be fantastic rather than true. Indeed, to believe in such things is a kind of blasphemy. In one sense, Bekker was preaching to the converted for the Dutch republic had long since ceased to prosecute witches. But his argument was already less about witchcraft than biblical interpretation, and so, not surprisingly perhaps, he made his principal impression on his fellow-clergymen. What struck them, however, was not the application of Cartesian reasoning to the problems inherent in demonological beliefs, but how close he, and through him his readers, must inevitably sail to what was then called 'atheism', that is, the acceptance of a mechanistic view of creation which, it might be argued, stood in no need of divine or non-human guidance, and in which neither God nor angel nor any other kind of spirit played an active part. Expulsion from his ministry in the Dutch Calvinist Church therefore quickly followed.[18]

This cry of 'atheism' raised against those who were pressing hard their doubts about demonology, however, is somewhat misleading. Bekker's arguments may have been radical in some of their conclusions, but he was essentially arguing for what Levack has called 'a proper historically contextualised interpretation of those Scriptural passages that referred to witches and devils',[19] and it was because he was at odds with many of his contemporaries in a long-standing debate within the Dutch Reformed Church over the nature of the Church and the proper interpretation of Scripture that he lost his job, not because he was some kind of champion of Enlightenment rationalism. Far from being anything of the sort, he was a religious controversialist quite in the traditional manner, whatever the effect of his arguments on demonological belief. In England, too, 'atheism' was used as a convenient stick by one set of political and religious propagandists wherewith to beat their political and religious opponents. As a Whig clergyman, Francis Hutchinson, put it in his *Historical Essay Concerning Witchcraft* (1718):

> These doctrines have often been made party-causes both in our own and other nations. One side lays hold of them as arguments of greater faith and orthodoxy and closer adherence to Scripture, and calls the other Atheists, Sadducees, and Infidels.[20]

Hutchinson himself was engaged in controversy of a relatively old-fashioned type with Richard Boulton who had produced *A Compleat History of Magick, Sorcery and Witchcraft* in 1715, reiterating and defending most of the standard concepts of witches and witchcraft, and the judicial methods which should be adopted to suppress them. Hutchinson's treatise on witchcraft repeats the usual objections raised by Weyer and other sceptics, but is less a theological disquisition and more a contribution to contemporary polemics, betraying as it does a preoccupation with certain immediate questions facing his own society. In his case, this meant the delicate problem of not upsetting the Scottish Presbyterian establishment so recently incorporated politically into England by the Act of Union (1707), since the Kirk was (or was thought to be) more wedded to acceptance of the reality of witchcraft than its English counterpart. More notably, however, Hutchinson tended to equate belief in witchcraft with religious 'enthusiasm' and factionalism, anathema to the good order and civility he identified with the new Hanoverian regime – demonology was, in effect, 'Tory ideology'; and by assuming that witchcraft and other types of superstition belonged to the lower classes, and were therefore causes of both anxiety and snobbish disdain in their betters and rulers, he also managed to repeat the old-fashioned but effective slur which suggested a close link between religious and intellectual backwardness and Catholicism.[21]

Hutchinson's appeal to history is a major feature of his book. Nine of his 15 chapters are concerned in one way or another with showing that witchcraft and related magical operations have been a thorn in society's side from the very earliest times. Thus, chapter 2 gives a chronological sequence of trials and executions of witches, starting with biblical history and then shading into modern examples, the last in 1707. Chapters 4–10 use specific historical cases – all except two (1593 and 1712) dating from the seventeenth century – as the bases of discussions between the three persons of his dialogue – an English clergyman and a juryman, and a Scottish advocate – in order to show the absurdity, fraudulence or incredibility of belief in witches. These chapters are salted with anti-Catholic remarks, and end with an almost Panglossian thankfulness that England has passed from the darkness of superstition into the light of scientific advance.

His format bears the mark of his immediate predecessors. Richard Bovet's *Pandaemonium* (1684) makes a similar appeal to very recent history – the second part of his book consists of 15 anecdotes about witches, demons and ghosts dating from 1667 to 1683 – but begins with the by now customary use of the Bible and Classical authors to indicate the antiquity of magic in human society. Like Hutchinson, he includes a good dose of anti-Catholicism – chapters 7 and 8 are, in fact, an example of history as rant – but unlike Hutchinson, he intends to show that those who deny the reality of witches and witchcraft are, in effect, denying God; so the appeal to biblical and recent history by both authors is made for entirely contrary ends, although both are making use of appeals to Scripture and history very much after the controversialist model of a hundred years earlier. Del Rio, for example, provides a similar chronological survey drawn from sacred and profane literature dating from the first to the seventeenth century, as part of an overt attack on those, such as heretics, who are impudent enough to doubt the existence of spirits and visions,[22] an additional subject which Bovet found important. So did Joseph Glanvill, the second part of whose *Saducismus Triumphatus* (1681) furnished 16 'proofs from Holy Scripture' of the reality of apparitions, spirits and witches, and 28 narratives dating from the second half of the seventeenth century for the same purpose, although 19 of these deal almost entirely with ghosts.

Glanvill, however, differs from Bovet and Hutchinson inasmuch as his book actually consists of more or less unedited materials, published after his death, which he had gathered for a scientific study. He was both a clergyman and a Fellow of the Royal Society, and was keen to have the Society investigate witchcraft and other occult phenomena by the most up-to-date scientific methods, just as though magic and so forth were subjects like any other. His book is thus an assemblage of data intended for the laboratory, so to speak, and as such has plundered Scripture and history after the style of (let us say) contemporary botanists scouring foreign countries for specimens which can be brought back for closer examination in the quiet of the study. Glanvill's aim, unlike that of Bovet or Hutchinson, was not controversy but investigation, analogous to the enquiries of the natural philosopher, from Aristotle and Roger Bacon to his own time – in Glanvill's case, an enquiry into the relationship between the spiritual and material worlds, and the interpenetration of the latter by the former.

A similar kind of purpose, although carried out in very different fashion, can be seen in the work of Christian Thomasius (1655–1728) who, very early in his university career, was accused of encouraging atheism among his students at Frankfurt-am-Oder on the grounds that he wanted to diminish the importance of theology. In 1701 he published 56 theses, *De Crimine Magiae*, largely devoted to legal procedure in connection with trials for witchcraft, but also inevitably considering such questions as whether or not the Devil actually exists. Thomasius had read both Bekker and Spee (whose arguments can clearly be seen informing his work), but whereas Bekker had more or less denied the reality of the Devil as an independent spirit, Thomasius preferred to straddle between those (such as himself) who accepted Satan's existence and those (such as Bekker) who were ready to argue him into oblivion. Thomasius was no Cartesian, either. He saw 'spirit' as a vivifying force without which matter is merely dead; but he also argued that, since the Devil is a spirit, he must keep to his own realm, and therefore cannot enter or have an effect upon the world of matter. Consequently, the diabolic pact, demonic copulation, and so forth, are fantasies, not realities, which means in turn that accusations against individuals, claiming that they had taken part in such actions, could not be sustained by the legal system. In 1705, Thomasius followed these arguments with a work in which he condemned the use of torture in witch interrogations, and in 1712 an expanded version of his thesis, *De Origine ac Progressu Inquisitorii contra Sagas*, particularly notable because he employed history to show that witches had not always been subject to legal prosecution of the type standard during the early modern period, and that their crimes were in fact 'invented' by those who examined them.

This kind of appeal to the past, however, was unusual and Thomasius had borrowed it, so to speak, from his reading of the works of a Dutch physician, Antonius van Dale, whose *De Origine ac Progressu Idolatriae et Superstitionum* had clearly influenced Thomasius' choice of title for his won work. Van Dale had argued therein that historical investigation could show that fear of the Devil had been sustained in the common people by priests and rulers in order to secure their own power and dominance. But the thrust and modus arguendi of witchcraft treatises tended to be that of the Renaissance. The recorded past was plundered for examples which would bolster the continuing truth of a present situation, and was interpreted, edited and revised to provide evidence of the point the author intended to argue. To that extent, witchcraft treatises existed in an enclosed world where Creation and the Last Judgement marked a known beginning and ending, and in which the material world shaded by known and perceptible degrees into the worlds of spirit and divinity. The developing intellectual love-affair with mathematics which, with increasing rapidity, informed other types of investigation into nature – physics, chemistry, cosmology, and so on – never touched witchcraft, however much it may have modified and advanced enquiry into other aspects of demonology. Witchcraft treatises, as opposed to disquisitions on magic or divination or possession, for example, concerned themselves essentially with practical questions: How should the law be applied to accused witches? In what points is my religious/political opponent mistaken or wrong in his opinions? Only when they concerned themselves with the basic question, Is witchcraft real or not?, did they begin to break out of their self-referential world, and then, as Stuart Clark has pointed out, they found that other modes of thought were offering what seemed to be more attractive ways of answering it.[23]

notes

1. Michel de Certeau, *The Writing of History*, English trans. (New York: Columbia University Press, 1988), pp. 2, 4.
2. Isaac de la Peyrère, *Praeadamitae*, (n.p. 1655), p. 19; Paracelsus, *Astronomia Magna* in *Sämtliche Werke*, ed. K. Sudhoff, Vol. 12 (Munich and Berlin, 1929), pp. 1–144.
3. Thomas Wright of Durham, *Second or Singular Thoughts Upon the Theory of the Universe*, ed. M. A. Hoskin (London: Dawsons of Pall Mall, 1968), pp. 72–3.
4. Simon Ravensberg, 'De magia', in G. Voetius, *Selectarum Disputationum Theologicarum Pars Tertia* (Utrecht, 1659), pp. 539–632 (his lecture was delivered on 2 April 1636); Thomas Vaughan, *Magia Adamica* (London, 1650), published under the pseudonym Eugenius Philalethes in *Works*, ed. A. Rudrum (Oxford: Clarendon Press, 1984), pp. 139–234.
5. René Descartes, *Discours de la Méthode* (Paris: Librairie Garnier, 1916), p. 5; Thomas Hobbes, *Leviathan*, ed. R. Tuck (Cambridge: Cambridge University Press, 1990), Part 1, chapter 7 and chapter 5.
6. *Conseils à un journaliste* (1737), 'Sur l'histoire'.
7. See further Eric Wilson, 'The Text and Context of the Malleus (1487)', unpublished PhD thesis (Cambridge, 1990), pp. 122–31. Wolfgang Behringer, G. Jerouschek, and W. Tschacher (eds), *Heinrich Kramer, Der Hexenhammer* (Munich: Deutscher Taschenbuch Verlag, 2000), pp. 35–7. On date of publication see pp. 22–7.
8. *Malleus*, Part 1, questions 1, 2 and 6; Part 3, question 1.
9. Pieter Spierenburg, *The Broken Spell* (Basingstoke: Macmillan, 1991), p. 117. As can be seen, Spierenburg accepts a dual authorship of the *Malleus*.
10. Hartmann Ammann, 'Der Innsbrucker Hexenprocess von 1485', *Zeitschrift des Ferdinandeums für Tirol und Vorarlberg* 34 (1890) 10, 12, 15, 18, 20–1, 22.
11. For this see C. D. Gunnoe, 'The Debate Between Johann Weyer and Thomas Erastus on the Punishment of Witches', in J. van Horn Melton (ed.), *Cultures of Communication from Reformation to Enlightenment* (Ashgate: Aldershot, 2002), pp. 257–85.

12. This last reference is ambiguous. *Veneficium* means 'poisonous substance', 'the act of poisoning' and 'a magical operation involving actual or potentially poisonous substances'. Wier discusses all of them. See Book 3, chapters 35–9; Book 4, chapters 29–30; Book 6, chapters 24, 26.

13. Cf. Johann Meyfahrt, a Protestant theologian, who also addressed his treatise, *Christliche Erinnerung* (1636), to the German princes.

14. On Boguet, see S. Houdard, *Les sciences du diable* (Paris: Les Editions du cerf, 1992), pp. 105–59.

15. See Houdard, *Les sciences du diable*, pp. 138–59, especially her closing paragraph.

16. See P. G. Maxwell-Stuart: *Witch Hunters* (Stroud: Tempus, 2003), pp. 32–57; Houdard, *Les sciences du diable*, pp. 161–216; J. Dusseau, *Le juge et la sorcière* (Luçon: Editions Sudouest, 2002); M. M. McGowan, 'Pierre de Lancre's Tableau de l'inconstance des mauvais anges et démons', in S. Anglo (ed.), *The Damned Art* (London: Routledge & Kegan Paul, 1977), pp. 182–201.

17. Stuart Clark, *Thinking With Demons* (Oxford: Clarendon Press, 1997).

18. See further G. Stronks, 'The Significance of Balthasar Bekker's The Enchanted World', in M. Gijswijt-Hofstra and W. Frijhoff (eds), *Witchcraft in the Netherlands from the Fourteenth to the Twentieth Century* (Rotterdam: Universitaire Pers, 1991), pp. 149–56.

19. Brian Levack, 'The Decline and End of Witchcraft Prosecutions', in Bengt Ankarloo and Stuart Clark (eds), *The Athlone History of Witchcraft and Magic in Europe: The Eighteenth and Nineteenth Centuries* (London: Athlone Press, 1999), p. 38.

20. Francis Hutchinson, *Historical Essay Concerning Witchcraft* (London, 1718), p. 181.

21. See further Ian Bostridge, *Witchcraft and its Transformations, c.1650–c.1750* (Oxford: Clarendon Press, 1997), pp. 143–54; Roy Porter, 'Witchcraft and magic in Enlightenment, Romantic and Liberal Thought', in Ankarloo and Clark, *Athlone History of Witchcraft and Magic in Europe*, pp. 205–8.

22. Del Rio, *Disquisitiones Magicae*, Book 2, question 26, section 5.

23. Clark, *Thinking With Demons*, p. 686.

3

science, medicine and witchcraft

peter elmer

In attempting to understand the origin of the witch trials and the emergence of a demonological tradition in early modern Europe, writers on witchcraft have consistently sought to ascribe a role for science and medicine in that process. Until fairly recently, two broad assumptions have informed this literature. Firstly, that the rise of the mass persecution of witches, and the accompanying set of ideas which underpinned those trials, were in part the product of the superstitious, backward-looking and erroneous state of post-medieval science and medicine. And secondly, that the demise of witch trials and beliefs can be largely accounted for by the overthrow of antiquated scientific and medical opinion following the 'Scientific Revolution' of the seventeenth century. Since the 1960s, aspects of this overarching explicandum have been slowly challenged and modified. It is only in the last decade, however, that a thoroughly revisionist account has emerged which has not only challenged the assumption that witchcraft was essentially a by-product of early modern science and medicine, but has also posed major objections to the idea that witchcraft was effectively argued out of existence by the onset of scientific, and to a lesser extent medical, innovation and change. This chapter seeks to chart the wider political and intellectual currents that have informed these interpretative traditions and to assess their contribution to our present understanding of the place of witchcraft in early modern history.

The idea that belief in witchcraft and the persecution of witches was largely a product of a faulty worldview, predicated on outmoded and discredited notions of the natural world, is an old one that dates back to the early eighteenth century and the final age of the European witch trials. In England, for example, where the 'new science' of Boyle and Newton epitomised the victory of the 'moderns' over the 'ancients', numerous spokesmen for the former were quick to assert that one of the positive achievements of this revolution was its role in undermining belief in witches, demons and devils.[1] The widespread victory of Enlightenment values, of which scientific and medical progress were staple elements, thus sealed the fate of antiquated superstitions like magic and witchcraft. This was the view propagated by the *philosophes* of the eighteenth century, most notably by Voltaire, and subsequently bequeathed to nineteenth-century writers on witchcraft, who were only too eager to consign such beliefs to the dustbin of history.[2]

During the course of the nineteenth century, such views were consistently repeated and developed by historians trained in the positivist methodology of Comte, who subscribed to the view that the age of the Industrial Revolution marked the peak of human development and a golden age for reason, science and progress. Typical of this 'triumphalist' approach to the past was the widely cited opinion of the English historian, William Lecky, who in 1865 depicted the 'witch craze' as a struggle between reason and science on the one hand, and intellectual obscurantism and religious bigotry on the other. Famously, he went on to announce that the destruction of witchcraft marked 'the first triumph of the spirit of rationalism in Europe', a view that echoed through the works of historians of witchcraft for the next 100 years at least.[3] It is particularly evident in the first attempts to investigate more fully European witchcraft using the empirical methodology of the new history, which were produced in the nineteenth and early twentieth centuries on both sides of the Atlantic. In Germany, for example, the pioneering scholarship of Wilhelm Gottfried Soldan and Joseph Hansen fits this pattern.[4] Soldan's work has been described as 'a standard rationalist account', in which the end of the trials is said to mark 'a vital stage in human progress' and an end to the oppressive and overweening authority of the Roman Catholic Church, which Soldan blamed for the superstition. Similar sentiments reappear in the ground-breaking archival work of Hansen, 'a bitterly anti-clerical archivist' from Cologne.[5]

The idea that witchcraft was essentially the product of the unscientific and religious temperament of the age was most fully developed, however, in the writings of American and French historians of this period. In both, liberalism, rationalism and anticlericalism combined to create a powerful tradition of historical writing in which opposition to witchcraft became indelibly linked with the forces of scientific and medical progress. In the United States of America, the focus lay on the historic battle between religious dogmatism and science – a conflict that mirrored continuing tensions in ante-bellum America surrounding the place of religion in the political and intellectual life of the nation. The conflict itself was largely played out in the popular press, but the battle lines were drawn in the campuses of America's universities, where a number of academic historians sought to enlist witchcraft as a stick with which to beat their fundamentalist opponents. James Russell Lowell, for example, a distinguished Harvard professor and editor of the *Atlantic Monthly*, published a lengthy essay on witchcraft in which he lauded the victory of rational science over irrational theology.[6]

More significant, perhaps, was the contribution of Andrew Dickson White, the first president of a non-sectarian college established by Ezra Cornell at Ithaca in up-state New York. White spent much of his time defending his new institution from the attacks of neo-conservatives and religious fundamentalists, who, he believed, posed a threat not only to Cornell, 'but to the very idea of free thought and its paramount expression, science, which he and most other liberal rationalists held to be central to their philosophical concerns'. His most celebrated work, *A History of the Warfare of Science with Theology in Christendom* (1896), not surprisingly contained a section on the witch craze, which he ascribed to the 'dogmatic theology' of the age. He also asserted that witchcraft's disappearance was a direct result of the Scientific Revolution, '[t]he newer scientific modes of thought, and especially the new ideas regarding the heavens' undermining 'the whole domain of the Prince of the Power of the Air'.[7] The legacy of this liberal rationalist tradition in America is most evident, however, in the work of two of the most influential historians of witchcraft of this era, George Lincoln Burr and Henry Charles Lea. Both men subscribed to the view that science and witchcraft were incompatible, at opposite ends of the spectrum from each other, and that the latter was ultimately undone by the advance of the former. They also did much to perpetuate another enduring belief among subsequent generations of witchcraft scholars, namely that participants in the contemporary debates surrounding witchcraft could be easily placed in one of two camps – the credulous or sceptical – depending on their scientific, intellectual and moral outlooks.[8]

The anticlericalism and secularism that formed such a vital ingredient of early American historical responses to witchcraft, and helped to propagate the idea that witchcraft was inimical to 'true' medicine and science, is equally evident in nineteenth-century France where, if anything, the political ramifications of such a stance were even more hotly contested. Here, the principal source of opposition to the pretensions of the Catholic Church came not so much from liberal academics, but from certain quarters of the medical profession. In particular, interest in historical witchcraft coalesced around a group based at the Salpêtrière asylum for women in Paris, led by the innovative neurologist, Jean-Martin Charcot (1825–1893). Under the inspiration of Charcot, a large body of work was produced comparing historical accounts of witchcraft, possession and ecstatic religious experience with contemporary cases of hysteria. Employing the concept of 'retrospective medicine', a term first coined by the physician and radical republican, Emile Littré, in 1869, Charcot and his colleagues began a vast historiographical project, which reached its peak in the years between 1882 and 1897 with the publication of eight well-known tracts of early modern witchcraft under the direction of Désiré-Magloire Bourneville.[9]

Among the texts reproduced by Bourneville was Johann Weyer's *De Praestigiis Daemonum* (*On the Tricks of Demons*) of 1563. Prior to the French edition of this work, which was published in 1885, Weyer's work, in which he provided a wide range of medical and other explanations for the behaviour of alleged witches, was little known among either physicians or historians. However, its reappearance at this time – probably inspired by a series of lectures given to the Paris Faculty of Medicine by Alexandre Axenfeld in 1865 – was destined to catapult Weyer to a position of some prominence, indeed hero status, in the eyes of those who contrasted his sceptical, humane and scientific approach to witchcraft with that of his credulous contemporaries. Moreover, as an enlightened physician, battling against the superstitions of his time, particularly those such as witchcraft that were widely credited as the brain-child of the clergy, Weyer stood out for the rationalist Charcot and his anticlerical colleagues as a model for his profession.[10]

During the first half of the twentieth century, little occurred to disturb this post-Enlightenment, rationalist account of witchcraft outlined so far. If anything, it gained wider credence in academic circles at this time as it was endlessly repeated and popularised in general histories of the early modern period. Belief in witchcraft was characteristically depicted as unscientific and irrational, the product of medieval superstition and obscurantism. Its decline was automatically assumed, though with little corroboratory evidence, to have proceeded from the advance of science and medicine in the second half of the seventeenth century. This triumphalist account received widespread and uncritical acceptance precisely because it accorded so closely with contemporary historiographical conventions that viewed human progress in all spheres of life as both natural and ineluctable.[11] It was nonetheless the case that the 'Whig' view of history was particularly applicable to the history of science and medicine, where man's discovery of the truths of nature was typically depicted as a heroic struggle in which individual geniuses such as Galileo and Harvey laboured bravely, and often against the odds, to uncover the secrets of the universe.

Historians of witchcraft reared in this tradition not unnaturally sought their own heroes, who fought against the tide in order to expose the hypocrisy and deceit of the priests and witchmongers. They soon found their man in the shape of the Dutch-born physician, Johann Weyer. Despite the interest of Charcot and his associates at Paris, Weyer's reputation as the exemplary sceptic did not gain wide credence until the publication of Gregory Zilboorg's *The Medical Man and the Witch* in 1935. Part of a wider history of medical psychopathology that was not published until 1941, Zilboorg focused on the period of the Renaissance in general, and on Weyer in particular, in order to elucidate the early origins of his own profession, psychiatry. Weyer, as a witchcraft sceptic, was accorded a key role in this process. In the words of the medical historian, Henry Sigerist, who wrote a preface to Zilboorg's first work, it was '[i]n the changing attitude towards witchcraft [that] modern psychiatry was born as a medical discipline'.[12]

Zilboorg's work, particularly his espousal of Weyer, demonstrates numerous facets of the 'presentist' approach to the past. This is most obviously apparent in his lauding of Weyer as the 'founder of modern psychiatry' or first true psychiatrist. Reading his works, 'one gains the impression that a new man, a new type of individual, has entered upon the scene of medicine'. He is curious, methodical and systematic in his medical practice, betraying all the distinctive features of the modern-day psychiatrist. He is also a great clinician, whose case studies are so insightful that Zilboorg claims it is possible to detect in his writings an unchanging psychopathy, including a host of recognisable mental illnesses such as depression, megalomania and schizophrenia. Weyer is thus credited with being the first man to diagnose accurately the mental illnesses that afflict early modern men and women, and to reject the theologians' view of them as proceeding from the Devil. Everywhere, he is seen to anticipate future developments in psychiatry, more often than not without being conscious of the fact:

> He knew that the answer to the problem was to be found in medicine; he knew that the doctors were woefully inadequate; he knew also that medicine itself was not yet ready to meet the problem. He felt that a psychiatry must be created, and he sensed that if it was to be created it would be brought about by medicine and not by theology, philosophy, or jurisprudence ... Thus, although he left us no system, no well-rounded theory – in fact no theory at all – he actually viewed the psychopathological problems of his day with an empirical matter of factness which proved revolutionary in the history of medical psychology.[13]

It is fairly clear with the privilege of hindsight that Zilboorg's hagiographic approach to Weyer says more about Zilboorg's anxiety and concern for his nascent profession, psychiatry, than it does about the historical reality of Weyer's medical practice and his attitude to witchcraft. Writing in the 1930s, Zilboorg was acutely aware of the uncertain status of his profession and the contempt in which it was held by vast swathes of the population.[14] In order to achieve respectability, it was crucial to create an image of the psychiatrist, and his history, that was heroic in scale and which exceeded that normally ascribed to the physician and scientist. Accordingly, Zilboorg characterised the work of the former as relatively straightforward in that the existence of physical illness and the corresponding need for medicine and the learned physician was self-evident. He then contrasted this with the problems faced by earlier psychiatrists like Weyer who had to convince the 'insane' that they were ill and so create the specialism of psychiatry 'uninvited and against the terrible odds, against the will of the public, against the will of established legal authority and against the will of a variety of established religious faiths'.[15]

Despite Zilboorg's claim to write with 'detachment' about his subject, and to eschew the 'undercurrent of condescending admiration' which he believed typified most history of medicine, Zilboorg was clearly working to his own agenda. It is a measure of his success that his claims for Weyer and psychiatry were rapidly assimilated by academic colleagues, both within his profession and outside, and soon took on the status of undisputed fact.[16]

It is difficult to underestimate the extent of Zilboorg's influence on subsequent generations of historians of medicine and psychiatry. The 'foundation myth' surrounding the role of Weyer in the field of psychiatry was endlessly repeated, particularly in manuals of abnormal psychology.[17] It also featured prominently in general histories of science and medicine, where Weyer's heroic stand against the theologians and demonologists fitted neatly into the whiggish, great man approach that informed so much writing in this historical subdiscipline. George Sarton, for example, one of the leading exponents of the new history of science, repeatedly invoked the notion of a conflict between religion and science in his discussion of the place of the 'witchcraft delusion'. For Sarton, Weyer was the first to introduce the 'medical point of view' in his treatment of witchcraft, and 'thanks to him and his followers, "witches" are now considered mental cases and treated as such'. Accordingly, the Renaissance represented an ambiguous phase in the development of science, one in which courageous iconoclasts such as Weyer fought against the tide in attempting to combat the general superstitions of the age. Typically, he is portrayed as a pioneer, exploring new ground for others to follow and expand.[18] Later writers working in this historical tradition sought to uncover Weyer's 'disciples' and followers, and to trace the heritage of Weyer's scepticism through subsequent generations down to the age of Enlightenment. In England, for example, Hunter and Macalpine discovered an English Weyer in the shape of the early seventeenth-century physician, Edward Jorden, whose work they claimed represented 'the first book by an English physician which reclaimed the demoniacally possessed for medicine'.[19] The American psychiatrist and historian, Oskar Diethelm, was yet more ambitious. In an article published in 1970, he attempted to break new ground by demonstrating the slow and corrosive influence of Weyer's ideas on the medical profession as a whole which, by 1700, he depicted as broadly sceptical and antagonistic toward belief in witchcraft and witch hunting.[20]

The example of the historical treatment of Johann Weyer epitomises the whiggish and optimistic tone of much of the writing on witchcraft produced by historians of science and medicine in the first half of the nineteenth and second half of the twentieth centuries. Weyer stands for progress, secularism and the rights of the individual in the eyes of his admirers. He was, in Zilboorg's memorable phrase, 'a true forerunner of what we would call today the conservative, neighbourly, middle class'. Weyer's appropriation in this way could also be used as a warning to those who took for granted the benefits that have accrued to modern societies as a result of the victory of science and medicine over dogmatic religion and superstition. The cautionary tone adopted by Zilboorg in defence of his beleaguered profession was echoed by some of his like-minded colleagues, who wished to adopt Weyer's exposé of the evils of witch hunting in order to prosecute and allay modern demons such as the holocaust and racial prejudice.[21] In the case of historians of science like Sarton, this could often reach apocalyptic levels; as, for example, in the following extract with which he concluded his section on the place of witchcraft in European history:

> Irrelevant learning is no longer a danger today, but the fight against superstition, intolerance, and endemic irrationalism must be continued … We cannot relax. And we must continue also to fight against lying and hypocrisy. Even now there are too many people, good citizens, pillars of society, who are afraid of science because they are afraid of truth …
>
> The Renaissance was an age of superstition, but so is our own, under the surface; science had made gigantic progress in certain fields, but in others, e.g. in politics, national and international, we are still fooling ourselves.
>
> The history of science is not simply the history of discoveries and new ideas that are closer to reality; it is also the history of the defence of these ideas against recurrent illusions, and lies. We must replace darkness with light; that is the main function of science.[22]

Writing in the mid 1950s, in the aftermath of the horrors of the Holocaust and the Second World War and the onset of the Cold War, Sarton's defence of science against the forces of superstition betrays a loss of that earlier confidence and optimism that characterised earlier accounts of the rise of science. The events of mid century cast a shadow over the western liberal democracies, and generated serious doubts as to many facets of social and intellectual life that were taken for granted by earlier generations. The expression of doubt took many forms in academia, and included a revival of interest in religion and its historic relationship to science. The work of scholars such as Alfred North Whitehead and Robin Collingwood, for example, now argued for what Andrew Fix has described as 'the constructive influence of religious ideas in the formation of many of the key concepts of modern science'.[23] Fix cites as other examples of this new approach the work of the American scholar, Robert Merton, who formulated an intriguing and highly influential hypothesis linking the rise of modern science in the West with the ethos of Puritanism. This, in turn, provoked a new generation of historians to investigate the links between religion and science, where emphasis was placed on the 'creative interaction' between the two.[24]

The impact of this new approach, however, on the study of witchcraft and its relationship to science and medicine was limited. Witchcraft continued to be widely seen as an aberrant 'superstition', though less as the natural product of religion and more as the by-product of the pseudo-scientific frame of mind spawned by the occultism of the Middle Ages and the backward-looking Scholasticism of the universities. Decline remained the prerogative of the Scientific Revolution. During the 1960s, however, the first major challenge to this traditional picture was initiated by the work of the English scholar, Frances Yates. Yates rejected the verdict of earlier historians of science, who routinely castigated the so-called 'occult sciences' (alchemy, astrology and natural magic) as deleterious to the advance of true science, and proposed instead that this ancient tradition of thought, newly revived by Renaissance Neoplatonists, provided a vital impetus to scientific innovation and laid the groundwork for the Scientific Revolution associated with men such as Boyle and Newton.[25]

Witchcraft scholars were not slow to pick up on the ramifications of Yates' work that implied a rational core to ideas previously regarded as inherently illogical and unscientific. The work of the British historian, Hugh Trevor-Roper, stands out in this respect, as his essay on the European witch craze was considered by many as an important new milestone in the history of the subject.[26] Unlike his predecessors, Trevor-Roper questioned some of the basic assumptions of traditional Whig history that had characterised European history as a period of unrelenting progress toward the present. In particular, he cited the recrudescence of new forms of barbarity in the middle decades of the twentieth century, particularly anti-Semitism, as a warning to historians to 'think again' about witchcraft and to seek to understand it from within the context of the peculiar social and intellectual changes initiated by the Renaissance and Reformation. Rejecting present-centred history and adopting a relativist pose, he thus argued, with witchcraft and the 'occult sciences' in mind, that 'mental structures differ with social structures, [and] that the "superstition" of one age may be the "rationalism" of another'. One fruit of this new approach was Trevor-Roper's suggestion, prompted by Yates' positive re-evaluation of the place of the occult or hermetic philosophy in Renaissance science, that demonological scepticism was most likely to flourish among 'natural magicians' like Agrippa and Cardano, and 'alchemists' like Paracelsus and his followers. Support for the witchcraft 'delusion' was, on the other hand, most likely to emanate from the opponents of the natural magical tradition, particularly those trained up in the traditional scholastic and Aristotelian curriculum of the universities.[27]

Among other things, Trevor-Roper provoked renewed attention to the problem of science's precise relation to the onset of the 'witch craze', and its contribution, if any, to the demise of the trials. Trevor-Roper himself remained firmly attached to the view that 'what ultimately destroyed the witch-craze ... was not the two-edged arguments of the sceptics', but rather 'a philosophical revolution which changed the whole concept of Nature and its operations'. In the short term at least, few challenged this orthodoxy, though they did seek to develop and critique specific aspects of Trevor-Roper's analysis. Keith Thomas, for example, in his ground-breaking study of English witchcraft, cited both Kocher and Trevor-Roper to the effect that the vogue for Neoplatonism encouraged a scientific and medical naturalism that was antithetical to belief in diabolic witchcraft. Like Trevor-Roper and others before him, he also placed the 'triumph of the mechanical philosophy' at the core of any explanation for the demise of the witch trials and the demonological system that underpinned the persecution of witches.[28] Others, such as Brian Easlea and Thomas Jobe, while subscribing in general terms to the schema of scientific change outlined by Trevor-Roper and Thomas, have offered variant readings of how developments in early modern science may have impacted upon belief in witchcraft.[29] Both Easlea and Jobe were troubled by specific aspects of the general theory offered for the decline of witchcraft in late seventeenth-century England. Easlea found 'problematic' the role allotted to the mechanical philosophy by Thomas, particularly in the light of the failure of Cartesian 'mechanists' like Joseph Glanvill and fellow members of the newly founded Royal Society to repudiate belief in witches. Jobe, who likewise focused on the figure of the scientist-demonologist, Joseph Glanvill, contrasted his support for witchcraft with the profound scepticism of the hermeticist Paracelsian and radical sectarian, John Webster. In both cases, the failure of the 'mechanist' Glanvill to acknowledge the logic of his scientific affiliations and reject the existence of witches, demons and spirits was put down to the subversive religious and political implications of such a stance and the need to accommodate the new mechanical philosophy to the status quo.[30]

By focusing on the religious and political allegiances of key figures in the witchcraft debates, as well as their specific scientific orientation, historians of witchcraft were becoming increasingly aware of the pitfalls of anachronism and the present-centred methodology that had dominated so much earlier writing in this field. This was especially true in relation to the work and thought of the sceptical physician, Johann Weyer, who, as we have already seen, was widely lauded in medical and psychiatric circles as a worthy forerunner of Freud as well as a lone advocate for the 'mad' witch. By the 1970s, however, this view of Weyer, and early modern approaches to mental illness, were undergoing a dramatic sea change that would radically reorientate historical understanding of the medical aspects of witchcraft.

The challenge to this orthodoxy came from two directions. Firstly, it arose among historians of witchcraft, who, imbued with relativist ideas about the past, sought to contextualise Weyer's views on witchcraft and subject them to detailed analysis. Placed under the historical microscope, a new view emerged of Weyer that was radically at odds with his former image. Instead of the thoroughgoing sceptic portrayed by Zilboorg and his successors, Weyer was now depicted as a medical conservative, whose views were neither novel nor convincing, and whose approach to witchcraft differed little from that of his opponents. Particularly damaging to his standing as the leading exponent of scepticism was the renewed emphasis that was now placed on his inveterate belief in the existence of the Devil, and the latter's ability to work evil in this world through the intercession of willing accomplices in the shape of sorcerers or black magicians. Weyer may have exculpated old and poor women accused of witchcraft by offering naturalistic medical explanations, but he nonetheless remained wedded to a system of ideas that allocated a central role in the world of man to the Devil, demons and sorcerers.[31]

The debunking of Weyer, however, was not the work of historians alone. A second source of opposition arose from within the same subdiscipline, the history of psychiatry, which had originally done so much to establish Weyer's vaunted position among the pantheon of medical greats. Here, a number of forces were at work which facilitated a revisionary approach to Weyer's erstwhile heroic status in the profession of psychiatry. During the course of the 1960s, the emergence of the 'anti-psychiatry' movement, spearheaded by figures such as Thomas Szasz, encouraged members of the profession to re-examine their past as well as their present. Szasz himself paved the way by tracing the 'mythological' history of psychopathology from Pinel, through Charcot and Freud, to Zilboorg and his associates. He also alludes to the key role played by Zilboorg in perpetuating the notion of the witch as mentally ill.[32] Others soon followed. In 1977, Thomas Schoeneman, a clinical psychologist, focused specifically on the work of Zilboorg in establishing this historiographical tradition as an orthodoxy within the psychiatric profession and the teaching of abnormal psychology. Citing much recent work by scholars of witchcraft, as well as that of the polemicist Szasz, Schoeneman demolished Zilboorg's thesis that all witches were mad and argued for a more historically sensitive approach to the question of what it meant to be mentally ill in early modern Europe. Much the same approach was taken by the psychologist Nicholas Spanos, who in 1978 produced a devastating attack on the 'presentist' assumptions in Zilboorg's work, rejecting among others things the claims of Zilboorg and his followers that it was possible to apply modern psychiatric diagnostic labels to the various ailments suffered by witches and their early modern contemporaries.[33]

An important by-product of the debate surrounding Zilboorg, and the role allotted to Weyer in the history of psychiatry, was the tendency among historians to question many of the aprioristic assumptions about madness, and by extension witchcraft, that featured so prominently in the work of earlier scholars. During the 1980s, a number of pioneering studies suggested a radically different approach to these issues. Foremost among these was the work of Michael MacDonald, whose micro-historical study of the casebooks of a provincial doctor, Richard Napier, demonstrated the complex variety of responses to the diagnosis and treatment of the mentally sick in early modern England.[34] Applying an eclectic combination of approaches (ethnographic, anthropological and historical) to an analysis of his material, MacDonald avoided all use of anachronistic psychological jargon and allowed the afflicted to speak for themselves. His results provide fascinating reading. Witchcraft and madness, it becomes apparent, were clearly not synonymous in the eyes of both Napier and his patients. Early modern English men and women possessed a rich vocabulary of terms and medical symptoms through which they were able to express and understand their day-to-day fears and anxieties. In short, MacDonald had produced the first evidence-based study of the social and cultural construction of madness in pre-industrial Europe.[35]

The idea that mental illness was socially constructed – in much the same way that all other facets of life such as morality, religion and the law were created by human artifice – has proved of great consequence for historians of witchcraft. Rather than seeking to 'explain away' the abnormal behaviour routinely ascribed to witches and their victims, witchcraft scholars have become increasingly sensitive to the idea that such afflictions are, in the words of Erik Midelfort, 'too idiosyncratic and culture-bound to fit well into our secular and psycho-pharmacological notions of neurosis and psychosis'.[36] The religio-centric nature of witchcraft thus defies modern-day categorisation. It also strongly suggests for many working in this field today that if we wish to understand the real motives of physicians like Weyer and Jorden for publishing 'sceptical' works on witchcraft, then we need to acknowledge their particular debt to religious and other non-medical influences. Midelfort, for example, while playing down the originality of Weyer's psychiatric thinking, has identified his Erasmianism as a central element in his assault on the crime of witchcraft. Likewise, Michael MacDonald has recently claimed that earlier writers such as Hunter and Macalpine have misconstrued Jorden's motives for entering the witchcraft debate on the side of the sceptics. Though his diagnosis of a form of hysteria in the case of a suspected witch in 1603 is still heralded as medically significant, MacDonald nonetheless argues strenuously that his published book on the subject was first and foremost 'a work of religious propaganda'.[37]

During the course of the 1980s, there appeared a number of general textbooks devoted to a synthesis of the burgeoning literature on witchcraft that had accrued in the previous two decades. In most cases, these were slow to reflect the radical implications of the work of scholars like MacDonald and Midelfort in the history of madness.[38] Moreover, they continued to perpetuate many of the assumptions of the older literature, in particular the idea that the fate of witchcraft was inextricably tied to the onset of the Scientific Revolution.[39] Other works, such as that of Brian Levack, acknowledged the problematic nature of the relationship between science and witchcraft, and the limited role of Neoplatonism in fomenting scepticism and laying the groundwork for the onset of the mechanical revolution. The traditional historiography was therefore left largely undisturbed; even Weyer's image as the 'father of modern psychiatry' remained intact.[40] At the same time, however, more fundamental developments were beginning to reshape the work of historians of early modern science, which in turn would prove invaluable to students of witchcraft, and its relation to medicine and science. By the 1990s, the traditional picture of the Scientific Revolution as a linear process whereby the outmoded scientific worldview associated with Aristotelianism was first challenged by the revival of Neoplatonism and then jettisoned altogether (with the occult sciences) following the success of the mechanical philosophy was itself under threat. The work of Charles Schmidt confirmed the continuing vitality and flexibility of Aristotelian scholasticism in sixteenth- and seventeenth-century science.[41] Others demonstrated the fundamental eclecticism of early modern science, which made it impossible to categorise scientific practitioners into neat, ideological pigeonholes.[42] Early modern Europe was now perceived as an age of 'radical epistemological instability', which required, according to a new school of historians of science, a new vocabulary or terminology with which to adequately describe the activity of scientists in this period. 'Science' itself was now seen as a distorting anachronism; 'natural philosophy' more accurately described what early modern practitioners such as John Dee or Robert Boyle were doing when they sought to analyse and explain the divine mysteries of the natural creation.[43]

These insights have opened up new ways of understanding the role which science and medicine played in witchcraft and demonology. Its impact is most evident in relation to the fate of witchcraft in Restoration England, and the role of many of the leading 'scientists' of the day in providing intellectual and scientific justification for belief in witches, demons and spirits. Steven Shapin, for example, has stressed the compatibility of the 'godly' mechanism and natural theology of Robert Boyle and his associates in the Royal Society with belief in witchcraft.[44] Allison Coudert has offered a similar analysis with respect to Henry More.[45] However, the most radical and original challenge to the traditional historiography, which incorporates much of the new history of science, is to be found in Stuart Clark's *Thinking With Demons* (1997), a major reassessment of the place of demonology in early modern European thought.[46] In a long section devoted to the scientific status of witchcraft beliefs, Clark turns on their heads all previous assumptions on this topic. Formerly depicted as antithetical and conflict-bound, Clark now claims that science and witchcraft (traditionally categorised as species of 'superstition') should be seen as complementary and mutually reinforcing. In stressing the preternatural (that is, natural, but unexplained or 'occult') status of the acts ascribed to the Devil, demons and witches – a point substantiated by reference to a vast literature on the subject – he demonstrates the centrality of demonology to early modern science. Demonologists investigated the preternatural precisely because it promised to yield further understanding of the natural world and its operations. In this sense, the student of witchcraft was not simply engaged in the pursuit of esoteric knowledge, but was performing 'cutting edge' science. Amidst the 'conceptual chaos' that reigned in early modern natural philosophy, investigation of the powers attributed to demons and witches offered what Clark terms, following Francis Bacon, 'prerogative instances'.[47]

In a separate chapter, Clark also addressed the thorny issue surrounding the role of science in the decline of witchcraft and, in line with much else in the book, proposed a radical new approach to the problem. Whereas earlier accounts sought to excuse or apologise for the appearance of some of the most celebrated figures of the Scientific Revolution on the side of the demonologists, Clark, invoking the latest work in this field, sought to demonstrate that the witchcraft beliefs of men like Glanvill and Boyle were perfectly in accordance with their wider natural philosophical preoccupations. Citing the work of scholars such as John Henry and Keith Hutchison, he affirmed that mechanism, which lay at the heart of the Scientific Revolution, was 'a radically supernaturalist philosophy' that was ideally suited to the aims of men like Glanvill and Boyle, who wished to promote belief in witches and spirits as a bulwark against the twin dangers of atheism and enthusiasm. It also held a special place for the notion of occult causation – a precondition of demonology. Demonism then, just like Newton's concept of gravity, occupied a central place in the thought of many of the natural philosophers associated with the new science in late seventeenth-century England. Like gravity, it 'was held to be intelligible in its effects but not in its causes, something real and manifest as an "experienced" matter of fact but as yet unexplained'. Accordingly, witchcraft was designated as a suitable topic of investigation by certain members of the Royal Society, who wished to construct a natural history of the demonic by amassing well-attested and verifiable (that is, Baconian) matters of fact. The process, as outlined by Glanvill, was collaborative and empirical, and like other forms of knowledge studied there, it was neither dogmatic nor doubting, but based on the precepts of probability and reasonable hypothesis.[48]

With Clark, we have come full circle in a radical reappraisal of the traditional historiography of this subject. Where to from here? Adopting some of the approaches outlined here in the last few paragraphs it is apparent that leading figures in the so-called 'witchcraft debate' between sceptics and believers will continue to attract close reading of their supposed intentions in invoking scientific support for their views. A good example can be found in the recent work of Andrew Fix, who has re-analysed the motives of the leading Dutch Cartesian, Balthassar Bekker, for publishing on witchcraft in the 1690s. Frequently cited as an example of the rational sceptic, imbued with the spirit of the new philosophy, Fix instead locates the heart of Bekker's scepticism in his liberal, non-confessional Calvinism and his disdain for his dogmatic, intolerant and literalist co-religionists. Without disputing his debt to Cartesianism, Fix claims that it was always 'more a weapon to be used against religious authority than a deeply held conviction'.[49]

In other respects, however, particularly in relation to the issue of decline, one might expect a growing disinclination to pursue the role of the Scientific Revolution as a factor in that process. In many respects, this approach is already well established among present-day historians of witchcraft.[50] The trend for some time has been to view the scientific changes of this period as part of a wider process of change, variously described as a 'shift of temperament' or revolution in general worldview. Multiple explanations, rather than monocausal ones, are now the order of the day if one wishes to get to grips with the various strands of thinking that brought about the 'disenchantment of the world'. Changes in the law and legal procedure and the emergence of religious and political pluralism both offer fruitful, alternative explanations for the decline in witch hunting.[51] Finally, if science did play an active role in undermining belief in witches and spirits – a view seemingly endorsed by those who lived in the age of Enlightenment – it is perhaps on this later period that historians of science should now focus since it now seems clear that

> the period from Boyle to Newton – the one on which the historiography of the decline of demonology has hitherto been premised ... pointed backwards to the preoccupations of the later medieval theologians ... rather than forwards to the eighteenth century.[52]

notes

1. This was the view, for example, of Richard Bentley and Francis Hutchinson, writing in 1713 and 1718 respectively; cited in Keith Thomas, *Religion and the Decline of Magic* (London: Weidenfeld & Nicolson, 1971), p. 579, n.3.
2. For a brief but instructive overview of Enlightenment reactions to magic and demonology, and the encroaching medicalisation of witchcraft, possession and related phenomena such as religious ecstasy and enthusiasm in the eighteenth century, see Roy Porter, 'Witchcraft and Magic in Enlightenment, Romantic and Liberal Thought', in Bengt Ankarloo and Stuart Clark (eds), *Witchcraft and Magic in Europe: The Eighteenth and Nineteenth Centuries* (London: Athlone Press, 1999), pp. 219–35.
3. William Edward Hartpole Lecky, *History of the Rise and Influence of the Spirit of Rationalism in Europe*, 2 vols (London, 1865), Vol. I, p. 90. This work was printed in countless later editions.
4. Wilhelm Gottfried Soldan, *Geschichte der Hexenprozesse aus dem Quellen Dargestellt* (Stuttgart: Cotta, 1843); Joseph Hansen, *Zauberwahn, Inquisition und Hexenprozess in Mittelalter und die Entstehung der Grossen Hexenverfolgung* (Leipzig: Oldenbourg, 1900); Hansen, *Quellen und Untersuchungen zur Geschichte des Hexenwahns und der Hexenverfolgung im Mittelalter* (Bonn: C. Georgi, 1901).
5. E. William Monter, 'The Historiography of European Witchcraft: Progress and Prospects', *Journal of Interdisciplinary History* 2 (1971–72) 435; Ronald Hutton, *The Triumph of the Moon: A History of Modern Pagan Witchcraft* (Oxford: Oxford University Press, 1999), p. 132. Chapter 8 of this work ('Finding a Witchcraft') contains a very useful summary of the post-Enlightenment and nineteenth-century historical reaction to witchcraft.
6. James Russell Lowell, 'Witchcraft', in Lowell, *Among My Books* (London: Trübner, 1870), pp. 81–150.
7. Andrew Dickson White, *A History of the Warfare of Science with Theology in Christendom*, 2 vols (London: Macmillan, 1896), Vol. I, p. 361. Chapter 15 deals with witchcraft. White originally published abridged versions of these views in the popular science journal, *Popular Science Monthly*, in 1889. For a good general analysis of nineteenth-century American historiography of witchcraft, see Leland L. Estes, 'Incarnations of Evil: Changing Perspectives on the European Witch Craze', *Clio* 13 (1984) 136–8.

8. George Lincoln Burr (ed.), *Narratives of the Witchcraft Cases, 1648–1706* (New York: Scribner, 1914); Burr, 'The Literature of Witchcraft', in *George Lincoln Burr: His Life by R. H. Bainton. Selections from his Writings*, ed. Lois Oliphant Gibbons (Ithaca: Cornell University Press, 1943), pp. 166–89; Henry Charles Lea, *Materials Toward a History of Witchcraft*, ed. A. C. Howland, 3 vols (London and New York: Thomas Yoseloff, 1957). Burr was the first librarian at Cornell and was largely responsible for collecting the vast archive of witchcraft tracts now held there. His fellow-bibliophile, Lea, was responsible for amassing a similar collection at the University of Pennsylvania.

9. For this, and what follows, see especially Jan Goldstein, 'The Hysteria Diagnosis and the Politics of Anticlericalism in Late Nineteenth-Century France', *Journal of Modern History* 54 (1982) 209–39; Goldstein, *Console and Classify: The French Psychiatric Profession in the Nineteenth Century* (Cambridge: Cambridge University Press, 1987), pp. 322–84, esp. pp. 369–73; Sarah Ferber, 'Charcot's Demons: Retrospective Medicine and Historical Diagnosis in the Writings of the Salpêtrière School', in Marijke Gijswijt-Hofstra, Hilary Marland and Hans de Waardt (eds), *Illness and Healing Alternatives in Western Europe* (London and New York: Routledge, 1997), pp. 120–40.

10. An abridged version of Axenfeld's lectures was appended as a preface to Bourneville's 1885 French edition of Weyer. Another colleague of Charcot and Bourneville, Charles Richet, pointedly wrote in 1880 that the sick women whom he treated at the Salpêtrière 'would have been burnt in another time'; Charles Richet, 'Les Démoniaques d'aujourd'hui', *Revue des Deux Mondes* 37 (1880) 340; cited in Ferber, 'Charcot's Demons', p. 137.

11. For one example among many that might stand as exemplary: 'The belief in witchcraft decayed with as little apparent reason as it arose. The civilized world gradually discovered that it had ceased to believe in the existence of witches even before it had given up the practice of burning them ... Clearly the change of attitude was due chiefly to the advance of science, which slowly defined the limits of man's mastery over nature, and disclosed the methods by which this mastery is attained'; Sir William Cecil Dampier, *A History of Science and Its Relations with Philosophy and Religion* (4th edn, Cambridge, 1966; 1st edn, 1929), p. 144.

12. Gregory Zilboorg, *The Medical Man and the Witch During the Renaissance* (Baltimore: Johns Hopkins Press, 1935), p. x. This work formed the central chapters of Gregory Zilboorg and George W. Henry, *A History of Medical Psychology* (New York: Norton, 1941).

13. Zilboorg and Henry, *A History of Medical Psychology*, pp. 207, 211, 221.

14. This is evident from Zilboorg's prologue to his *History of Medical Psychology*, where he refers, among other things, to the fact that the profession abounds in 'quacks' and 'various nonmedical amateurs, lay and clerical', who 'are more readily accepted by patients and their relatives than are psychiatrists, and mental hospitals' (p. 25).

15. Zilboorg and Henry, *A History of Medical Psychology*, pp. 24–5.

16. Zilboorg and Henry, *A History of Medical Psychology*, pp. 17–18, 20.

17. These are discussed and cited most fully in Thomas J. Schoeneman, 'The Role of Mental Illness in the European Witch Hunts of the Sixteenth and Seventeenth Centuries', *Journal of the History of the Behavioral Sciences* 13 (1977) 337; Nicholas P. Spanos, 'Witchcraft in Histories of Psychiatry: A Critical Analysis and an Alternative Conceptualization', *Psychological Bulletin* 85 (1978) 417.

18. George Sarton, *Six Wings: Men of Science in the Renaissance* (London: J. Lane, [1957] 1958), pp. 212–18.

19. Richard Hunter and Ida Macalpine, *Three Hundred Years of Psychiatry, 1535–1860* (London: Oxford University Press, 1963), p. 68. They would appear to have provided the chief source and inspiration for R. E. Hemphill, 'Historical Witchcraft and Psychiatric Illness in Western Europe', *Proceedings of the Royal Society of Medicine* 59 (1966) 891–902.

20. Oskar Diethelm, 'The Medical Teaching of Demonology in the 17th and 18th Centuries', *Journal of the History of the Behavioral Sciences* 6 (1970) 3–15. Diethelm extends the medical roll call of honour to include Jean Fernel, Felix Platter and Daniel Sennert in addition to Weyer. He breaks new ground, however, in citing the evidence of medical dissertations to uphold his argument that the gradual acceptance of medical psychiatry, or primitive psychotherapy, set in motion a period of advance in relation to the medical diagnosis of witchcraft. This in turn, he argues, gained further momentum following the publication of the work of Descartes, Willis and Stahl in the late seventeenth and early eighteenth century. In tracing the creation of a tradition in this way Diethelm's work contrasts with that of Zilboorg, who was scathing in his condemnation of the contribution of later medical and scientific figures such as Willis to the progress of medical psychology; Zilboorg and Henry, *A History of Medical Psychology*, pp. 247–53, 256, 263, 277. Zilboorg's antipathy toward the neurological approach of iatromechanists like Willis is probably best explained by his commitment to an autonomous psychiatric profession, practising Freudian psychotherapy, which eschewed neurological explanations for mental illness and a therapeutics based on drugs.

21. See, for example, Joost A. M. Meerloo, 'Four Hundred Years of "Witchcraft", "Projection" and "Delusion"', *American Journal of Psychiatry* 120 (1963) 85–6. Meerloo describes Zilboorg as 'one of our best historians of medical psychology' (83).

22. Sarton, *Six Wings*, p. 218.

23. Andrew Fix, *Fallen Angels: Balthassar Bekker, Spirit Belief, and Confessionalism in the Seventeenth Century Dutch Republic* (Dordrecht, Boston and London: Kluwer, 1999), p. 6; citing Alfred North Whitehead, *Science and the Modern World* (New York, 1925) and Robert G. Collingwood, *The Idea of Nature* (Oxford, 1945).

24. Robert K. Merton, *Science, Technology and Society in Seventeenth-Century England* (New York: Harper & Row, 1970); originally published in *Osiris* 4 (1938). As further examples of those working in this tradition, Fix cites the work of Eiejer Hookyaas, Paul Kocher, Richard Westfall, Theodore Raab, Barbara Shapiro, J. R. and Margaret Jacob, and Margaret Osler.

25. Frances A.Yates, *Giordano Bruno and the Hermetic Tradition* (London: Routledge & Kegan Paul, 1964). Yates was building on the earlier work of scholars such as Lynn Thorndike and D. P. Walker, who did much to reawaken academic interest in this otherwise esoteric subject; see Thorndike, *A History of Magic and Experimental Science*, 8 vols (New York, 1923–58); Walker, *Spiritual and Demonic Magic from Ficino to Campanella* (London: Warburg Institute, 1958).

26. Hugh Trevor-Roper, 'The European Witch-Craze of the Sixteenth and Seventeenth Centuries', in Trevor-Roper, *Religion, the Reformation and Social Change* (London: Macmillan, 1967), pp. 90–192.

27. Trevor-Roper, 'The European Witch-Craze', pp. 90–1, 99–100, 132. For an earlier attempt to hint at a similar division of interests in relation to scientific orientation and attitudes to witchcraft, see Paul H. Kocher, *Science and Religion in Elizabethan England* (San Marino, Calif.: Huntingdon Library, 1953), pp. 70–1.

28. Keith Thomas, *Religion and the Decline of Magic* (London: Weidenfeld & Nicolson, 1971), pp. 578–9, 643–4. Thomas did, however, acknowledge that the problem of decline represented 'the most baffling aspect of this difficult subject' (p. 570).

29. Brian Easlea, *Witch-Hunting, Magic and the New Philosophy: An Introduction to Debates of the Scientific Revolution 1450–1750* (Brighton: Harvester Press, 1980); Thomas Harmon Jobe, 'The Devil in Restoration Science: The Glanvill-Webster Witchcraft Debate', *Isis* 72 (1981) 343–56.

30. Easlea, *Witch-Hunting*, pp. 196–220; Jobe, 'The Devil in Restoration Science'.

31. This is the view developed by Sidney Anglo in his 'Melancholia and Witchcraft', in A. Gerlo (ed.), *Folie et Déraison à la Renaissance* (Brussels: University of Brussels, 1976), pp. 209–22, esp. pp. 210–12. Others, such as Christopher Baxter, have stressed the greater significance attached to his innovative religious beliefs (radical Protestant) in shaping his approach to witchcraft, while Erik Midelfort has argued that Weyer's legacy was greater in the field of forensic medicine, where he pioneered the insanity defence; see C. Baxter, 'Johann Weyer's *De Praestigiis Daemonum*: Unsystematic Psychopathology', in S. Anglo (ed.), *The Damned Art: Essays in the Literature of Witchcraft* (London: Routledge & Kegan Paul, 1977), pp. 53–75; H. C. Erik Midelfort, 'Johann Weyer and the Transformation of the Insanity Defense', in R. Po-Chia Hsia (ed.), *The German People and the Reformation* (Ithaca: Cornell University Press, 1988), pp. 234–61.

32. Thomas Szasz, *The Manufacture of Madness* (New York: Harper & Row, 1970), p. 75. At the same time as Szasz was launching his crusade against the psychiatric profession of psychology as a form of social and political control, the French philosopher and historian, Michel Foucault, was turning on its head traditional Whig historiography by postulating, among other things, that the eighteenth century had witnessed a 'great confinement' across Europe which stigmatised the mentally ill as antisocial outcasts and deviants; see esp. M. Foucault, *Madness and Civilization*, trans. Richard Howard (New York: Pantheon Books, 1965). A revised French edition of this work appeared in 1972.

33. Schoeneman, 'The Role of Mental Illness'; Spanos, 'Witchcraft in Histories of Psychiatry'. See also Irving Kirsch, 'Demonology and the Rise of Science: An Example of the Misperception of Historical Data', *Journal of the History of the Behavioral Sciences* 14 (1978) 149–57, who argues for a relativist approach in understanding the early history of psychiatry and its relation to witchcraft. Spanos rejected retrodiagnosis partly on the grounds that it was ahistorical, and partly because of the 'ambiguity and unreliability' surrounding the diagnosis of many contemporary medical conditions, such as schizophrenia; Spanos, 'Witchcraft in Histories of Psychiatry', pp.418–19. Clearly divisions within the psychiatric profession in the 1960s and 1970s were partly responsible for the revisionist approach to Zilboorg adopted by 'insiders' such as Schoeneman and Spanos. For a more recent critique of Zilboorg and his role in creating a foundational myth for psychiatry centred on Weyer, see Patrick Vandermeersch, 'The Victory of Psychiatry over Demonology: The Origin of the Nineteenth-Century Myth', *History of Psychiatry* 2 (1991) 351–63.

34. Michael MacDonald, *Mystical Bedlam: Madness, Anxiety, and Healing in Seventeenth-Century England* (Cambridge: Cambridge University Press, 1981).

35. For a good overview of these developments, and the application of social constructionism to the history of madness, see Roy Porter, *Mind-Forg'd Manacles: A History of Madness in England from the Restoration to the Regency* (London and Cambridge, Mass.: Harvard University Press, 1987), pp. 1–18.

36. H. C. Erik Midelfort, *A History of Madness in Sixteenth-Century Germany* (Stanford: Stanford University Press, 1999), p. 49. Midelfort's work probably represents the best example to date of an attempt, using a wide range of archival sources, to describe the cultural construction of mental illness in early modern Europe. His earlier ideas on this subject are summarised in his 'Madness and the Problems of Psychological History in the Sixteenth Century', *Sixteenth Century Journal* 12 (1981) 5–12. For an earlier case study seeking to refute the idea that modern concepts of mental illness first formed in reaction to witch hunts and cases of demonic possession, 'so that scientific beliefs triumphed over superstition', see David Harley, 'Mental Illness, Magical Medicine and the Devil in Northern England, 1650–1700', in Roger French and Andrew Wear (eds), *The Medical Revolution of the Seventeenth Century* (Cambridge: Cambridge University Press, 1989), pp. 114–44.

37. Midelfort, *A History of Madness*, p. 205; Michael MacDonald, *Witchcraft and Hysteria in Elizabethan London: Edward Jorden and the Mary Glover Case* (London and New York: Routledge, 1991), pp. viii–ix.

38. A notable exception is G. R. Quaife, *Godly Zeal and Furious Rage: The Witch in Early Modern Europe* (London: Croom Helm, 1987), pp. 205–6.

39. See, for example, Joseph Klaits, *Servants of Satan: The Age of the Witch Hunts* (Bloomington: Indiana University Press, 1985), pp. 161–8. Outlining the traditional view of the onward march of science in the seventeenth century, Klaits argues that it was the 'new ways of thinking inspired by science and Cartesian philosophy' that informed the judgements of the French judicial elite, who were responsible for bringing witch trials to an end in France. In doing so, Klaits relies almost exclusively on the earlier work of the French witchcraft scholar, Robert Mandrou; see Mandrou, *Magistrats et Sorciers en France aux XVIIe Siècle: Un Analyse de Psychologie Historique* (Paris: Plon, 1968), pp. 539–64.

40. Brian P. Levack, *The Witch-Hunt in Early Modern Europe* (London and New York: Longman, 1987), pp. 54–5, 56–7, 218–22. Levack's generally optimistic assessment of the impact of medicine on retarding the witch hunts is also noteworthy. It runs counter to the somewhat idiosyncratic and largely unsubstantiated view expressed by Leland Estes that the witch hunts were principally the fault of early modern physicians, who in their ignorance routinely deferred cases of suspected bewitchment and possession to witch-hunting clerics; see Leland L. Estes, 'The Medical Origins of the Witch Craze: A Hypothesis', *Journal of Social History* 17 (1983) 271–84. In related fashion, Garfield Tourney had earlier attributed the prevalence of witchcraft in late-seventeenth-century England to the 'superstitious' nature of contemporary medical practice and thinking; Tourney, 'The Physician and Witchcraft in Restoration England', *Medical History* 16 (1972) 143–55.

41. Charles Schmitt, *Aristotle and the Renaissance* (Cambridge, Mass.: Harvard University Press, 1983).

42. See, for example, Michael Hunter, *Science and Society in Restoration England* (Cambridge: Cambridge University Press, 1981).

43. There is now a vast literature on this subject. One of the most original and influential texts in this field, however, remains Steven Shapin and Simon Schaffer, *Leviathan and the Air Pump: Hobbes, Boyle, and the Experimental Life* (Princeton: Princeton University Press, 1985). For a broad overview of these developments, see Simon Shapin, *The Scientific Revolution* (Chicago: Chicago University Press, 1996); Margaret J. Osler, 'The Canonical Imperative: Rethinking the Scientific Revolution', in Osler, *Rethinking the Scientific Revolution* (Cambridge: Cambridge University Press, 2000), pp. 3–22.

44. Shapin, *The Scientific Revolution*, pp. 153–5. For Boyle on witchcraft, see also Michael Hunter, 'Alchemy, Magic and Moralism in the Thought of Robert Boyle', *British Journal for the History of Science* 23 (1990) 387–410.
45. Allison Coudert, 'Henry More and Witchcraft', in Sarah Hutton (ed.), *Henry More (1614–1687): Tercentenary Studies* (Dordrecht, Boston and London: Kluwer, 1990), pp. 115–36.
46. Stuart Clark, *Thinking With Demons: The Idea of Witchcraft in Early Modern Europe* (Oxford: Oxford University Press, 1997). Clark also provides in Part 2 of this work a useful introduction to the historiography of science and medicine in relation to witchcraft.
47. Clark, *Thinking With Demons*, esp. pp. 251–80. Clark identifies the contemporary preoccupation with monsters and a host of similar preternatural phenomena as related to the early modern interest in witchcraft. In both cases, they are said to provide exemplary material that, it was widely believed, might help to resolve some of those 'boundary disputes' spawned by the 'epistemological uncertainty' of the time.
48. Clark, *Thinking With Demons*, esp. pp. 299–309. Among other works cited by Clark, see especially Keith Hutchison, 'What Happened to Occult Qualities in the Scientific Revolution?', *Isis* 73 (1982) 233–54; Hutchison, 'Supernaturalism and the Mechanical Philosophy', *History of Science* 21 (1983) 297–333; John Henry, 'Occult Qualities and the Experimental Philosophy: Active Principles in Pre-Newtonian Matter Theory', *History of Science* 24 (1986) 335–81.
49. Fix, *Fallen Angels*, pp. 11–12. For the traditional view of the debt of Bekker's scepticism to Cartesian mechanism, see Robin Attfield, 'Balthassar Bekker and the Decline of the Witch-Craze: The Old Demonology and the New Philosophy', *Annals of Science* 42 (1985) 383–95.
50. Geoffrey Scarre, *Witchcraft and Magic in Sixteenth- and Seventeenth-Century Europe* (Basingstoke: Macmillan, 1987), p. 56. Roy Porter makes no mention of the role of science in his treatment of the decline of the witch trials in Enlightenment England; R. Porter, *Enlightenment: Britain and the Creation of the Modern World* (London: Penguin Press, 2000), pp. 219–24. Jim Sharpe is equally reticent to accord science a prominent place in any explanations for the demise of witchcraft; James Sharpe, *Instruments of Darkness: Witchcraft in England 1550–1750* (London: Hamish Hamilton, 1996), pp. 256–75, esp. p. 26.
51. See, for example, B. P. Levack, 'The Decline and End of Witchcraft Prosecutions', in Ankarloo and Clark, *Witchcraft and Magic in Europe*, pp. 3–93; I. Bostridge, *Witchcraft and its Transformations c.1650–c.1750* (Oxford: Clarendon, 1997).
52. Clark, *Thinking With Demons*, p. 299.

4

the nineteenth century: medievalism and witchcraft

christa tuczay

This chapter will explore how two apparently very different traditions developed from the rediscovery of witchcraft as an integral part of a medieval world. On the one hand, Romantics, such as Walter Scott (1771–1832) in Scotland and Jules Michelet (1798–1874) in France, rediscovered witch beliefs as an integral part of a lost society, potentially offering a critique both of the medieval world, as in Michelet's rebel peasant sorceress, and also of modernity. On the other hand, the liberal Enlightenment tradition, strongest in Germany and the US, and exhibited in the works of such influential figures as Joseph Hansen (1862–1943) and Henry Charles Lea (1825–1909), charted in meticulous detail the relationship of witchcraft to medieval intolerance and inquisitorial techniques. Both movements drew on the rise of folklore and early anthropology, with its assumptions that folk culture and 'primitive societies' preserved traditional beliefs in witchcraft, which were inevitably eroded by the rise of civilisation.

the romantics

Witchcraft may have been finally decriminalised across Europe by the late eighteenth century, and witches no longer held much fear for educated society, but the fascination with witches and witchcraft certainly did not end. The notion of witchcraft as an integral part of the medieval word seems to have fascinated as well as horrified nineteenth-century educated society and pandered to the 'age of sensation'.[1]

The first profound, comprehensive delineation of the topic was provided by the Protestant minister Georg Conrad Horst (1769–1832) of Lindheim, a town where three major witch trials had taken place in 1663–64. When the proprietor of the 'witches' tower' in Lindheim asked for a historical outline of the trials, Horst plunged into researching the subject, which led to a lifelong fascination. In 1818 he published *Dämonomagie, oder Geschichte des Glaubens an Zauberei und dämonische Wunder*, aiming for a complete encyclopaedia of sorcery and witchcraft persecution.[2] He presented his study in a detached, scientific spirit and interpreted his abundant material in an unprejudiced manner. Bearing in mind that the last execution of a witch had taken place in Switzerland less than 40 years before, Horst felt that witchcraft was situated firmly in the past and therefore a fitting subject for detached scrutiny. His history of witchcraft persecutions begins with Pope Innocent VIII (1484–1492) and closes with a summary of the witch trials in his hometown of Lindheim. Horst was one of the first historians of witchcraft to use archival sources, and several years later he performed a valuable service to future historians by putting together the famous seven-volume *Zauber-Bibliothek* or 'Library of sorcery', which contained a compendium of rare manuscripts and printed texts on magic.[3]

In Horst's work we can detect a nascent nationalistic bias in witchcraft studies, for he stressed that it had been enlightened Germans who fought against the 'witch craze'. We also find the same tendency in the work of the liberal-minded Jules Garinet (1797–1877) of Chalons-sur-Marne, who in the same year that Horst's *Dämonomagie oder Geschichte* was published, produced a weighty history of French trials.[4] His research reached a wider English audience through the use Charles Mackay made of it in his chapter on 'The Witch Mania' in his bestselling *Extraordinary Popular Delusions and the Madness of Crowds* (1841).

The other major study in this first phase of nineteenth-century attempts to historicise the witch trials was Walter Scott's *Letters on Demonology and Witchcraft* (1830).[5] Published two years before his death, *Letters* was the product of his lifelong interest in what would later be called 'folklore'. His knowledge of Scottish magical beliefs had coloured his novels, and the witches' Sabbath, more at home in the continent, did not play an important part in his fiction because he was more interested in the day-to-day aspects of the supernatural that had concerned the Scottish peasantry. He thought of witchcraft as characteristic of periods and places and so his witches range from the rather exotic figures in *Ivanhoe* to the hags of his Scottish novels.[6] While Scott situated his fictional witches in medieval settings he was aware of the continued belief in such figures into his own time and ended his ninth 'Letter' with a 'Case of supposed Witchcraft, related from the Author's own knowledge, which took place as late as 1800'.[7] As a consequence, he was not as supremely confident as some of his contemporaries in the idea of an epochal Enlightenment leap from the days of the witch trials to the present, at least with regard to the labouring classes. 'There remains hope', he concluded, 'that the grosser faults of our ancestors are now out of date; and that whatever follies the present race may be guilty of, the sense of humanity is too universally spread to permit them to think of tormenting wretches till they confess what is impossible, and then burning them for their pains.'[8]

Scott was by no means unique in his interest in the supernatural beliefs of the common people. In Germany, for example, Johann Gottfried Herder (1744–1803), stimulated by the *Sturm und Drang* literary movement and Romantic stirrings, reversed traditionally negative attitudes towards the populace and turned the '*Volk*', its legends, legacy, spirit and, above all, its language into objects of esteem. Furthermore, the founders of German philology, the Brothers Grimm – and especially Jacob Grimm (1785–1863) – were of utmost relevance, influence and impact, particularly with regard to the development of witchcraft studies.[9] Jacob had an excellent grasp of the literary sources on the subject from antiquity onwards and a huge number of records. Scrutinising the studies of his contemporaries he considered Scott's efforts as imprecise and negligent, and even called Horst tasteless. Long before Michelet, Grimm was confident he had identified in the witch trial material elements of a pre-Christian Germanic pagan cult, though he did not go so far as to say that those prosecuted as witches were the surviving followers.[10] He was also first to identify those persecuted as witches as wise women, an equation that was to be reformulated by radical feminists 140 years later.[11] In both these ways Jacob Grimm's influence on witchcraft studies has been long-lasting if not always recognised.

However, Grimm was not the only person at the time identifying the survival of traces of pre-Christian cults into Christian times. In 1828, Karl Ernst Jarcke (1801–1852), a converted Catholic and professor of criminal law at the University of Berlin, argued that witches were the followers of a Germanic nature religion, which the Church wrongly interpreted and persecuted as Devil worship.[12] Franz Josef Mone (1796–1871), professor of history in Heidelberg, proposed that during a great migration of Germanic tribes south-eastwards they came into contact with the pagan cults of Dionysos and Hecate. When the Germanic tribes moved westwards again they brought with them their hybrid, debased religion centred on the worship of a goat-like god. These pagan cults survived among the peasantry well into the Christian era and constituted a secret society, which was unveiled through the witch trials.[13] Neither of these advocates of an unbroken tradition took the pains to prove their assertions. Both being devout Catholics they were apologists for the Church's role in the witch trials, asserting that it was engaged in a godly struggle to root out wicked anti-Christian secret societies. Catholic conservatism also emerged in excrescences like those of Josef Görres (1776–1848) who echoed the early modern demonologists in his *Die christliche Mystik* ('The Christian Mystic').[14] During the many years he held the first professorship for history in Munich he exploited his position to inhibit the establishment of modern historical science. His initially Romantic interpretation of witchcraft became increasingly dogmatic, as is evident in the third and fourth volumes of *Die christliche Mystik* (1840–42). In his writings on demoniacs and witchcraft he deliberately chose authors that fitted his own conservative opinions, like Delrio and Grimm, but ignored Protestants like Thomasius, Hauber and Horst. Like Pierre de Lancre he interpreted details of the flight of the witches and Sabbaths as real, and like Jarcke and Mone he expressed no sympathy for those persecuted as witches. His stance would be echoed nearly a century later in the writings of Montague Summers, who was familiar with Görres' work, and who considered the witch 'a social pest and parasite; the devotee of a loathly and obscene creed ... a member of powerful secret organization inimical to Church and State; a blasphemer in word and deed; swaying villagers by terror and superstition'.[15]

While German philologists, folklorists and historians were amongst the most active witchcraft researchers – partly because of the academic search for a Germanic national identity in the folk cultures of the past, partly because of the large amount of source material – it was an anticlerical French historian who achieved the most widespread popular influence with a powerfully politicised reading of the witch persecutions. Jules Michelet, a fine historian who was forced out of academia in 1851 by a conservative administration concerned by his radicalism and lack of historical neutrality, believed the practitioners of witchcraft were medieval peasants rebelling against oppressive feudalism. This secret rebel society adopted the festivities of a pre-Christian fertility religion as a symbol of defiance against the intolerant, suffocating medieval Church. For Michelet these witches were admirable holders of the French radical tradition, the forerunners of the revolutionaries who overthrew absolutism in 1789. Michelet defined history more generally as a class and gender competition, as a struggle between brutal overlords and intuitive serfs. His romantic, positive reinterpretation of witchcraft in this respect was laid out in *La sorcière* ('The Witch'), which was first published in 1862, and although it made little academic impact, it became a bestseller.[16] In it he described how at the centre of this witch society was a priestess, and how the female members were possessed of great healing skills and were the physicians of the people. It was Michelet, then, who formulated the hypothesis of the alleged ancient wisdom of the 'wise women' that still has much potency today.

Despite their contrasting viewpoints, the radical Michelet and the conservative Catholic German theorisers on the 'witch cults' shared one fundamental trait – none of them conducted detailed research on the trials. The danger of this reliance on printed sources is evident in the dubious contribution made to witchcraft studies by the French Baron Etienne-Léon de Lamothe-Langon (1786–1864), an impoverished aristocrat who earned his living by writing, and who published a huge number of books and manuscripts over 50 years. He lacked the time and inclination for detailed research and profound analysis, so if necessary he made up sources.[17] In 1829 he published his *Histoire de l'Inquisition en France*[18] ('History of the Inquisition in France'), in which he described large-scale witch trials which had supposedly taken place in southern France in the early fourteenth century. Instigated by Inquisitions in Toulouse and Carcassonne, these trials led to the execution of hundreds of people, and in the purported trial transcripts printed by Lamothe-Langon we find the fully formed stereotype of the diabolic witch. We can see how these fake trials fostered the conception elaborated on by Romantics such as Michelet and Görres that the witch trials were a primarily medieval phenomenon.

The pernicious historiographical impact of the Lamothe-Langon forgeries was to be long-lasting. In the early twentieth century, the prominent historian Joseph Hansen included large sections of Lamothe-Langon's work in his compendium on medieval witchcraft.[19] Later historians cited Hansen's citations, apparently without closely examining Lamothe-Langon's credentials. Non-academic writers then regurgitated the writers who cited Hansen, and thus Lamothe-Langon's dramatic French trials became a standard part of the popular view of the witch hunt. However, as more research was done, Lamothe-Langon's fabricated trials began to look odd to historians. No sources mentioned them, and they were completely different from all other fourteenth-century trials. There were no other mass trials of this nature until 1428, no major outbreaks of witch persecution until the sixteenth century. This puzzling anomaly led Norman Cohn and Richard Kieckhefer to investigate Lamothe-Langon's background, and they soon found out that the great trials of the *Histoire* had never occurred. Cohn and Kieckhefer published their findings in the mid 1970s, and since then, academics have avoided this forged material, though unfortunately by this time, Lamothe-Langon's lurid trials had entered into the popular mythology of witchcraft.[20]

Lamothe-Langon's trials, however, were not only forged but they were the *terminus ante quem*: they had kept the theory of medieval witch hunting alive. Once these trials were exposed as non-existent, the development of witchcraft stereotypes became much clearer. Before Lamothe-Langon's forgeries were discovered, the earliest great hunts appeared to come from southern France in an area once the home of the Cathar heresy. This led some historians to suggest a link between Catharism and witchcraft, that witches were the remnants of an old dualism. Take away the forged trials, however, and the centre of the early witch prosecution shifts from Cathar territory to the Alps.

the historians respond

There was bound to be a backlash against the poor historical basis for the Romantics' interpretations. The hugely influential historian Leopold von Ranke (1795–1886) of the University of Berlin, who is credited with shaping modern historiography and 'scientific' history, demanded that history must rest on the strict, meticulous study of archival sources.[21] Small wonder that a topic like witchcraft, associated with myth, intuition and fantasy, was sidelined from the curricula of universities. Furthermore, historians distanced themselves from the interests of the philologists and folklorists in the cultures of the lower classes. While the subject of the witch trials remained a general source of fascination during the second half of the nineteenth century, attracting the attention of theologians, psychologists, medical men and folklorists, only a few professional historians like the Bavarian Sigmund von Riezler (1843–1927), a critic of Ranke's style of history, turned their attention to the subject. His ground-breaking study *Geschichte der Hexenprozesse in Bayern* ('History of the Bavarian Witch Trials'), published in 1896, has been praised by the modern expert on the Bavarian witch trials, Wolfgang Behringer, who describes him as an excellent scholar, even though some of his interpretations of the archival sources were incorrect.[22] Although not a study of the witch trials directly, it is also worth noting Gustav Roskoff's (1814–1889) still readable book on the history of the Devil, *Geschichte des Teufels* (1869), which for the time was both ground-breaking and profound in its discussion of the increasing importance of the Devil in Christianity from the thirteenth century onwards.[23]

The most historiographically influential study published during the second half of the nineteenth century was, however, the original and revised version of Wilhelm Gottlieb Soldan's *Geschichte der Hexenprozesse* ('History of the Witch Trials'), which, on the basis of Lamothe-Langon's forgeries, rooted the history of the witch trials erroneously in the late medieval period.[24] Soldan (1803–1869), a Protestant theologian from Hessen, was a liberal political opponent of Jacob Grimm and a widely known personality among his contemporaries. His major history of witchcraft based on archival sources was first published in 1843, and in 1880 his son-in-law Heinrich Heppe, a Protestant minister and professor of theology in Marburg, produced a revised and expanded edition. Although never translated, the Soldan–Heppe edition was nevertheless broadly acknowledged. Like Thomasius, Soldan considered witchcraft a mere chimera and regarded witch trials as criminal proceedings without a crime. Consequently the Church was deemed culpable of a grave error. His interpretation of the topic created what William Monter labelled the 'rationalist paradigm', which was to become the orthodoxy in witchcraft studies, though it was later threatened by Margaret Murray and Montague Summers.[25]

Conservative Catholic historians did not take the anticlerical attacks of Soldan and Riezler quietly, and something of a confessional 'witch war' broke out. The young Catholic historian Paul Maria Baumgarten (1860–1948), for example, polemically referred to witchcraft as the favourite subject of a biased anti-Catholic historiography, and was so carried away by his anger that he even justified the witch persecutions. Johannes Diefenbach, a Catholic priest, also criticised the witchcraft historiography as a one-sided and incriminating attack on the Catholic Church. After ten years studying the sources Diefenbach concluded that the trials were largely the fault of the secular authorities rather than the Catholic Church, and asserted that, in fact, the Protestant contribution to the progression of the witch trials was more influential.[26]

While Germany was certainly at the centre of witchcraft research, similar debates about the pernicious role of religion in promoting the witch trials were occurring overseas. In the third volume of George Bancroft's ten-volume *History of the United States* (1834–74), the Puritans were blamed for the Salem trials, and considering Bancroft's account of the episode became the standard history in schools and universities, its anti-Puritan stance proved widely influential.[27] One of those who borrowed from Bancroft's interpretation was the Irish historian and essayist William Edward Hartpole Lecky (1838–1903).[28] In his oft-printed *History of the Rise and Influence of the Spirit of Rationalism in Europe* (1865) he outlined the history of the 'witch craze' as a fundamental battle between the darkness of religious bigotry and the light of science. Just as Bancroft pointed an accusatorial finger at the New England Puritans, so Lecky blamed Calvinists for the witch trials in Scotland, lashing 'himself into paroxysms of rationalist rage' while so doing, as Christina Larner memorably put it.[29] Lecky sought to create a 'science of history', just as the new 'science' of economics was then arising. He looked upon human progress as the gradual ascendancy of rationalism over superstition. His dogmatic rationalism, which led him to somewhat simplify his overall approach by seeing whole periods of complex historical developments as simple instances of the warfare between reason and theology, was a key work of nineteenth-century scholarship and some of Lecky's points remain valid.

Of all the historians of the liberal-rationalist tradition the most widely read student of the sources on the European witchcraft and heresy trials was Henry Charles Lea (1825–1909). Born into a Philadelphian family of wealth and influence, he read and wrote classical Greek, Latin, French and English, Spanish and Italian and, later in life, German and Dutch. The head of a great publishing company, overwork led to a nervous breakdown in 1847 and he abandoned his intellectual and scientific work for some time. During his convalescence, he read several French memoirs of the medieval period. This changed the direction of his studies by creating a lifelong interest in medieval history, and he set about accumulating one of the best libraries in America containing many primary-source documents on medieval history. He retired in 1880 and devoted his time to academic studies. In his seminal books on the Inquisitions, Lea presented copious material culled from papal bulls, the proceedings of Church councils, letters between ecclesiastical authorities and trial records.[30] He approached the topic with an objective but critical eye. He once wrote in a letter that he was supportive of any effort to 'emancipate human intelligence from the trammels of medievalism'.[31] As Henry Kamen has observed, 'Though Lea had strong prejudices that he expressed uncompromisingly, his work once and for all rescued the tribunal from the make-believe world of invented history, and placed it firmly in the arena of documented fact.'[32] With regard to witchcraft, Soldan had already partly made that breakthrough, but Lea set about amassing a huge collection of material on the witch trials for a potentially ground-breaking history on the subject. We can gain an insight into how he would have approached the witch trials from his essay on 'Ethical Values in History' written in 1903. In it he promoted the role of historians in the onward progress of humanity. In discussing Philip II of Spain he described how to understand rather than merely condemn the 'distorted ethical conception' of religious persecution 'ennobles the historian's labours by rendering them contributory to that progress which adds to the sum of human happiness and fits mankind for a higher standard of existence'.[33] He died, however, before being able to complete his witchcraft project, though his notes and sources were published posthumously.[34] Although Lea's book never came to fruition, in Germany the Cologne archivist Joseph Hansen (1862–1943), an admirer of Lea's work and translator into German of Lea's *History of the Inquisition*, produced his own impressive investigation in two major books. The first was *Zauberwahn, Inquisition und Hexenprozeß im Mittelalter und die Entstehung der großen Hexenverfolgung* (1900) and the second was a source book, *Quellen und Untersuchungen zur Geschichte des Hexenwahns* (1901).[35] Hansen's thesis was a full-blooded reaffirmation of the anticlerical view that the witch trials were inspired by the medieval Church.

While the historians buried themselves in the archives, another group of scholars turned to a living source material – the beliefs of the labouring classes. Though adherents of historical 'progress' like Horst thought that witchcraft had largely withered away, reports from country parishes across Europe proved otherwise. Events like the epidemic possession in Morzine in the Haute-Savoie, at the peak of which, in 1861, nearly 200 villagers, mostly female, acted out the classic symptoms of convulsions, blaspheming and speaking in foreign tongues. The origins of the outbreak lay in accusations of witchcraft, like most cases of possession in the early modern period, though as Ruth Harris points out, the outbreak was also symptomatic of distinct nineteenth-century tensions over the relationship between state and community. For some intellectuals the Morzine affair was a lamentable recrudescence of medieval 'superstition', while for the nascent spiritualist movement it was a confirmation of the vibrancy of the spirit world. The uncomfortable truth for liberal historians was that such events as Morzine dented their progressive ideal which consigned witchcraft to the past.[36]

While the historians used the witch trials to build their models of societal progress from the dark ages of medievalism onwards, in the second half of the nineteenth century folklorists and anthropologists such as Edward Tylor and James Frazer, influenced by the evolutionary models formulated by Darwin and Spencer, attempted to chart human progress by collecting and examining the 'vestiges' of pagan and magical traditions of the ancient past, which were thought to have survived in corrupted and debased forms in the mental and cultural world of the peasantry. Karl Simrock, influenced by the Grimms, asserted, for example, that the folk legends they collected were ancient, pagan and of unbroken continuity, and Franz Josef Vonbun believed he had found references to heathen gods in both oral legends and the archival sources regarding witch trials in the alpine Vorarlberg region, though others like Wilhelm Schwartz were more circumspect.[37] As part of the methodology of charting the evolutionary decline of pagan beliefs, folklorists and anthropologists turned to the witch trials as a means of gauging the process. Thus in 1858 the Innsbruck philologist Ignaz Zingerle (1825–1892), a follower of Grimm, produced a study about two sensational witch prosecutions.[38] Fritz Byloff (1875–1940) of Graz started his scholarly career with a dissertation on the *crimen magiae*. As well as a couple of articles on the topic, he also wrote two major works on Austrian witch trials. Influenced by his reading of Frazer he had a distinctive insight into the nature of magic and witchcraft in the Austrian Alps.[39] In a similar vein were the writings of Ludwig Rapp (1828–1910), who analysed the Tyrolean witch trials, and Hartmann Ammann who scrutinised the distinctive traits of Tyrolean folk beliefs and the historic opposition to witch trials in the region.[40]

the witch figure in nineteenth-century literature[41]

The nineteenth-century Romantic and rationalist scholarly interpretations of the witch trials, particularly the emphasis on their supposed medieval genesis, was also mirrored in the fiction of the period. As Michel Zink has observed, scholars from many different disciplines 'found in medieval literature … a mirror that sent back a reflection of their own illusions. True illusions, however, since these false perspectives are truly set up by that literature.'[42] This is true in novels by the likes of Felix Dahn, Adalbert Stifter and Gustav Freytag. They all deliberately returned to an imaginary medieval period in order to break through the perceived barrenness of their own age in search of new visions. Freytag's bestselling series of historical novels, *Die Ahnen* (1872–80), justified Germany's unification by tracing German history back to the early Middle Ages. His compendium *Bilder aus der deutschen Vergangenheit* ('Pictures from the German Past'), in which he discussed the role of the Devil and witchcraft, although by his own definition historiographical, is a wild essayistic mixture of genres.[43]

At the end of the eighteenth century the novelist Johann Paul Friedrich Richter (1763–1825) had treated the witch trials as a source of humour in his *Auswahl aus des Teufels Papieren* (1789), but the European Romantic novelists of the early nineteenth century took a more earnest view regarding witches, presenting them variously as socially deviant women, femmes fatales, power-obsessed furies and even vampires.[44] Ludwig Tieck's novel *Hexensabbat* (1832), set in Burgundy around 1460, portrayed the witch figure as a fictitious construction. Tieck acknowledged the *Malleus Maleficarum*, highlighting its misogynistic content, and provided an insightful account of the mechanics of denunciation and torture. His vivid description of the dynamics of a village is punctuated by his enlightened comments and statements of disagreement with the trial procedure. Tieck's aim was to prove that prejudicial persecutions not only emerged in a period of religious fanaticism and aggressive 'superstition' but could also break out in a peaceful society.[45]

The dichotomy of the witch figure as either beautiful young woman or unpleasant old crone, which had been apparent in fifteenth- and early sixteenth-century imagery, re-emerged as a strong theme in nineteenth-century novels. The synthesis between 'old hag' and 'witch', in particular, was remodelled in the Romantic literature and heavily influenced by the folk tale as a literary form. The process can be seen, for example, in the different versions of the story of Hänsel and Gretel.[46] In Grimm's various versions we can detect the gradual development of the figure of an old woman to a very old, ugly woman and finally to a very old, ugly, abhorrent witch.[47] Ludwig Bechstein (1801–1860), a successful German romantic novelist, poet and folktale collector further reinforced the image of the witch as old crone in his popular collection *Hexengeschichten* ('Tales of Witches'), published in 1854.[48] Bechstein skilfully combined information from chronicles, trial records and pamphlets from the medieval and early modern periods with legends and tales from his own collection and the likes of Grimm's *Deutsche Sagen*. In his humorous and sometimes macabre refurbishment of the material he achieved an artistic peak compared with his otherwise rather conventional work. The cruellest of the tales was the story 'Furia infernalis' set on a Russian manor. A Tatar Muslim woman, stigmatised as a heathen, binds herself to the Devil and murders her enemies with the help of a mythical animal. Her appearance corresponds with the western stereotype of the old hag. The conflict with society escalates when she believes that her son is beaten to death as punishment for having slaughtered some doves. Ignorant of his recovery she conjures up the evil 'furia infernalis', a compound animal of scorpion and tarantula, to wreak revenge. Too late she learns that her son has survived and she is doomed.

The powerful re-emergence of the historical novel, a genre initially launched by Sir Walter Scott's Waverley novels, in which he made frequent use of 'superstitions', was one of the most remarkable phenomena in nineteenth-century literature.[49] From its beginnings the genre was attacked by both historians and literary critics as a 'bastard art' for its supposedly 'impure' mix of fiction and history, reason and romance. Yet it is exactly this mix and the contested relationship between the two that is the central concern of the historiography, as literature likewise mirrors the attitude of the time towards witchcraft and witchcraft persecutions.[50] Such criticisms were, perhaps, unfairly levelled at Scott, for most of his historical novels were not based on real events. They can be more appropriately levelled at one of Scott's admirers, the novelist William Harrison Ainsworth (1805–1882). Ainsworth fictionalised real events and people, turning the Lancashire witch trials of 1612 into a gothic romance, *The Lancashire Witches* (1849). The dual impulse – common to much nineteenth-century witch fiction – to highlight the chimera of witchcraft and condemn the persecutors of witches was compromised by the fictional device of the representation of witchcraft as a reality. In the process he distorted the facts to meet his dramatic needs.[51] His similar treatment of the highwayman Dick Turpin, in an earlier successful novel, *Rookwood* (1834), generated a bogus history of the criminal that soon became popularly accepted as historical truth.[52]

Across the Atlantic the popularity of the Waverley novels, coupled with the growing desire of American authors to find literary worth in their own country's brief history, as an expression of independence from Europe, inspired numerous fictional accounts of the New England witch trials, those at Salem in particular.[53] First to appear was *Salem, an Easter Tale*, published in three instalments in the *New York Literary Journal* in 1820. Others soon followed, such as the Scott-influenced but rather poor *The Witch of New England: A Romance* (1824), John Neal's *Rachel Dyer* (1828), William Leete Stone's *The Witches: A Tale of New England* (1837) and *The Salem Belle, a Tale of Love and Witchcraft in 1692* (1842). As the latter title suggests, its anonymous author presented the witch as attractive heroine. The same stereotype, along with that of the wizened old witch, was also at the centre of *Delusion, or the Witch of New England* (1840). Its author, Eliza Buckminster Lee, observed in the preface that 'If it is objected that the young and lovely are seldom accused of any witchcraft except that of bewitching hearts, we answer that of those that were actually accused, many were young.'[54] The romantic if tragic portrayal of the witch trials continued to be drawn upon by American authors throughout the rest of the century. The famed novelist Nathaniel Hawthorne (1804–1864), who was born in Salem, drew upon the trials in several works, most notably *The Scarlet Letter* (1850), and even the realist novelist John De Forest (1826–1906) could not break out of the European Romantic mould when it came to witchcraft, as is evident in his serialised novel *Witching Times*, published in *Putnam's Magazine* (1856–57).[55]

Back across the Atlantic again the witchcraft novels of the German writer and former pastor Wilhelm Meinhold (1797–1851) also proved tremendously popular in England. His *Maria Schweidler, die Bernsteinhexe* (1843) was translated by Lady Duff-Gordon, who was friends with Dickens and Thackeray, and published as *The Amber Witch* in 1844. Meinhold's *Sidonia von Bork, die Klosterhexe* (1847) was translated by Oscar Wilde's mother, Jane Elgee, and appeared in 1849 with the title *Sidonia the Sorceress*.[56] Both *Sidonia* and *The Amber Witch* were Wilde's favourite boyhood reading and influenced his own work.[57] Meinhold's *Bernsteinhexe* purports to be an authentic account written by a seventeenth-century pastor about a young female healer who helps the rural community against malevolent sorcery. When her powers cease, the villagers suspect her not only of having lost her virginity but also of having intercourse with the Devil. A 'real' witch and a bailiff plot against the innocent maiden and she is tried for witchcraft. The young woman is found guilty and sentenced to death but is rescued by her nobleman lover. In *Sidonia* the eponymous central character is a beautiful but sadistic sorceress in the court of seventeenth-century Pomerania who uses her witchcraft and her beauty to manipulate and murder courtiers.

The underlying message in Meinhold's witch novels, which mirrored long-held patriarchal conventions that were still prevalent in the nineteenth century, was that female craving for power was disruptive of the natural order and that 'good' women required male protection. Still, as Maureen Moran notes, in Victorian witchcraft fiction one also finds portrayals of the male victimisation of women. This is most apparent in a serialised story published in 1844 called *The Witchfinder*, in which the psychotic Black Claus sets out on a misogynistic mission to seek and destroy witches, only to realise later that his mission is a folly born of madness. As Moran suggests, 'Victorian writings – whether by men or women – expose fears of female power, of social disorder, and of cultural mobility and invasion.'[58]

Others tried to emulate the success of Meinhold with varying results. The translation of Paul von Heyse's *Die Hexe von Corso* ('The Witch of Corso'), published in 1882, received little acclaim. Wilhelm Heinrich Riehl took the trial of Maria Holl of Nördlingen as a model for his novel *Jörg Muckenhuber* (1860), and Wilhelm Raabe's successful novel *Else von der Tanne* was set during the Thirty Years' War. This time of great upheaval serves as a metaphor for the German decline of morals and social regression. Perpetual fear, famine and squalor brutalise the rural communities. In this atmosphere of suspicion the beautiful refugee Else attracts hostility through her mere otherness. She is accused of having bewitched the village minister, and the infuriated villagers stone her to death. Raabe highlights the dynamics and interdependence of fanaticism and intolerance. His general point is sociological: fear of witches, collective panic and mimetic violence are modes of crisis regulation.[59]

Theodor Storm's *Renate* (1878) harks back to the time and place of the witch trials in northern Germany. Storm collected fairy tales and demonological literature, and used both these sources to construct the story of *Renate*, which recounts the unfortunate love of a young theologian for the daughter of a peasant who is suspected of black magic. Like Meinhold and Raabe, Storm emphasises the authenticity of his fictional figures by incorporating into the story such real personages as the cleric Petrus Goldschmidt, whose involvement in the novel is based on his north German parentage. He portrays the preacher as a reactionary and embittered believer in a diabolic world who urges the persecution of witches and sows mischief and suspicion in the village. Storm recounts the doom of the theologian with noticeable relish. It symbolised the ultimate fate of what liberal historians saw as the benighted grip of medieval superstition over the population.

notes

1. I am much indebted to the following studies: Leland L. Estes, 'Incarnations of Evil: Changing Perspectives on the European Witch Craze', *Clio* 13 (1984) 133–47;

Jonathan Barry, 'Keith Thomas and the Problem of Witchcraft', in Jonathan Barry, Marianne Hester and Gareth Roberts (eds), *Witchcraft in Early Modern Europe: Studies in Culture and Belief* (Cambridge: Cambridge University Press, 1996), pp. 1–48; Wolfgang Behringer, 'Erträge und Perspektiven der Hexenforschung', *Historische Zeitschrift* 249 (1988) 619–40; Behringer, 'Witchcraft Studies in Austria, Germany, and Switzerland', in Barry, Hester and Roberts, *Witchcraft in Early Modern Europe*, pp. 64–95; Behringer, 'Zur Geschichte der Hexenforschung', in Sönke Lorenz (ed.), *Hexen und Hexenverfolgung im deutschen Südwesten* (Ostfildern: Cantz, 1994), pp. 93–146; revised and enhanced as 'Geschichte der Hexenforschung', in Sönke Lorenz and Jürgen Michael Schmidt (eds), *Wider alle Hexerei und Teufelswerk. Die europäische Hexenverfolgung und ihre Auswirkungen auf Südwestdeutschland* (Ostfildern: Thorbecke, 2004), pp. 485–668; Jon Butler, 'Witchcraft, Healing, and Historians' Crazes', *Journal of Social History* 18 (1984) 111–18; Ulrich von Hehl, 'Hexenprozesse und Geschichtswissenschaft', *Historisches Jahrbuch* 107 (1987) 349–75; H. C. Erik Midelfort, 'Witchcraft, Magic, and the Occult', in Steven Ozment (ed.), *Reformation Europe: A Guide to Research* (St. Louis, Minn.: Center for Reformation Research, 1982), pp. 183–209; William E. Monter, 'The Historiography of European Witchcraft: Progress and Prospects', *Journal of Interdisciplinary History* 2 (1972) 435–51; Donald Nugent, 'Witchcraft Studies, 1959–1971: A Bibliographical Survey', *Journal of Popular Culture* 5 (1971) 710–25; Bengt Ankarloo and Stuart Clark (eds), *Witchcraft and Magic after the Witch Trials. The Eighteenth and Nineteenth Centuries* (London: Athlone Press, 1999); Nils Freytag and Benoît van den Bossche, 'Aberglauben, Krankheit und das Böse. Exorzismus und Teufelsglaube im 18. und 19. Jahrhundert', *Rheinisch-westfälische Zeitschrift für Volkskunde* 44 (1999) 67–93.

2. Georg Conrad Horst, *Dämonomagie oder Geschichte des Glaubens an Zauberei und dämonische Wunder: mit besonderer Berücksichtigung des Hexenprocesses seit den Zeiten Innocentius des Achten; nebst einer Beschreibung des Hexenthurms zu Lindheim in der Wetterau* (Frankfurt am Main, 1818).

3. Georg Conrad Horst, *Zauber-Bibliothek oder von Zauberei, Theurgie und Mantik, Zauberern, Hexen und Hexenprocessen, Dämonen, Gespenstern und Geistererscheinungen* (Mainz, 1821–1826).

4. Jules Garinet, *Histoire de la magie en France, depuis le commencement de la Monarchie jusqu'à nos jours* (Paris: Chez Foulon, 1818).

5. Walter Scott, *Letters on Demonology and Witchcraft Addressed to J. G. Lockhart* (London: Murray, [1830] 1884).

6. See Coleman O. Parsons, *Witchcraft and Demonology in Scott's Fiction* (Edinburgh: Oliver and Boyd, 1964); Daniel Cottom, 'The Waverley Novels: Superstition and the Enchanted Reader', *English Literary History* 47, 1 (1980) 80–102.

7. Scott, *Letters on Demonology*, p. 228.

8. Scott, *Letters on Demonology*, p. 320.

9. Lothar Bluhm, *Die Brüder Grimm und der Beginn der deutschen Philologie: eine Studie zu Kommunikation und Wissenschaftsbildung im frühen 19. Jahrhundert* (Hildesheim: Weidmann, 1997).

10. Jacob Grimm and Wilhelm Grimm, *Deutsche Sagen* (Berlin: In der Nicolaischen Buchhandlung, 1816–18); Jacob Grimm, *Deutsche Mythologie* (Göttingen: In der Dieterichschen Buchhandlung, 1835).

11. Grimm, *Deutsche Mythologie*, pp. 872–960.

12. Karl Ernst Jarcke, 'Ein Hexenprozess', *Annalen der Deutschen und Auslandischen Criminal-Rechts-Pflege* 1 (1828) 450.

13. Franz Josef Mone, 'Über das Hexenwesen,' *Anzeiger für Kunde der teutschen Vorzeit* 8 (1839), cols 271–5.
14. Joseph Görres, *Die christliche Mystik*, 4 vols (Regensburg, 1836–42). See Mary Gonzaga, 'The Mysticism of Johann Joseph Görres as a Reaction against Rationalism'. Dissertation submitted to the Faculty of the Catholic University of America, Washington, 1920.
15. Montague Summers, *The History of Witchcraft* (London: K. Paul, Trench, Trubner & Co., 1926), p. xvi.
16. It was initially translated into English under the title *The Sorceress* (Paris: Carrington, 1904) and later as *Satanism and Witchcraft* (New York: Citadel, 1965). On Michelet see Norman Cohn, *Europe's Inner Demons: An Enquiry Inspired by the Great Witch Hunts* (London: Pimlico Press, 1975), pp. 105–7; Oscar A. Haac, *Jules Michelet* (Boston: Twayne, 1982); Paule Petitier, *'La sorcière' de Jules Michelet: l'envers de l'histoire* (Paris: Champion, 2004); Ceri Crossley, *French Historians and Romanticism: Thierry, Guizot, the Saint-Simonians, Quinet, Michelet* (London: Routledge, 1993); Linda Orr, 'A Sort of History: Michelet's *La sorcière*', *Yale French Studies* 59 (1980) 119–36.
17. See Cohn, *Europe's Inner Demons*, chapter 7.
18. Étienne Léon de Lamothe-Langon, *Histoire de l'inquisition en France depuis son établissement au XIIIe siècle, à la suite de la croisade contre les Albigeois, jusqu'en 1772, époque définitive de sa suppression* (Paris, 1829).
19. Joseph Hansen, *Zauberwahn, Inquisition und Hexenprozeß im Mittelalter* (Munich: Oldenbourg, 1900); Hansen, *Quellen und Untersuchungen zur Geschichte des Hexenwahns und der Hexenverfolgung im Mittelalter* (Bonn: C. Georgi, 1901).
20. Cohn, *Europe's Inner Demons*, chapter 7; Norman Cohn, 'Three Forgeries: Myths and Hoaxes of European Demonology', *Encounter* 44 (1975) 11–24; Richard Kieckhefer, *European Witch Trials: Their Foundation in Popular and Learned Culture, 1300–1500* (Berkeley: University of California Press, 1976).
21. See B. Stuchtey and P. Wende (eds), *British and German Historiography, 1750–1950* (Oxford: Oxford University Press, 2000); George Iggers and James Powell (eds), *Leopold von Ranke and the Shaping of the Historical Dimension* (Syracuse: Syracuse University Press, 1990).
22. Sigmund von Riezler, *Geschichte der Hexenprozesse in Bayern: im Lichte der allgemeinen Entwicklung dargestellt* (Stuttgart: Cotta, 1896); Wolfgang Behringer, *Witchcraft Persecutions in Bavaria* (Cambridge: Cambridge University Press, 1997; first published in German, 1987), pp. xiii, 23, 25.
23. Gustav Roskoff, *Geschichte des Teufels: eine kulturhistorische Satanologie von den Anfängen bis ins 18. Jh.* (Leipzig: Brockhaus, 1869).
24. Wilhelm G. Soldan, *Geschichte der Hexenprozesse* (Stuttgart: Cotta, 1843).
25. Monter, 'The Historiography of European Witchcraft'.
26. Paul Maria Baumgarten, *Die deutschen Hexenprozesse* (Frankfurt, 1883); Johann Diefenbach, *Der Hexenwahn vor und nach der Glaubensspaltung in Deutschland* (Mainz, 1886).
27. Chadwick Hansen, *Witchcraft at Salem* (London: Hutchinson, 1970), p. 12.
28. J. J. Auchmuty, *Lecky: A Biographical and Critical Essay* (Dublin: Hodges, Figgis & Co., 1945); H. Montgomery Hyde (ed.), *A Victorian Historian: Private Letters of W. E. H. Lecky 1859–1878* (London: Home & Van Thal, 1947); Donal McCartney, 'Lecky's Leaders of Public Opinion in Ireland', in *Irish Historical Studies* 14, 54 (1964) 119–94; Donal McCartney, *W. E. H. Lecky, Historian and Politician 1838–1903* (Dublin: Lilliput Press, 1993).

29. Christina Larner, *Enemies of God: The Witch-Hunt in Scotland* (Oxford: Blackwell, 1983), p. 34.

30. Henry Charles Lea, *Superstition and Force: Essays on the Wager of Law – The Wager of Battle – The Ordeal – Torture* (Philadelphia, 1866); Lea, *A History of the Inquisition of the Middle Ages*, 3 vols (New York: Harper & Brothers, 1887); Lea, *History of the Inquisition of Spain* (New York: Macmillan, 1906–08).

31. John M. O'Brien, 'Henry Charles Lea: The Historian as Reformer', *American Quarterly* 19, 1 (1967) 106.

32. Henry Kamen, *The Spanish Inquisition: An Historical Revision* (London: Weidenfeld & Nicolson, 1997), p. 312.

33. Henry Charles Lea, 'Ethical Values in History', *American Historical Review* 9, 2 (1904) 245.

34. Henry Charles Lea, *Materials Toward a History of Witchcraft*, 3 vols (Philadelphia: University of Pennsylvania Press, 1939).

35. Hansen, *Zauberwahn, Inquisition und Hexenprozeß im Mittelalter*; Hansen, *Quellen und Untersuchungen zur Geschichte des Hexenwahns und der Hexenverfolgung*. The work of Moriz Ritter (1840–1923) should also be mentioned. His three-volume *Deutsche Geschichte im Zeitalter der Gegenreformation und des Dreißigjährigen Krieges* (Stuttgart: Cotta, 1889–1908) for many years provided a valuable synopsis of the development and geographical distribution of the German witch trials.

36. On Morzine, see Ruth Harris, 'Possession on the Borders: The "Mal de Morzine" in Nineteenth-Century France', *Journal of Modern History* 69 (1997) 451–78; Jacqueline Carroy, *Le mal de Morzine: De la possession à l'hystérie* (Paris: Solin, 1981); Laurence Maire, *Les possédées de Morzine, 1857–1873* (Lyon: Presses Universitaires de Lyon, 1981).

37. Franz Josef Vonbun, *Beiträge zur deutschen mythologie. Gesammelt in currhaetien* (Chur: L. Hitz, 1862), p. 94. See also Peter Strasser, *Ein Sohn des Thales: Franz Josef Vonbun als Sammler und Editor Vorarlberger Volkserzählung* (Frankfurt am Main: P. Lang, 1993); Manfred Tschaikner, *'Damit das Böse ausgerottet werde': Hexenverfolgungen in Vorarlberg im 16. und 17. Jahrhundert* (Bregenz: Vorarlberger Autoren Gesellschaft, 1992); Karl Simrock, *Handbuch der deutschen Mythologie mit Einschluß der nordischen* (Bonn: A. Marcus, 1855); Friedrich Leberecht Wilhelm Schwartz, *Der heutige Volksglaube und das alte Heidentum mit Bezug auf Norddeutschland und besonders die Marken* (Berlin, 1849); Wilhelm Mannhardt, *Germanische Mythen*, 2 vols (Berlin: Schneider, 1858).

38. Ignaz Zingerle, *Barbara Pachlerin die Sarnthaler Hexe und Mathias Perger der Lauterfresser: Zwei Hexenprozesse herausgegeben* (Innsbruck: Wagner'schen buchhandlung, 1858).

39. Fritz Byloff, *Hexenglaube und Hexenverfolgung in den österreichischen Alpenländern* (Berlin: De Gruyter, 1934); Byloff, *Das Verbrechen der Zauberei (crimen magiae): Ein Beitrag zur Geschichte der Strafrechtspflege in Steiermark* (Graz: Leuschner and Lubensky, 1902); Byloff, *Volkskundliches aus Strafprozessen der österreichischen Alpenländer mit besonderer Berücksichtigung der Zauberei- und Hexenprozesse 1455 bis 1850* (Berlin: De Gruyter, 1929).

40. Ludwig Rapp, *Die Hexenprozesse und ihr Gegner aus Tirol* (Innsbruck: Wagner, 1874); Hartmann Ammann, 'Der Innsbruker Hexenprocess von 1485', *Zeitschrift. des Ferdinandeums für Tirol und Vorarlberg* 34 (1890) 1–87.

41. There is no space to consider the presentation of witchcraft on the stage, but see, for example, Jennifer Götz, 'Hexenvorstellungen im Theater des 17.–19. Jahrhunderts eine Untersuchung anhand ausgewählter Stücke,' Magisterarbeit, München,

2003; Christa Tuczay, 'Hebbels Agnes Bernauer – Rezeption der Hexenthematik', in *Hebbel, Mensch und dichter im Werk*, series 8, der Schriftenreihe der Friedrich Hebbel Gesellschaft (Berlin, 2004), pp. 71–87. Suffice it to say that apart from a few works like Friedrich Hebbel's *Agnes Bernauer*, the subject was generally treated in a frivolous fashion.

42. Michel Zink, *The Enchantment of the Middle Ages*, trans. Jane Marie Todd (Baltimore: Johns Hopkins University Press, 1998), p. 13.

43. Gustav Freytag, *Bilder aus der deutschen Vergangenheit*, 2 vols (Leipzig: Hirzel, 1859), esp. p. 465. See also Lynne Tatlock, 'Realist Historiography and the Historiography of Realism: Gustav Freytag's *Bilder aus der deutschen Vergangenheit*', *German Quarterly* 63, 1 (1990) 59–74.

44. Andrea Rudolph, 'Elisabeth v. Maltzahns Roman "Ilsabe" (1897). Die Hexenkeller der Alten Burg Penzlin im Horizont von Kulturkampf und Reichskrise', in Dieter Harmening and Andrea Rudolph (eds), *Hexenverfolgung in Mecklenburg. Regionale und überregionale Aspekte* (Dettelbach: J. H. Roll, 1997), pp. 57–77; Inge Schöck, 'Hexen heute. Traditionelle Hexenglaube und aktuelle Hexenwelle', in Richard van Dülmen (ed.), *Hexenwelten. Magie und Imagination vom 16.-20. Jahrhundert* (Frankfurt am Main: Fischer Taschenbuch Verlag, 1987), pp. 283–305; Ulrike Stelzl, *Hexenwelt. Hexendarstellungen um 1900* (Berlin: Frolich & Kaufmann, 1983); Hilde Schmölzer, *Phänomen Hexe. Wahn und Wirklichkeit im Laufe der Jahrhunderte* (Wien: Herold, 1986), pp. 74–83; Julia Bertschik, 'Von der Hexe zum Vampir: Dämonologische Geschlechterbilder in den historischen Erzähltexten Wilhelm Raabes', in Marion George and Andrea Rudolph (eds), *Hexen-Sorcières* (Dettelbach: Roll, 2004), pp. 271–85.

45. See Marion George, 'Vom historischen Fakt zur Sinnfigur der Moderne. Zur Gestalt der Hexe in Tiecks Novelle Hexensabbat (1832)', in George and Rudolph, *Hexen-Sorcières*, pp. 271–85.

46. See Jörg Kraus, *Metamporphosen des Chaos. Hexen, Masken und verkehrte Welten* (Würzburg: Königshausen & Neumann, 1998), p. 215.

47. Walter Scherf, 'Die Hexe im Zaubermärchen', in van Dülmen, *Hexenwelten*, pp. 219–53. Leander Petzoldt parallels the narrative elements in the *Malleus maleficarum* with nineteenth-century legends in 'Das Bild der Hexe in der populären narrativen Tradition des 19. Jahrhunderts. Zur Wirkungsgeschichte des Malleus maleficarum', in George and Rudolph, *Hexen-Sorcières*, pp. 75–91.

48. See Burghart Schmidt's latest well-researched monograph, *Ludwig Bechstein und die literarische Rezeption frühneuzeitlicher Hexenverfolgung im 19. Jahrhundert* (Hamburg, 2004).

49. Kurt Gamerschlag, *Sir Walter Scott und die Waverley novels: eine Übersicht über den Gang der Scottforschung von den Anfängen bis heute* (Darmstadt: Wissenschaftliche Buchgesellschaft, 1978).

50. See, for example, Wolfgang Müller-Funk, *Die Kultur und Ihre Narrative* (Wien: Springer, 2001), p. 102. See also Sonja Ausserer, *Hexe, Nymphe, Zauberin: Das Motiv der dämonischen Frau in der deutschsprachigen Prosaliteratur des 19. Jahrhunderts* (Universitätsbibliothek Innsbruck, 1996); Markus Kippel, *Die Stimme der Vernunft über einer Welt des Wahns: Studien zur literarischen Rezeption der Hexenprozesse (19.-20. Jahrhundert)* (Münster: LIT, 2001).

51. See Jeffrey Richards, 'The "Lancashire Novelist" and the Lancashire Witches', in Robert Poole (ed.), *The Lancashire Witches: Histories and Stories* (Manchester: Manchester University Press, 2003), pp. 166–87.

52. See James Sharpe, *Dick Turpin: The Myth of the English Highwayman* (London: Profile, 2004).

53. Heike Hartrath, *Fiktionalisierungen der Salemer Hexenverfolgung in amerikanischen Romanen vor 1860* (Frankfurt am Main: P. Lang, 1998); Bernard Rosenthal, *Salem Story: Reading the Witch Trials of 1692* (Cambridge: Cambridge University Press, 1993); G. Harrison Orians, 'New England Witchcraft in Fiction', *American Literature* 2, 1 (1930) 54–71.

54. Cited in Orians, 'New England Witchcraft', 61, n.35.

55. On Hawthorne's use of history see Deborah L. Madsen, 'Hawthorne's Puritans: From Fact to Fiction', *Journal of American Studies* 33 (1999) 509–17.

56. See Kippel, *Die Stimme der Vernunft über eine Welt des Wahns*. Patrick Bridgewater compares the Medusa-figure Sidonia with Burne-Jones' Lucretia Borgia, William Morris' Medea, Rosetti's Sister Helen, Swinburne's Catherine di Medici and Oscar Wilde's Salome; Bridgewater, 'Who's afraid of Sidonia von Bork', in Susanne Stark (ed.), *The Novel in Anglo-German Context. Cultural Cross-Currents and Affinities*. Papers from the Conference held at the University of Leeds from 15 to 17 September 1997 (Amsterdam: Rodopi, 2000), pp. 213–28. The portrait of Sidonia von Bork by Edward Burne-Jones is in the Tate Gallery. The importance of pictorial images for historiography, gender research and witchcraft historiography is emphasised in Sigrid Schade, 'Hexen und Hexenkünste. Hexen in der bildenden Kunst vom 16. bis 20. Jh.', in van Dülmen, *Hexenwelten*, pp. 170–218.

57. Barbara Belford, *Oscar Wilde: A Certain Genius* (London: Random House, 2000), chapter 1.

58. Maureen F. Moran, '"Light no Smithfield fires": Some Victorian Attitudes to Witchcraft', *Journal of Popular Culture* 33 (2000) 126.

59. See Jennifer Cizik Marshall, *Betrothal, Violence, and the 'Beloved Sacrifice' in Nineteenth-Century German Literature* (New York: P. Lang, 2001), pp. 137–43.

5

the reality of witch cults reasserted: fertility and satanism

juliette wood

There is room, there always will be, for studies of witchcraft, of haunting, of the occult. We only ask that these books should be written seriously ... The amateur and alas there are all too many of them who invade the occult are awaking forces of which they have no conception.

Montague Summers, *The Galanty Show*, p. 164.

In the first half of the twentieth century, two writers challenged the view that witchcraft had never been a 'real' cult. Margaret Murray (1863–1963), an Egyptologist and member of the Folklore Society, applied James Frazer's ideas on the universality of fertility cults at the primitive stage of cultural development to re-create an ancient religion. She put forward the idea that what Christianity demonised as witchcraft was a medieval survival of an ancient Palaeolithic/Neolithic religion based on the agricultural cycle. In contrast, Montague Summers (1880–1948), a convert to Catholicism, claimed that genuine satanic cults had existed in previous centuries, and that witches did indeed practise satanic rites. At the time there was widespread interest in occultism, and the modern disciplines of anthropology and psychology were emerging. The very different conclusions reached by Murray and Summers shared the assumption that the language of accusation could be matched to a real phenomenon, rooted in the past, but still affecting the present. Researchers have again become interested in witches' assemblies, although there are still widely expressed reservations about the way Murray and Summers fused popular motifs and historical material to produce their respective benign or satanic cults. Despite this, both writers have been influential in the development of modern popular religion and lifestyle movements.

Murray and Summers both based their view of witchcraft primarily on records of witch trials and on the writings of demonologists. They substantiated their findings using, in Murray's case, descriptions of folk customs and, in Summers' case, reports about satanic practices. They saw the relationship between witchcraft and the dominant culture as one of conflict, although they differed sharply on the nature of this relationship. For Murray, the Christian Church attempted to obliterate the already fading remnants of an ancient cult. For Summers, the Church was defending itself against a conspiracy of evil. Both focused on the ritual as presented in the trials and demonological literature, the Sabbat or Black Mass, seeing it as a pagan celebratory sacrifice or a perversion of Christian ritual. Neither attributed any inherent power to the rituals practised by either the latter-day pagans or the satanists. Murray saw witchcraft as a force for change which transformed a medieval world, dominated by feudal social structures and religious orthodoxy, into modern culture, an idea favoured in post-Enlightenment models of history. Despite her preface to Gerald Gardner's book on Wicca, one of the founding texts of modern Neo-Pagan witchcraft and the only context in which she discussed contemporary witchcraft, she does not seem to support witchcraft as a force in the modern world, and her later work suggested that it died out about the eighteenth century. Summers viewed witchcraft as an attempt to disrupt the social and moral order. For him, modern forms of witchcraft, namely the occult practices of spiritualism and satanism, remained a threat, although he adopted a more positive view of mystical activity. Murray's theories, through their influence on Gerald Gardner and his followers, have become part of the belief system of modern Wicca. Montague Summers' influence is more diffuse. By the 1960s, Summers' notion of a heretical satanic religion which challenged Christianity was no longer taken seriously in the field of witchcraft studies. However, his particular brand of Romantic satanism has had an enduring effect on horror films and novels, especially those of Dennis Wheatley, while his early studies on the English gothic novel and his interest in werewolf and vampire traditions have important echoes in modern 'goth' culture.

Of the two, Margaret Murray is more easily situated in the historiography of twentieth-century witchcraft studies. Despite her lack of formal qualifications, she worked at University College London with the distinguished Egyptologist, Sir William Flinders-Petrie. Although her books on witchcraft, the work for which she is best known, were written in the latter part of her life, she published several articles before *The Witch Cult in Western Europe* in 1921. Murray believed that what the medieval Church called witchcraft was 'an old religion' based on the natural cycle of vegetation. This ancient religion, whose rituals were dominated by a horned hunting god and his consort, had been practised in Neolithic Europe, and despite medieval persecution, traces of the cult, which died out around the eighteenth century, could still be found in seasonal folk practices. Murray refined these ideas in articles and in two further books, *The God of the Witches* (1933) and *The Divine King in England* (1954). She expanded Frazer's ideas about the dying god and the divine king into a theory of periodic voluntary human sacrifice linked to covens and Sabbats. By the time she wrote *The Divine King,* her notion of voluntary royal sacrifice extended to the Stuart kings whose position on the divine right of kings she interpreted, rather eccentrically, as belief in divine kingship. This model of ritual renewal underpinned her research throughout her life, and her last book, *The Genesis of Religion* (1963), attempted to trace the dominance of the male over a female deity in religion. With the rise of modern paganism, Murray's ideas have gained new, and even greater, currency. *The Witch Cult* was reprinted in 1952, appeared in paperback in 1962, and is now available on the internet.[1] Her work has been an important influence on Wicca since the 1950s, although many of its practitioners today readily point out that their doctrines are not dependent on the historical reality of a cult as re-created by Murray.[2]

Murray was active in the Folklore Society from the late 1920s until her death in 1963. Her ideas about the survival of a primitive religious cult have become problematic for professional folklorists, embodying as they do assumptions associated with J. G. Frazer's three-stage model of rising human progress (magic > religion > science), which was outdated in relation to academic folklore studies even by Murray's time. Nevertheless, this approach, which sees 'survivals' in traditional custom and uses 'folk memory' to justify links between past and current practices, still characterises the popular image of what folklore is. The topic of Murray and her relation to folklore studies was reopened recently in a Folklore Society presidential address which emphasised Murray's selective use of material in creating her 'old religion'. Although the sexual and anti-religious content of Murray's work caused some public controversy at the time of publication,[3] reservations expressed by historians and anthropologists since the 1970s focus on her idea that witchcraft originated as an ancient fertility cult and on her failure to consider the phenomenon in the wider context of sympathetic magic.

Murray regarded the members of her ancient fertility cult as victims of bigotry and never made a secret of her antagonism to organised religion, but neither did she suggest that they conserved a tradition of ancient wisdom. Her main concern was the

existence through the Middle Ages of a primitive religion in Western Europe. No doubt that the cult was spread in early times through central and Eastern Europe and the Near East. The Old Religion was seldom recorded for Paganism belonged there as here to the inarticulate uneducated masses who remained for many centuries untouched by the new religion.[4]

The link between witchcraft and a 'secret tradition' appealed to the occult movement. Substantial extracts from *The Witch Cult* appeared in the *Occult Review* and provided a degree of popularisation for Murray's work.[5] However, when the book was re-published after the witchcraft laws had been repealed in 1951, its influence was felt on the pagan revival through the writings of Gerald Gardner.[6] In 1954, Murray wrote a preface to his book, and there is a suggestion that she herself was a witch, but she vigorously denied this.[7] Current criticisms of Murray's work focus on her methodology and, more crucially, on her chosen cultural model, problems already identified by earlier critics. George Lyman Kittredge and G. L. Burr criticised her overly literal interpretation of trial material, as did Cecil L'Estrange Ewen whose work was savaged by Murray, and even the review in *Folklore* noted a few problems.[8] A review in the Roman Catholic *Tablet* noted that a misheard version of a common Catholic prayer was insufficient grounds to suggest that Marian devotion was based on goddess worship. Murray refused to accept criticism on this point which involved her publishers in a costly reprint.[9] Montague Summers also caught a piece of circular reasoning, namely Murray's suggestion that similarities between the witches' Sabbat and the Christian Mass arose because the pagan ritual influenced the Christian one, and thus demonstrated the antiquity of the former.[10]

A more fundamental problem with Murray's work is the model itself, an ancient fertility cult of the type suggested by Frazer. For Murray, the Devil at witch assemblies, a prominent feature in trial accounts and demonological literature, was a man (priest) in a mask. The masked man-priest in Murray's witch cult replaced an ancient priest-king who was sacrificed periodically as the human embodiment of a dying and reviving horned god of vegetation. This was seemingly substantiated by various factors; a symbolic marriage between the king and the land, horned deities, a Palaeolithic drawing of a figure with horns, and seasonal customs using horned masks and mock deaths. These were taken as proof that a pan-European cult of the dying god survived as a medieval folk religion. Unfortunately, Murray's position involved a priori assumptions of considerable proportions imposed on evidence which is fragmentary, ambiguous and, as her critics are quick to point out, often manipulated or poorly argued. There is, for example, little evidence that an actual human sacrifice was ever part of European kingship rituals or that seasonal customs are very ancient. The fertility aspect of the British/Gaulish horned deity is not so clear as Murray assumed, nor is the identity of the *Trois frères* cave painting.[11] The consensus today is that this picture of a nature/fertility cult centred on a male and female who represent god and goddess or divine king and consort is a modern construct rather than a record of history. Few scholars today give serious consideration to her suggestion that historical figures like William Rufus, Joan of Arc, Thomas Becket and Gilles de Rais were willing victims in a periodic human sacrifice needed to ensure agricultural fertility and social stability.[12]

Murray's autobiography, written as she approached her hundredth birthday, reveals a clear mind convinced that her witchcraft research, 'the interpretation of beliefs and ceremonies of certain ancient forms of religion', was her most important contribution.[13] Her ideas about the origin of witchcraft developed through articles on topics such as the number 13, child sacrifice, animal transformation, the Devil's mark, the witch coven and the familiar which appeared in *Folklore* and the journal *Man* before her first book was published. Journals like these played host to lively debates on the nature of culture, and most influential was the theory of cultural evolution that influenced the development of social sciences at this period. James Frazer identified references to the death and resurrection of divine and semi-divine figures in the context of seasonal activity and the return of vegetation in ancient texts. He interpreted these as survivals of a universal ritual aimed at ensuring the continuance of the fertile seasonal cycle which propitiated natural forces through the actions of significant individuals such as priests, kings and, eventually, their substitutes. Frazer's rationalist stance saw primitive man as a kind of proto-rationalist who, lacking a scientific understanding of cause and effect, attempted to control the forces of nature by means of ritual. The sacrifice of the king in the primitive world of magic was replaced by substitutes and symbols as magic gave way to religion and eventually disappeared, except in folklore practices, as man's rational faculties developed.

The limitations of Frazer's theories about the evolution of religious behaviour have been treated elsewhere.[14] The concept of the 'divine king' as a representative of a vegetation god influenced Murray's approach to Egyptian archaeology as well as witchcraft, and their methods have much in common.[15] Both attempted to reconstruct the mental world of primitive man from texts, such as classical sources, witchcraft trials and demonological tracts. Both assumed that man's primitive psyche evolved towards rationality, and thus they shared a mistrust of organised religion. Frazer eventually included the Christian Resurrection story among his examples of the dying god myth, whereas Murray differed from Frazer in placing her cult in conflict with Christianity. Instead of a cultural dynamic which moved from magic (which attempts to control natural forces) to religion (which focuses on supernatural forces) to civilisation (which depends on reason), Murray suggested that a native European magical cult continued after the Christian religion had been introduced and created tensions which could only be resolved violently. She endowed it with the hallmarks of a counter-culture movement which could bring about social change, but also caused social unrest.

The formative period for Murray's ideas, fin-de-siècle Europe, was a time of both secularisation and re-spiritualisation. Secular and radical modern thinkers viewed old forms of authority as oppressive, and the idea of a counter-culture working in secret, but espousing remarkably modern, liberal ideas, became popular. There was widespread interest in the occult and in mysticism (the terms often used interchangeably), a growing interest in mystery religions, and in religion itself. New forms of spirituality appeared, and new disciplines, such as anthropology, helped to understand the past and the wider world.[16] Such broad cultural concerns formed the background to Murray's thinking, and the institution where she worked, University College London, was strongly associated with this new agenda. In an essay entitled 'Woman as Witch: Evidence of Mother-Right in the Customs of Mediaeval Witchcraft', another colleague at University College London, the statistician and polymath, Karl Pearson, anticipated substantial elements of her theory.[17] The essay interpreted features of medieval witch gatherings such as communal feasting, dancing and a sacrifice under a sacred tree, together with modern agricultural folklore, peasant festivals and dances as fossils of an 'old mother age' of prehistoric civilisation. Inheritance during the period of 'mother-right' (that is, matriarchy) descended through the female. Thus, the presiding spirit of witchcraft was a female served by a male deity. Gradually, the male deity became prominent, eventually becoming the 'devil' of the witch trials. Pearson's essay drew heavily on ideas about medieval culture put forward by the French historian, Jules Michelet, and on Erich Neumann's concept of matriarchy. Again following Michelet, Pearson characterised Joan of Arc as a white witch or folk healer and suggested that medieval female witches were followers of an old folk religion. In other words, witchcraft was a predominantly female counter-cultural movement forced into confrontation with authority.

The possible survival of ancient civilisation and the idea that a secret cult offered an alternative history to the oppressive forces of feudalism and the Church was a potent image in an age of rationalism and revolutionary politics. Few writers stated this as dramatically and persuasively as Jules Michelet (1798–1874).[18] Even though his sources have been largely discredited, his work had an immense influence in creating the notion that witchcraft was persecuted because Christianity regarded it as a rival religion. Murray may very well have come to Michelet's ideas through Pearson's work. However, other aspects of her witch cult resemble Charles Leland's *la vecchia religione*. Leland's description of an organised witch cult in medieval Italy was heavily influenced by Michelet and appeared in the 1890s.[19] Her mentor in archaeology, Flinders Petrie, speculated about the significance of popular rites associated with megalithic monuments and may have reinforced Murray's interest in European Neolithic culture. The eminent religious historian, E. O. James, is another possible influence. His study of mystery religion, *The Cult of the Mother Goddess*, complements *The Witch Cult*.[20] Both present rituals as the product of the needs of an agricultural world in which dying god, divine marriage, and the rebirth of vegetation re-enacted in a great annual festival are central concepts. However, Frazer's agricultural rituals and James' goddess cult were situated in the context of ancient civilisations or remote primitive ones. Murray's work sought evidence closer to home, namely a European Neolithic religion whose rituals centred on magical maintenance of the vegetative cycle and survived as a misunderstood and persecuted cult called witchcraft. She presented her cult as a real and living part of a relatively recent medieval European past, made it a force for social conflict and change, and, by extension, a factor in the emergence of modernity. By the time she wrote *The Divine King* in the 1950s, Murray's ideas about this ancient cult had developed even further. Christianity became a foreign religion, and witchcraft a 'fiercely insular' native cult which resisted subjugation for centuries. The idea of English resistance to foreign domination was perhaps understandable in the wake of two world wars, but even her staunchest supporters found it difficult to accept that the principle of the divine right of kings invoked by Stuart monarchs could be equated with a divine king.

In her autobiography, Murray gives an account of her perception of the nature of the central figure in the witch meeting. Her claim that while working 'only from contemporary records' she 'had the sort of experience that sometimes comes to a researcher' (namely that the Devil was a disguised man, not a demon) has a heightened dramatic tension often found in auto-biographical discourse. This convinced her that witches were members of a primitive form of religion, and a male deity, represented by a masked priest, continued to be the focal point of her witch cult.[21]

Historians of witchcraft, from L'Estrange Ewen onwards, suggested that her failure to use newly available records undermined her theories. She insisted that she consulted original documents, but these were mostly printed trials records and always mined for the information that she expected to find there. Much the same can be said of the 'fieldwork' that supported her findings. Her working methods were typical of a time when collecting meant finding a 'good' informant who would have access to other sources. This meant, in practice, educated men and women collecting from servants, employees and rural workers. Murray's work addressed a number of contemporary concerns such as changing sexual identities, apparent in the contrasting roles of god and goddess, and the attraction in an industrial society for the idea of a religion based on nature. Murray transformed the rituals of sacred kingship, as described in *The Golden Bough,* into a religion which existed within the context of a later and, according to this anthropological model, more developed culture. Here it was not merely a fragmentary survival, but a living entity which could challenge the hegemony of authority.

Her work influenced the development of the Wicca movement. Gerald Gardner, a central figure in this revival, had contacts with both the Folklore Society and Margaret Murray. The concept of a priest and priestess, which lies at the centre of Gardner's writing, was certainly influenced by her theories. The articulation of Frazer's ancient vegetation cult as a European, and more specifically English, cult gave Gardnerian magic a unique nativist character. Although Gardner claimed that his sources antedated Murray, he acknowledged a debt of gratitude for her sympathetic treatment of witch beliefs.[22] The very flamboyance of Gardner's character makes it difficult to evaluate some of his assertions, but it is now clear that his sources included the 'high magic' traditions of practitioners like Aleister Crowley and local Masonic groups.[23] These traditions are themselves not very ancient, but they place Gardner more firmly in the occult context of this period, a context relevant to Murray and even more so to Montague Summers.

Vivianne Crowley, one of the earliest Wicca historians, acknowledged Murray's importance, but staunchly defended the prior existence of an ancient cult. Her chapter on 'The Witch God' mirrored Murray's argument and use of material such as the Palaeolithic 'shaman' cave drawing about which Murray wrote an enthusiastic note.[24] In the first edition of Crowley's book, Murray is given pride of place in a discussion of Wicca's predecessors and is mentioned several times in connection with king sacrifice and initiation rituals. However, the revised edition underplays this role. Although Crowley stresses Murray's testimony to Gardner, she omits reference to the (by now far-fetched) self-sacrifice of English kings, and mentions Frazer only (that is, omits the earlier references to Murray) in connection with ritual king murders.[25] Murray's thesis affected the development of Wicca in America as well. The writings of Leo Louis Martello (1930–2000) parallel Gerald Gardner's work in many ways. Like Gardner, Martello translated his own heritage as an Italian-American immigrant into a magical life and became the central figure in a modern witch cult. The 'old religion' in Martello's background was a combination of Italian witchcraft traditions, in which the influence of Leland is strong, and Murray's witch god cult.[26]

Unfortunately for Murray's theory, the idea of a universal agricultural cult centred on a dying god did not survive in modern anthropology. The practice of modern Wicca has also changed and is broader based, and more influenced by feminist principles and ecological thinking. Its increased confidence as a movement makes it less dependent on arguments about continuity with an ancient period, and the reality or lack thereof, of the model of witchcraft described by Murray. The use of depositions and accusations and the writing of micro-history have shifted the focus away from elements that Murray associated with paganism, the coven, the Sabbat and onto beliefs associated with healing and cursing in a more local context. These popular beliefs were not associated with a specific cult and they certainly did not relate to the worship of a horned god. However, five years after the publication of *The Witch Cult*, the same sources were to be given an entirely different spin.

Montague Summers' (1880–1948) scholarly interests included theatre and literature as well as the supernatural.[27] He undertook pioneering research on the gothic novel, edited the plays of early dramatists and supported new productions of Restoration plays for London audiences in the 1920s. His work on the reality of satanic witchcraft caused interest when it first appeared, but was largely overlooked by witchcraft scholarship in the 1960s. However, his ideas that ancient evils could erupt into modern urban life were already entrenched in the genre of horror film and fiction and with the development of alternative lifestyles were moving into a different arena.

Augustus Montague Summers was born in Clifton near Bristol into a comfortable middle-class Victorian family. He attended Trinity College, Oxford and lived there during the 1930s where his knowledge of classical and modern languages enabled him to make good use of the Bodleian Library. His early European travels exposed him to Roman Catholicism, and he became a convert in 1909. Self-image was crucial to Summers' life and work, and he affected the 'clerical attire reminiscent of some exotic abbé of the time of Louis Quatorze'.[28] His personal tastes looked back to a romanticised eighteenth-century world, one without revolutionary change in which behaviour and imagination were liberated and belief was orthodox. One commentator described him as a mixture of 'spooks and sex and God', and, given his temperament, his lifestyle and his homosexuality, rumours inevitably clung to him.[29] As a student, he had an interest in elaborate Church ritual and an intense belief in the supernatural. His early poetry, in the style of Wilde and Swinburne, included decidedly baroque religious poems, ones with homoerotic overtones and perhaps a hint of satanism. An actor friend claimed that he and Summers were actively involved with homosexuality and the Black Mass from 1918 to 1923, but the truth behind this is difficult to assess.[30] Summers' career in the Anglican Church certainly ended after an unproven charge of sexual misconduct. His conversion to Catholicism in 1909 has been likened to that of 'the actor who had found a new role more suited to his talents, or the magician who had found a more authentic and powerful magic'.[31] He claimed that he had been ordained in Italy and was authorised to live privately and devote himself to scholarship, His biographers, in general, accept that Summers' accounts of his clerical status were evasive and inconsistent.[32] In keeping with his sincere, but theatrical, piety, he maintained the image of an ordained priest and said Mass in private oratories fitted out with relics and saints' portraits. Many of his books contained attacks on spiritualism whose popularity worried some Victorian and Edwardian intellectuals as strongly as it attracted others.[33] Summers set spiritualism in the context of historical witchcraft and necromancy. His position as a cleric certainly added credence to his writing on the dangers of satanism, and this may explain his reluctance to clarify the circumstances of his ordination. He insisted that only specialists in theology and psychology (another of his interests) were adequately qualified to discuss the 'very dark and terrible aspect' of the subject.[34]

Both *The History of Witchcraft and Demonology* and *The Geography of Witchcraft* appeared in a history of civilisation published by Kegan Paul whose intent was to examine changes in post-First World War attitudes to culture. The series editor, C. K. Ogden, commissioned Summers to do two volumes on witchcraft and out of series volumes on werewolves and vampires. Summers' stance in these volumes, and in all his writings, was that modern society had, to its detriment, abandoned the supernatural for science. Ogden, himself a Roman Catholic convert, may have favoured Summers' approach for this reason, and Summers certainly felt he was less favourably treated by Ogden's successor.[35]

The first book on witchcraft appeared in 1926, just five years after Murray's *Witch Cult*. Like her, he had been studying the subject for many years, and it too provoked controversy. In his autobiography Summers recounts with relish that the 'vulgarians snapped and snarled'.[36] The prevailing historiographic model at the time proposed rational explanations such as superstition or political unrest for the supernatural phenomena associated with witch hunts. Summers, by contrast, accepted the reality of the supernatural and attempted to explain its hold on the human psyche. He never tried to explain it away and always assumed that the practices of witchcraft were linked to beliefs not delusions. The ideas on witchcraft and satanism set out in this first book did not change much in anything he wrote subsequently. A second volume, *The Geography of Witchcraft,* followed as well as several later, more popular, treatments, and related studies on the vampire and the werewolf brought together an even greater variety of related material.[37]

According to Summers, the cult of magic was not an outcome of primitive/peasant mentality. For him, evil was real, and witches were members of a 'powerful secret organization inimical to church and State'. Although not paranoid, he accepted the reality of conspiracy and quoted early writers on conspiracy theories such as Nesta Webster.[38] On the surface this seems odd in someone who relished the artistic liberality of the late seventeenth and eighteenth centuries, but is perhaps easier to understand in a man who deplored 'the heavy and cross materialism' of post-Enlightenment England which 'intellectually disavowed the supernatural'. For Summers, rationalism meant abandoning the spiritual and political certainties of the past. He mistrusted the secular tendency in modern society where a phenomenon such as spiritualism, which he equated with necromancy, could be tolerated. However, he felt that

> rationalistic superstition is dying fast. The extraordinary vogue for and immense adherence to spiritism would alone prove that, whilst the widespread interest that is taken in mysticism is a yet healthier sign that the world will no longer be content to be fed on dry husks.[39]

Summers also took up the idea of mystical transformation, but this, unlike his work on the darker side of the supernatural which combined historical reference and contemporary anecdotes in a persuasive and attractive manner, has not survived changing scholarly fashions.[40]

Murray's 'Dianic cult' was among the rationalistic explanations which he felt merely tried to explain away the existence of diabolic rites: 'amongst English writers' wrote Summers,

> witchcraft in Europe has not of recent years received anything like adequate attention from serious students of History, ... Magic ... [is] the subject of vast and erudite studies mostly from an anthropological and folkloristic

point of view, but the darker side of the subject the History of Satanism seems hardly to have been attempted.

Like later historians, he criticised Murray's a priori model and selective use of material. He also distrusted the 'Dianic cult' hypothesis because of its perceived link with developments such as the French Revolution, 'illuminism' and rationalism. These were factors that led to the secularisation of culture and undermined Summers' romantic view of a stable past. He referred to her interpretation of William Rufus as a king sacrifice 'a curious and irresponsible fantasy'. His own suggestion that the king's death was 'due to homosexual jealousy' is, however, equally coloured by concerns with his own sexual identity.[41] Nevertheless, he recognised similarities between Murray's 'Dianic cult' and ideas presented by Leland and his sources, similarities which have only recently been discussed in witchcraft scholarship.[42] The idea that witches were devotees of an ancient goddess cult remains an important element in modern witch belief and in scholarly discourse, and Summers was among the first to notice the lines of influence. The eighteenth-century writer and sceptic, Girolamo Tartarotti, dismissed the supposed diabolism of witch Sabbats as nothing more than meetings of latter-day devotees of Diana. Summers notes how Tartarotti's ideas influenced both Leland and Michelet, and through them Margaret Murray.[43]

Like writers such as Margaret Murray, Jules Michelet and Charles Leland, Summers characterised witchcraft as a covert rebellion against authority. Indeed one of the reasons for the popularity of the occult at the turn of the twentieth century (and of paganism today) is exactly that it offers an alternative to more standard forms of spirituality. The cult these two writers describe is similar in many ways and includes a Sabbat assembly with the appearance of a 'devil', as well as dancing, feasting and sacrifice. Summers regarded accounts of magic flight in confessions as comparatively rare and assumed that the participants were deluded in some way. Like Murray, he underplayed certain aspects of the descriptions by assuming a 'rational' stance.[44] The Devil at the Sabbat presented just such a problem of credibility. While Summers favoured the theological position of Augustine and Aquinas that witches had no intrinsic power to harm, he also accepted the real existence of the Devil, an entity who could therefore appear at a Sabbat. Nevertheless, in his opinion, the Sabbat leader was an ordinary man attempting to disguise his real identity at an elaborate burlesque of the 'ritual of Holy Mass'.[45] Although Murray and Summers both assumed that witches were real people doing real things, overtly supernatural occurrences had no place in human ritual. Murray typically ignored or dismissed them as exaggeration. This may have been a factor in Murray's selectivity with regard to her sources, a feature of her methodology, which, not surprisingly, antagonised her critics. Summers, on the other hand, believed in supernatural phenomena, although his theological stance on satanism denied it any intrinsic power. In this context, the idea of exaggeration in the original sources provided a very useful filter to eliminate extraordinary elements and keep material within the parameters of logic and credibility.

Summers edited important works on demonology and on high-profile witch trials, frequently presenting them to English-speaking audiences for the first time. His main interest lay in material from the seventeenth century onwards which suited his classical education and knowledge of European languages. This considerable editorial feat provides the means to evaluate his attitudes to his material and his influence on later perceptions of it.[46] The best known of these editions is the fifteenth-century witchcraft treatise, *Malleus Malificarum*. J. Allen Ashwin provided the translation for this, and for several other works on demonology, while Summers did the editing, notes and bibliography. Although an able classicist, Summers was by no means a professional editor or Latin scholar. Modern editors, even of his Restoration works, can find him frustrating, and he was not above enhancing his material, the very shortcoming he seizes on in his criticism of Murray. For example, in a quote from a demonological tract translated by Ashwin, Summers substituted the word 'Devil' for 'demon', thereby creating a more intensely dramatic effect.[47] Summers' influence is also felt through other works he edited. His depiction of Matthew Hopkins as 'a charlatan and a deceiver' has contributed to the negative attitudes surrounding this figure and to Hopkins' continued prominence in the popular mythology of witch hunts.[48] His editions of Madelaine Bevant's *Confessions* and Remy's *Demonolatry* caused problems under the obscenity laws and reflect the 'blend of eroticism and religion' which feature strongly in his writing.[49] While this was no doubt a factor in his choice of subjects, he had a genuine interest in psychology, especially deviant psychology. He was a member of the British Society for the Study of Sex Psychology, although he left abruptly in 1923. Two of his talks, one of the first English studies of the Marquis de Sade and a discussion of Richard Barnfield's poems, appeared in print.[50]

He described his work on the existence of a satanic cult as an impartial investigation by the 'keenest minds of the centuries' whose writings 'were here in England practically unknown, or at any rate unread'.[51] He certainly helped to make works like the *Malleus* more widely known, and his extensive bibliographies provide additional insights into his use of material and his attitudes to his subjects. He glossed the sources in *History of Witchcraft and Demonology* (1926) with revealing annotations favouring the work of believers. For example, he added a caveat to his citation of Henry Charles Lea, the author of a pioneering study in the rationalist disbelief in the reality of witchcraft; while the work of Dr Frederick George Lee, a writer on occult subjects who believed the supernatural experiences he collected, is glossed as scholarly and valuable. He praised Wallace Notestein's *History of Witchcraft in England* (1911) for his facts, but not for his interpretation that confessions were the result of physical and mental pressure. He drew on the important two-volume collection of Thomas Wright, *Narratives of Sorcery and Magic* (1851), and on

Joseph Hansen's *Quellen und Untersuchungen* (1901), and included substantial listings of pamphlet literature.

From this distance in time, Summers' work stands out perhaps more than it should. He was not the only writer to consider the realities of satanism, and his research might usefully be seen in the context of a wider interest in the nature and reality of psychic and spiritual phenomena at the beginning of the twentieth century.[52] If Murray was seeking the popular roots of a counter-religion behind an elite distortion, Summers was looking for a distorted mirror image of a stable worldview. The core of his research drew on French material relating to the supposed existence of elite black magic cults, and he is an important conduit for ideas about the Black Mass. Descriptions of these satanic practices, supposedly active in France from the sixteenth to eighteenth centuries, were based on material from the historian Jules Michelet (1798–1874), the novelist/occult historian Joris-Karl Huysmans (1824–1893) and writers like Baron Etienne Lemothe Landon and Jean-Antoine Boullan.[53] These sources reconstruct the past from a post-Enlightenment position with the result that rumour and outright fabrications abound. The effects of this filtering are particularly strong in Summers' account of the Chambre Ardente affair. The air of delicious decadence which makes Summers so popular today, and also probably caused him to be dropped by serious witchcraft scholars, derives in large measure from the romanticism of writers like Huymans, Michelet and their sources. Summers' prose invites his readers to revisit the *ancien régime* and drift along the corridors of power overhearing racy gossip about covert magical activities involving Louis Quatorze, Cardinal Richelieu, latter-day Templars and assorted Renaissance princes. Unfortunately this moves the discussion away from serious examination of witchcraft. Significantly, it comes closer to gothic fantasy and to glimpses of Summers' social life in Paris when, for example, he alludes to a Folie Bergère review in 1925, called 'Le Sabbat and la Herse Infernale'.[54] This may explain his attraction for modern 'goth' culture. Summers presents an apparently real world, unlike the fantasy realms of writers like Lovecraft or Tolkien, which offers a venue for alternative identities.

Descriptions from various printed sources are supplemented by seemingly first-hand examples, such as a report of a Black Mass in the ruins of Godstow Nunnery in Oxford and one in a basement of a house in Merthyr Tydfil.[55] Elsewhere, Summers hinted that a satanic cult, centred on the figure of Francis Barrett, operated in Cambridge. Barrett's compilation of magical writing, *The Magus*, was published in 1801 and reprinted in 1875. An advertisement in the original edition offered lessons in magic by the author, and the eighteenth-century cunning man, John Parkins, was reputed to have been one of Barrett's pupils.[56] However, there is no hard evidence for a Cambridge coven, and this sounds suspiciously like a bit of Oxford versus Cambridge one-upmanship. Summers' methodology comes quite close to Murray (and Leland) on the occasions when he assumes the existence of organised cults behind genuine folk practices. This occurs in his description of the 'Mass of St Secaire', actually a cursing ritual first described in a nineteenth-century book of Gascon folklore, which has become standard fare for descriptions of the Black Mass by subsequent popular writers.[57]

Much of the material which Summers incorporated into his books is perhaps best understood in the context of contemporary legend, namely, rumours which express fears and anxieties and contain just enough detail to make them possible, but not enough to verify them incontrovertibly.[58] His influence on modern magical traditions is not as widespread as Murray's. However, like her he was inclined to see organised cults of long standing behind the folk practices he described. His interest in ritual magic cults, which he locates in late medieval and Renaissance contexts, do bring to mind the 'high magic' traditions which underpin the modern revivals associated with Crowley's Thelemic magic, John Dee's Enochian magic and, increasingly, it seems, the groups founded by Gerald Gardner. He is concerned with a learned magical tradition, not with a primitive cult dependent on seasonal changes. However, the two meet in Dennis Wheatley's novel *The Devil Rides Out* which links the horned god of Murray's witch cult with the demonic sacrifice in Summers' satanic one.

Summers did discuss witchcraft outside Europe, such as the Salem witch trials, and he was aware of American research. He knew, for example, that the American scholar, George Lincoln Burr, had reservations about his overly theological approach. His range of sources contracted dramatically in relation to magic and witchcraft outside Europe.[59] Here, Summers relied on surveys, often authoritative enough for their day, but lacking the breadth of material which he brought to the subject in Europe. William Crooks' *Folklore of Northern India* (1896) provided the bulk of the Indian material. Summers was dismissive of African magic, seeing it as a source of voodoo and witchcraft. His attitude is almost entirely dependent on studies which present voodoo as a travesty of Catholicism rather than a syncretic religion.[60]

Summers' influence on popular culture, in particular gothic horror, is undoubtedly more significant than his influence on witchcraft historiography. His personal reputation and his lifestyle overlap in a way that Murray's do not. His conversion to Catholicism, flamboyant dress style and eccentric habits of piety reflect his interests in the theatre and street life of eighteenth-century London, in gothic literature, elite satanic cults and the nightwalker folklore of vampires and werewolves in a way which ensured that he would become part of the tradition himself. The testimony of friends reveals a man childishly fond of ceremony and romantic posturing. He was a homosexual in a difficult time after the Oscar Wilde trials of the 1890s, an aspect of his life which his friends treat very tactfully. In actual fact he had a long and stable relationship with Hector Stuart Forbes, who inherited his papers and considerable library, and died two years after him in 1950. Just as Murray's reputation for the actual practice of witchcraft seems stronger in the United

States, so too does Summers' reputation as an occultist. The American publisher of his works, Felix Morrow, suggested that he was a member of some occult organisation. References to youthful involvement with satanic rites are at best illusive, and despite the macabre descriptions in his writing, there is no practical information on the occult.[61] While he condemned magic in the strongest terms, he does not advocate any evangelical attempts to eradicate it in the modern world. His attitude is that the application of purely rational categories, which characterise the modern age, cannot in the end explain away the existence of the supernatural or the need for a faith that had an external reference and a claim to truth.[62]

Montague Summers presented satanism as a ceremonial cult in which human sacrifice and orgiastic behaviour parodied the Christian Eucharist. His views achieved wider popular appeal through the novels of writers like Dennis Wheatley (1897–1977) and horror films like those produced by Hammer Film Studios. Many of Wheatley's novels have occult or supernatural themes.[63] The elaborate ceremonials, sacrifices and orgies owe much to the fevered decadence of events such as the Chambre Ardente affair as transmitted through the books of Montague Summers. Wheatley incorporated many other aspects of the occult revival, and he certainly never took Summers' theological line. His sensational descriptions with their emphasis on heterosexual sex and sadism owe much to Aleister Crowley as well. Nevertheless, the historical entries in Wheatley's picture-book encyclopaedia of satanism echo Summers' work even though no specific references are given.[64] Wheatley wrote introductions for over 40 mass-market horror and occult titles published by Sphere Books in the 1970s. Some items in this series, such as J. K. Huysman's novel *La Bas* ('Down There') and *Malleus Malificarum,* are directly dependent on Summers' editions. Others range from novels with occult themes by Charles Williams, John Cowper Powys, F. Marion Crawford and others, which demonstrate the wide interest in the occult among literary writers, to studies like Alfred Metraux's work on voodoo.[65]

Whatever the ultimate sources, the images of satanic rituals presented in Wheatley's novels have contributed to popular ideas about the nature of modern witchcraft and pagan cults. This is most notable in the cult classic *The Devil Rides Out,* which was published in 1934 and filmed in 1968. The novel helped fuse the notion of a pagan fertility cult as presented by Margaret Murray with the idea of satanic sacrifices as described by Montague Summers. The Hammer film version downplayed the xenophobia and jingoism of Wheatley's novel, but kept the idea of wealthy, sinister devil worshippers who were foiled by an exotic demonologist helped by a number of very English, slightly dim assistants. This served to reinforce the notion of satanism in the midst of middle-class England. Montague Summers' books have also influenced the perception of vampirism in contemporary popular culture both through his books, which added an element of satanic behaviour to the traditional *topos,* and were an important source for the Hammer vampire films of the 1970s.

These films revived as a trend which had begun with Bram Stoker and other Victorian novels by reinterpreting a Slavic folklore figure in a British context of gothic country houses and fog-obscured streets. Other Hammer films that popularised the figure of the werewolf and witch hunters like Matthew Hopkins also echo Summers' work.[66]

Like Murray, Summers became popular again in the 1960s, and also like her, he is the subject of an ever-expanding number of internet sites. His work on Restoration drama and the gothic novel anticipates contemporary interests in the psychological aspects of these genres and the way they treat taboo social subjects. He identified the appeal of the gothic novel with typically florid prose. 'The romantic writer ... created ... a domain which fancy built and fancy ruled', a land of mystery, beauty and wonder,

> this longing for beauty intermingling with wonder and mystery ... will express itself perhaps exquisitely in the twilight moods of romantic poets ... perhaps a little crudely and even a little vulgarly in tales of horror and blood.[67]

Summers' own fascination for the occult paraphernalia of witchcraft has been compared to the wizardry and bawdy that informs the texts of the gothic novel and late seventeenth-century drama.[68] Like the gothic and horror literature which fascinated him, he blurred the boundary between fictional and real in order to create an urbanised realm with vaguely medieval overtones where the Inquisition lurked in dark corners and nightstalkers walked the surrounding woods. With great verve, Summers re-located this gothic world in the apparently real context of Europe during the witch trials. This concern with the macabre aspects of the supernatural has a very modern feel, and the links between vampires and satanic masses, so beloved of horror films and popular exorcisms, owe much to his particular body of work. Perhaps his real legacy is that he combined all the elements of the gothic novel into an allegedly real satanism that creates a tension between reality and fiction that appeals so strongly to postmodern imagination.

For all their differences, Murray and Summers operated in the context of a post-Enlightenment understanding of historical process in which statements about witchcraft either from participants or accusers could be tested by reference to an empirical reality. For Margaret Murray, what was being described as a satanic meeting was in reality a pagan ritual. The practices at the witches' Sabbat described by prosecutors and demonologists were not simply an elite fiction but harked back to a pre-Christian popular thinking. Montague Summers was more romantic, and his attitude to historical witchcraft is underpinned by nostalgia for pre-Enlightenment social and religious stability which Murray's suffragette modernism rejected. For Summers, trial records and demonological tracts described not an 'old religion' but an inverted one for which the Christian Eucharist ritual still held power and meaning central.

Thus the Black Mass is a recurrent theme. Writing in the first half of the twentieth century, in a period interrupted by two world wars, they shared an interest in the meaning and future of society and a belief that scholarly writing could influence it. They differed on the motivation for the persecution, whether the authorities were attempting to obliterate a distinctive pagan sect or whether a genuine satanic cult was attempting to undermine Christian civilisation. Witchcraft scholarship has undoubtedly moved away from both these extremes and sees diverse activities being treated as antisocial under particular circumstances and labelled as pagan or satanic. However, Murray's bucolic pagan religion and Summers' gothic urban cult continue to have an enormous impact on the public understanding of the nature of witchcraft.

notes

1. Margaret Murray, *The Witch Cult in Western Europe* (Oxford: Clarendon Press, 1921; reprinted 1952, paperback 1962); Murray, *The God of the Witches* (London: Sampson Low, 1933); Murray, *The Divine King in England* (London: Faber and Faber, 1954); Murray, *My First Hundred Years* (London: W. Kinber, 1963); Murray, *The Genesis of Religion* (London: Routledge and Kegan Paul, 1963); see also 'Murray's Unlikely History', <www.wicca.timerift.net/murray.html>; <www.Sacred-texts.com/pag/wcwe>; <www.Sacred-texts.com/pag/gow>.
2. Norman Cohn, *Europe's Inner Demons: An Enquiry Inspired by the Great Witch Hunts* (London: Pimlico Press, 1975); Ronald Hutton, *Triumph of the Moon: A History of Modern Pagan Witchcraft* (Oxford: Oxford University Press, 1999), pp. 377–9; Hutton, *The Pagan Religions of the Ancient British Isles: Their Nature and Legacy* (Oxford: Blackwell, 1991), all articulated the problem of Murray's thesis in relation to historical witchcraft; Jacqueline Simpson, 'Margaret Murray: Who Believed Her and Why', Presidential Address of 1992, *Folklore* 105 (1994) 89–96; Caroline Oates and Juliette Wood, *A Coven of Scholars: Margaret Murray and Her Theories of Witchcraft* (London: Folklore Society Archive Series, 1998).
3. Oates and Wood, *Coven of Scholars*, p. 14. In the preface to her next book, *The Divine King* (1954), Murray protested against the number of anonymous letters she received.
4. Murray, *God of the Witches*, preface. For the key modern reassessment see Cohn, *Europe's Inner Demons*, pp. 108–15.
5. Oates and Wood, *Coven of Scholars*, pp. 13, 96, n.23.
6. Gerald Gardner, *High Magic's Aid* (London: Michael Houghton, 1949); Gardner, *Witchcraft Today* (London and New York: Rider, 1954); Gardner, *The Meaning of Witchcraft* (London: Aquarian, 1959).
7. Gardner, *Witchcraft Today*, Murray's introduction, pp. 15–16; Leo Louis Martello, *Witchcraft: The Old Religion* (Secaucus, NJ: Citadel Press, 1969), p. 59; unsourced internet anecdote; Oates and Wood, *Coven of Scholars*, p. 93 n.29; 3rd *Stone, Archaeology, Folklore and Myth* 34 (1999) 18–22.
8. Oates and Wood, *Coven of Scholars*, pp. 27–8, 98, n.93; Cecil L'Estrange Ewen, *Some Witchcraft Criticism* (privately printed, 1938); W. R. Halliday, review in *Folklore* 33 (1922) 228 and note.
9. Oates and Wood, *Coven of Scholars*, pp. 12–13.
10. Montague Summers, *History of Witchcraft and Demonology* (London: Kegan Paul and Knopf, 1926), pp. 42–3.

11. Christopher Cawte, 'It's an Ancient Custom but How Ancient?', in Theresa Buckland and Juliette Wood (eds), *Aspects of British Calendar Custom* (Sheffield: Sheffield Academic Press, 1993), pp. 37–76 ; Hutton, *Triumph of the Moon*, pp. 196–7.

12. Murray, *Divine King*, pp. 15–16; Oates and Wood, *Coven of Scholars*, p. 28, n.97. Even her friend and supporter E. O. James felt that this book was fundamentally unsound.

13. Murray, *My First Hundred Years*, p. 204; Oates and Wood, *Coven of Scholars*, pp. 27–8, 98, n.92.

14. James G. Frazer, *The Dying God*, Part 3, *The Golden Bough* (London: Macmillan, 1911).

15. Murray thought Frazer was behind a negative review of *The Witch Cult* in the *Scotsman*. Oates and Wood, *Coven of Scholars*, pp. 16–17.

16. Holbrook Jackson, *The Eighteen Nineties: A Review of Art and Ideas at the Close of the Nineteenth Century* (New York: Capricorn Books, [1913] 1966); Alex Owen, *The Place of Enchantment: British Occultism and the Culture of the Modern* (Chicago: University of Chicago Press, 2004): for a discussion of the distinctions between occult and mystic see pp. 22–9.

17. Karl Pearson, 'Woman as Witch: Evidence of Mother-Right in the Customs of Mediaeval Witchcraft', in Pearson, *The Chances of Death*, Vol. II (London: Edward Arnold, 1897), pp. 1–49; Oates and Wood, *Coven of Scholars*, pp. 14–22.

18. Jules Michelet, *La sorcière* (1862); Michelet, *Satanism and Witchcraft: A Study in Medieval Superstition*, trans. A. R. Allison (New York: Citadel Press, 1939).

19. Charles Leland, *Aradia: Gospel of the Witches* (London: D. Nutt, 1899); Hutton, *Pagan Religions*, pp. 301–6.

20. E. O. James, *The Cult of the Mother Goddess* (London: Thames & Hudson, 1959). James edited *Folklore* from 1932 to 1958 and wrote Murray's obituary; see *Folklore* 74 (1963) 568–69.

21. Oates and Wood, *Coven of Scholars*, pp. 32–90; Murray, *My First Hundred Years*, pp. 103–4.

22. Gardner, *Witchcraft Today*, p. 149.

23. Philip Heselton, *Wiccan Roots: Gerald Gardner and the Modern Wiccan Revival* (Chieveley, Berks: Capall Bann, 2000); Heselton, *Gerald Gardner and the Cauldron of Inspiration: An Investigation into the Source of Gardnerian Witchcraft* (Chieveley, Berks: Capall Bann, 2003); Aiden A. Kelly, *Crafting the Art of Magic* (St. Paul, Minn.: Llewelyn, 1991); Bill Ellis, *Raising the Devil* (Lexington: University of Kentucky Press, 2000), pp. 148–56.

24. Margaret Murray, 'The Witch-Cult in Palaeolithic Times', *Man* 22 (1922) 3.

25. Vivianne Crowley, *Wicca: The Old Religion in the New Age* (London: Aquarian, 1989), pp. 33, 45–6, 55, 196; new edition revised and updated (London: Thorsens, 1996), pp. 39, 164.

26. Leo Louis Martello, *Weird Ways of Witchcraft* (New York: H. C. Publishers, 1969); Martello, *Witchcraft*; Martello, *Black Magic, Satanism and Voodoo* (New York: Castle Books, 1973).

27. Montague Summers, *The Galanty Show: An Autobiography*, with an introduction by Brocard Sewell (London: C. Woolf, 1980); see 'Hauntings', pp. 131–53. For a complete listing of Summers' works, see Timothy D'Arch Smith, *Montague Summers: A Bibliography* (Wellingborough: Aquarian Press, 1983) and *Montague Summers: A Bibliographical Portrait*, ed. Frederick S. Frank (Metuchen, NJ: Scarecrow Press, 1988); Robertson Davies, 'Augustus Montague Summers', *Oxford Dictionary of National Biography* (Oxford: Oxford University Press, 2004), Vol. 53.

28. Joseph Jerome, *Montague Summers: A Memoir*, with a foreword by Dame Sybil Thorndike (London: C. Woolf, 1965), p. 14; Brocard Sewell, Introduction to Summers' *The Galanty Show*, p. 6.

29. Timothy D'Arch Smith, *Montague Summers: A Talk* (Edinburgh: Tragara Press, 1984), p. 22. There is no in-depth biography, but comments from his wide acquaintance-ship are consistent in presenting a man well-liked by friends who were aware of his considerable learning and tolerant of his homosexuality, theatricality and self-drama. Brocard Sewell, *Tell Me Strange Things*, introduction by Sandy Robertson and Edwin Pouncey (Upton: Aylesford Press, 1991) drew on the unpublished reminiscences of his friends, Redwood-Anderson and C. R. Cammell. Montague Summers' papers were dispersed after his death, but letters appear in the following collections: *Letters to an Editor*, Montague Summers to C. K. Ogden, introduction and notes by D. E. Wickham (Edinburgh: Tragara Press, 1986); The British Institute of Florence, Harold Acton Archive, Edward Hutton collection, 33 letters from Summers between 1928 and 1948; Cambridge University Library, Kings JDH/24/95, Papers of John Davey Hayward, Guide to Louis F. Peck Papers, collection includes letters from Summers to Frank Algar concerning the gothic novel; Chapel Hill Library, North Carolina, Sadlier Black Collection, Micheal Sadleir Papers contain gothic novel material; British Library, B.L. Add Mss 5772 ff. 88, 89, Ashley B5568 ff. 92, 93, B3183 ff. 194–5, 5755 f. 139, B3180 ff. 186, 187, B5500 ff. 53–4, B700 ff. 163–4.

30. Montague Summers, *Antinous and Other Poems* (1907), reprinted with an introduction by Timothy D'Arch Smith (London: C. Woolf, 1995), pp. 9–10, 18–19; Gareth J. Medway, *Lure of the Sinister: The Unnatural History of Satanism* (New York: University Press, 2001), pp. 382–3.

31. Jerome, *A Memoir*, p. 22.

32. Sewell, *Tell Me Strange Things*, pp. 9, 15, 18, 29.

33. Owen, *Place of Enchantment*.

34. Summers, *Witchcraft and Black Magic*, p. 111: 'Priest and psychologist stand out as those best qualified to investigate the subject of witchcraft and Satanism which as a political and social factor permeates all History and is the undercurrent influencing and polarizing events today in its hell born eternal impulse to precipitate the world into the abyss of utter perdition'; Montague Summers, *The History of Witchcraft and Demonology* (London: Kegan Paul and Knopf, 1926), 2nd impression with a foreword by Felix Morrow (New York: University Books, 1956), p. xiii; Summers, *The Galanty Show*, pp. 154–64.

35. Summers, *Letters to an Editor*, introduction, p. 5.

36. Jerome, *A Memoir*, pp. 18–19; media interest never bothered Summers, nor did he waver in his conviction that witchcraft was real; *The Galanty Show*, pp. 154–60.

37. Summers, *The Galanty Show*, p. 52, n.1, 2; Summers, *The History of Witchcraft and Demonology*; Summers, *The Geography of Witchcraft* (London: Kegan Paul and Knopf, 1927), 2nd impression (New York: University Books, 1958); Summers, *A Popular History of Witchcraft* (London: Kegan Paul, 1937); Summers, *Witchcraft and Black Magic* (London: Rider, 1946); Summers, *The Physical Phenomena of Mysticism: With Especial Reference to the Stigmata, Divine and Diabolic* (London: Rider, 1950); Summers, *The Werewolf* (London: Kegan Paul, 1933); Summers, *The Vampire, his Kith and Kin* (London: Kegan Paul, 1928); 2nd edn with article on Summers by Felix Merrow (New York: University Books, 1960); Summers, *The Vampire in Europe* (London: Kegan Paul, 1929).

38. Nesta Webster, *Secret Societies and Subversive Movements* (1929) (Palmdale, Christian Book Club of America reprint); Summers, *Witchcraft and Black Magic*, p. 111.

39. Summers, *History of Witchcraft and Demonology*, pp. viii–ix, xiv; chapter VI, 'Diabolic Possession and Modern Spiritism'. To give just one example of his rolling prose; 'the clairvoyance [sic] is merely playing with fire – I might say – hell fire' (p. 237).

40. St Alphonses Liguouri, *The Glories of Mary*, trans. with a foreword by Montague Summers, 2 vols (London: Fortune Press, 1938 and 1948); Summers, *Physical Phenomena*.

41. Summers, *History of Witchcraft and Demonology*, pp. viii–xii, 33–6, 40–3, 45–6; Summers, *Popular History*, pp. 119, 120–1.

42. Summers, *History of Witchcraft and Demonology*, p. 91; Summers, *Witchcraft and Black Magic*; Girolamo Tartarotti, *Del congresso Notturno delle Lamia* (Rovereto, 1749).

43. Summers, *Witchcraft and Black Magic*, pp. 112–14; Summers, *Popular History*, pp. 103–4; Nicholas Remy, *Demonolatry*, trans. E. A. Ashwin, ed. with introduction and notes (London: John Rodker, 1930), foreword.

44. Cohn, *Europe's Inner Demons*, pp. 120–1.

45. Summers, *History of Witchcraft and Demonology*, p. xii, 110, Sabbat section, pp. 122–30, 133, 151; Summers, *Popular History*, pp. 109–16.

46. Ludovico Maria Sinistrari, *Demoniality*, trans. from the Latin with an introduction and notes (London: Fortune Press, 1927); Heinrich Kramer and Jacob Sprenger, *Malleus Maleficarum* (London: John Rodker, 1928); *The Discovery of Witches: A Study of Matthew Hopkins* by the Rev. Montague Summers, together with a reprint of the discovery of witches from the rare ordinal of 1647 (London: Cayme Press, 1928): Henry Boguet, *An Examen of Witches*, trans. E. Allen Ashwin with an introduction (London: John Rodker, 1929); *Reginald Scot's The Discoverie of Witchcraft*, introduction (London: John Rodker, 1930); Noel Tallepied, *A Treatise of Ghosts*, introduction and commentary (London: Fortune Press, 1933); Richard Bovet, *Pandaemonium*, introduction and notes (Aldington, Kent: Hand and Flower Press, 1951); Francesco Maria Guazzo, *Compendium Maleficarum*, Montague Summers edition trans. E. A. Ashwin (London: John Rodker, 1929).

47. Guazzo, *Compendium Maleficarum*, quoted in Summers, *History of Witchcraft and Demonology*, p. 144; see Jerome, *A Memoir*, p. 55.

48. Hammer Studios depicted Matthew Hopkins as an obsessed fanatic and charlatan in the 1968 film *Witchfinder General*.

49. Jerome, *A Memoir*, pp. 52–7; *The Confessions of Madeleine Bavant* translated from the French of 1652 with introduction, notes and bibliography (London: Fortune Press, 1933); Remy, *Demonolatry*; D'Arch Smith, *Montague Summers*, p. 7.

50. Jerome, *A Memoir*, pp. 18–19; Montague Summers, *Essays in Petto* (London: Fortune Press, 1928); Summers, *Richard Barnfield: The Poems*, introduction (London: Fortune Press, 1936).

51. Summers, *The Galanty Show*, p. 158.

52. A. E. Waite, *Devil Worship in France* (London: Redway, 1896); H. T. F. Rhodes, *The Satanic Mass: A Criminological and Sociological Study* (London: Rider, 1954); Medway, *Lure of the Sinister*, pp. 1–8, 70–100, 380–9.

53. Summers, *History of Witchcraft and Demonology*, pp. 321, 338–43; Summers, *Letters to an Editor*, pp. 9, 14; Jerome, *A Memoir*, p. 22; Michelet, *Satanism and Witchcraft*; Joris-Karl Huysmans, *Las Bas: A Novel*, additional bibliography and notes by Montague Summers (London: Fortune Press, 1943). Summers was the only English member of the Societé J-K Huymans in Paris.

54. Summers, *History of Witchcraft and Demonology*, pp. 150–1.

55. Summers, *Witchcraft and Black Magic*, chapter 7.

56. Francis Barrett, *The Magus or Celestial Intelligencer* (London, 1804).

57. Jean-Francois Bladé, *Contes populaires de la Gascogne* (Paris, 1886); Medway, *Lure of the Sinister*, pp. 87–8.
58. Ellis, *Raising the Devil*, pp. 144–8; Jeffrey S. Victor, *Satanic Panic: The Creation of a Contemporary Legend* (Chicago: Open Court Books, 1993); Bill Ellis, 'The Highgate Cemetery Vampire Hunt: The Anglo-American connection in Satanic Cult Lore', *Folklore* 104 (1993) 113–39.
59. Summers wrote a foreword to only one work on non-European magic: Frederick Kaigh, *Witchcraft and Magic of Africa* (London: Lesley, 1947); Summers, *Witchcraft and Black Magic*, p. 10; Summers, *History of Witchcraft and Demonology*, p. xv.
60. Summers, *History of Witchcraft and Demonology*, p. 163; Robert H. Nassau, *Fetichism in West Africa* (London, 1904); Summers, *Popular History*, pp. 246–50.
61. Jerome, *A Memoir*, pp. 59, 67–8. Summers spoke with distaste of Crowley but he did know him and kept a file on him; for Felix Morrow's remarks about Summers, see p. 57.
62. See Owen, *Place of Enchantment*, pp. 22–9.
63. *Such Power is Dangerous* (1933); *The Devil Rides Out* (1934); *Strange Conflict* (1941); *The Haunting of Toby Jugg* (1948); *To the Devil a Daughter* (1953); *The Ka of Gifford Hilary* (1956); *The Satanist* (1960); *Gateway to Hell* (1970); *They Used Dark Forces* (1964); *The White Witch of the South Seas* (1968); *The Irish Witch* (1973).
64. Dennis Wheatley, *The Devil and All His Works* (London: Hutchinson and Co., 1971).
65. <www.denniswheatley.info>; Ellis, *Raising the Devil*, pp. 156–8.
66. <www.hammerfilms.com>.
67. Jerome, *A Memoir*, quoting introduction to Summers' edition of *Castle of Otranto*.
68. Summers, *The Galanty Show*, p. 158; Summers, *Antinous and Other Poems*, p. 10.

6
the witch-craze as holocaust:
the rise of persecuting societies

raisa maria toivo

In the aftermath of the Second World War, with the experiences of systematic genocide and totalitarianism, new developments in sociological and political analysis began to emerge, such as the logics of persecution and victimology. The history of ethnic minorities and of ideological crimes also assumed greater prominence. As a consequence there was a general reassessment of the historical significance of the witch trials. The study of the persecutions both before and during the war and the study of early modern witch hunts seemed mutually supportive. Furthermore, in the post-war period the term 'witch hunt' became a common descriptor for contemporary persecutions, such as the McCarthyite campaign against suspected communists and political purges in the Soviet Union.

This new academic interest in the European trials was initially most evident in the research of German law students and in the work of German émigré sociologists,[1] but the writings on the subject that achieved the widest appeal and greatest influence were by two British historians, whose perspectives were strongly influenced by the experience of war and their understanding of the Holocaust. The first was Hugh Trevor-Roper's essay 'The European Witch-Craze of the 16th and 17th Centuries', which was first published in a collection of essays in 1967 and then published in its own right two years later.[2] Then in 1975 Norman Cohn, a founder of the Centre for Research in Collective Psychopathology and the Centre for the Study of Persecution and Genocide, produced *Europe's Inner Demons: An Enquiry Inspired by the Great Witch-Hunt*.[3] In these works, the 'witch craze' was presented as the result of elite fears of ideological or social enemies, conspiring 'others', projected onto witches more or less consciously. On the one hand, the irrationality of persecutions – the unfoundedness of the suspicions of conspiracy – was as prominent in these interpretations as it had been in previous studies of witchcraft. On the other hand, 'mass irrationality', which was once believed to belong firmly in the past but which, it seemed after the wars, had returned, drew attention to the notion that conceptions of the rational and irrational were defined by time and context: the persecutions of witches could be explained rationally in the context of early modern religious strife and power struggles. The witch hunts were, as Geoffrey Scarre, a philosopher who has written on the witch trials and the Holocaust, put it, a phenomenon that historians dared to explain rather than apportion blame. Terrible as the trials were supposed to have been, the judges and the demonologists could still be given the moral benefit of the doubt, in the sense that because of their cultural and religious circumstances, they earnestly believed they were doing right in the eyes of God.[4]

stereotypes of conspiracy: witchcraft as anti-society?

The main focus of Trevor-Roper's and Cohn's interpretations was on the stereotyping of fear and conspiracy, with persecution being seen as expressing social and cultural fears, particularly of the 'enemy within'. As Cohn explained:

> The essence of the fantasy was that there existed, somewhere in the midst of the great society, another society, small and clandestine, which not only threatened the existence of the great society but was also addicted to practices which were felt to be wholly abominable, in the literal sense of anti-human.[5]

The fear of a hidden 'other' created frustration and tensions, which generated mythologies, or theories, concerning that 'other'. These mythologies were often alike. Ritual murder and cannibalistic feasts, for example, had long been attributed to perceived conspiracies or secret societies hungry for power, whether Christians in the early first millennium, Jews in the medieval period or witches in the early modern.[6] In the introduction to *Europe's Inner Demons*, Cohn also compared the witch persecution to the late medieval and early modern chiliastic movements he had examined in his earlier work, *The Pursuit of the Millennium*. In both phenomena, the aim was to purge and so save society through the annihilation of those who were though evil and destructive. In one case, it was the establishment judging marginal groups; in the other, the fantasies of those in the margins were central.[7]

Whether or not originally so, the mythology regarding witches soon became widely shared and believed. As soon as the authorities in the trials had internalised the theory, each witch would have to not only comply with it in his or her confession but also add something personal to it in order to make the confession credible. Once the mythology had been established, all individual deviations could be interpreted through it and so become evidence for it.[8] Such ideas regarding the fear of conspiracy have great dramatic potential, and were perceptively exploited in Arthur Miller's famous portrayal of the Salem trials in his play *The Crucible*. Although Miller, according to the ideals of realism, claimed historical accuracy, the play is the work of an artist, not an academic historian, and should be read as such. It is relevant here, because it has profoundly influenced the popular conception of witchcraft in America, as have its Trevor-Roperian counterparts in Europe. One of the reasons for the vast publicity attracted by Miller's play was obviously that it linked the Salem religious persecution to the 'American' identity, the declared basis of which had been the religious and ideological freedom sought by the Puritans. The play was also influenced by Miller's own experience of the McCarthyite hunt. But it is also an investigation into the stereotypes of fear and hate, their psychology and their capacities for identity-building that were later examined by scholars like Cohn and Kurt Baschwitz, the German journalist and sociologist who fled Nazi persecution in 1931, and who wrote an influential German text on the witch trials in 1963.[9] Understanding the nature of the fear of conspiracy, furthermore, made it possible to show the terror of both the persecuted and the persecutor. In his introduction to an edition of his collected plays, for example, Miller highlighted how it was worse to 'have nothing to give' persecutors than to be able to relieve the collective guilt and fear by confessing.[10]

Cohn's view that witchcraft, and especially the Sabbat, represented an anti-society has been influential on subsequent witchcraft historiography. The Sabbat is often interpreted as an anti-fertility rite where the witches caused hailstorms and frost or made babies ill in festivities that were the reverse of the semi-Christian village festivals and rituals to ensure good crops.[11] Descriptions of Satan and Hell and the activities of witches were characterised by ritual backwardness. Witchcraft presented a social and hierarchical inversion of the whole cosmic order. Contemporary rhetoric legitimised the power of the ecclesiastics and secular magistrates by claiming these authorities had been selected by God. Stuart Clark has linked this notion of the world upside-down in learned demonology to a general conception of a polar world, which can be found in many areas of elite culture from literature and theatre to science and theology. In this understanding of the world, order was dependent on both the opposite polarities. Witchcraft was seen as a force that disrupted the balance.[12] The witches' Sabbat as an anti-society also helped to establish a polar relationship between God and Satan, something that seems to have been part of the elites' educational projects in late medieval and early modern Europe.[13] Through these interpretations, scholars emphasised a more or less conscious blackening of the persecuted. The imagery was also used to reinforce the Christian faith by showing not only its opposite but also what happened to the subversionary opposition: Jews and witches could, in fiction, be made to defile sacred objects and consequently be punished.[14]

We can see the language of subversion in numerous trials, most obviously those involving stories of the Sabbat – the core of Trevor-Roper's 'crazes' – but also in more mundane charges of *maleficium* and even benevolent magic, where popular mentalities are revealed more clearly than in the Sabbat trials. Take, for example, the following questioning of Valpuri Kyni, an accused Finnish witch in Finland, tried in the Turku High Court in November 1649:

> After a few questions concerning her name, parents, children and place of domicile, she was asked if she had previously been in court accused of harming the Pastor's cattle or any other witchcraft that she had committed? Answer: No, I have neither bewitched the Pastor's cattle nor anything else. God takes cattle and humans as He pleases. They can say whatever they want of me, but it is all just envious talk and I am innocent before God and man.
>
> Question: Why has she allied herself with Satan and hurt the Pastor's cattle by her witchcraft? And why had she taught the Pastor's maid to practise witchcraft and harm her master's cattle? Answer: God preserve me from any connections with Satan, I intend to be God's child just as well as anyone. I have not done or taught anything wrong. She had only taught the maid how to stay in service and eat what her master and mistress would give her as her food, be it gruel, cabbage or whatever.
>
> Question: She had indeed given the Pastor's maid a snake's skin, with which she was to bewitch the milk so it would turn into blood? Answer: Denied this altogether.[15]

After a few questions regarding her previous trials and other harm she was accused of by her fellow-villagers, the court 'pointed out to her that her mother's father and her mother had indeed been burnt as witches when the late Nils Bielke of noble birth had been judging and they had indeed taught her the same arts'. Answer: 'She knew not nor remembered that her parents were burned; they had taught her nothing unnecessary.'

Valpuri's case is one that totally lacks any notion of a Sabbat, but it serves to illustrate the use of the conspiracy theory. The court seems to have tried to present Valpuri joining Satan, conspiring with him and other witches and luring the maidservant to hurt and ruin the pastor. Valpuri used the same polar rhetoric when she repeatedly mentioned God in her defence. Later on, the court pursued a rather unexpected line as Valpuri was asked to recite the Creed and her inability to remember the third article was thought suspicious.[16] In the rest of the record, destructive traits towards the good of the community were as clear as in any *maleficium* charges. Valpuri was said to have harmed not only the pastor but also the crops of some villagers, and was subsequently accused of having bewitched the father of her maid and of having brought forth wolves to attack people. Conspiracy against society is hinted at, but is it the only explanation?[17]

Cohn's identification of the stereotypes of conspiracy and their appearance in certain kinds of witch trials has not received as much criticism as other aspects of the comparisons between the witch hunts and other persecutions, although Ginzburg countered the criticism Cohn levied on his theories of the connection between ancient fertility cults and night-flying witches, which are discussed in another chapter of this book.[18] It is important to stress that Cohn never meant to explain *all* witch trials. He was primarily interested in the mechanisms of persecution, though his ideas about stereotyping, anti-societies, inversion and, to some extent, conspiracy, have influenced broader scholarly interpretations of witchcraft, borrowing from psychology and literature, which move beyond the setting of elite and state fears.[19]

confessional conflict, newly established powers and 'the persecuting society'

Whereas Norman Cohn's conception of the general fear of conspiracies throughout history eventually became – for good reason – highly influential in academic circles, Trevor-Roper's earlier and slightly different perspective had more popular appeal. Trevor-Roper presented the 'witch craze', as he called the phenomenon, as a persecution of social and cultural deviants, of scapegoats, by Catholic and Protestant elites competing for power. Whereas Cohn situated his analysis of the witch trials within a broad comparative context of persecutions across hundreds of years, Trevor-Roper's understanding was shaped by the focus on a very specific time-frame. He wrote his essay on the witch craze as a part of a larger collection of older essays published under the title *The Crisis of the Seventeenth Century*.[20] Trevor-Roper argued that there was not only a political and economic crisis in early modern Europe, but also an ideological and intellectual one, of which the Reformation, Counter-Reformation and the witch trials were an integral part. The collection, in Trevor-Roper's own terms, was a favourable reconsideration of the sociologist Max Weber's thesis that the Calvinist ethic was crucial to the birth and development of the 'Spirit of Capitalism'.

According to Weber (1864–1920), the formation of new states and the struggle of new rulers to establish their power was likely to create a state monopoly of violence – local power would be channelled and enforced via the judicial system. In addition to the mundane crimes concerning the interests of individuals in local communities, more abstract crimes against the ruler, the state and religion would emerge. These were crimes that seemed to undermine the newly established powers, but their prosecution actually served to enhance the influence of the judicial system, and thus the ruling order.[21] Trevor-Roper's description of the witch craze was very similar. The newly established powers were the Protestant and post-Tridentine Catholic Churches and their secular supporters. Persecution, he claimed, attained new vigour wherever either side had won a major ideological or actual battle: demonising and persecuting the opposition helped enforce the new theologies, ideologies and regimes. Weber's ideas were already influential in his lifetime and some of their popularity in continental Europe may have been related to the changing political world order after the First World War. From the post-war point of view, competing and emerging new powers might appear again and again in history, and initiate similar persecutions.

This scenario of the behaviour of new powers in times of social, political and cultural crises is evident in the comparisons made between the witch persecution and the twentieth-century persecution of the Jews. The Nazi regime used anti-Semitism and judicial persecution to establish itself against a background of dire economic problems and a cultural crisis born of the terms of the Versailles Treaty, while some historians have likewise linked the witch panics to economic and agricultural crises, as well as the Trevor-Roperian model of cultural upheaval.[22] Such comparisons are obviously tempting, but can lead to over simplification; likewise, the comparisons made by Gerhard Schormann, for example, between authoritarian witch-hunting rulers and Nazi leaders. Although Schormann is aware of significant differences, he nevertheless sees close parallels between the witch 'extirpation' programme undertaken by Elector Ferdinand of Cologne in the 1620s and 1630s and the Third Reich's 'final solution' for Jews. Lucy Dawidowicz, in her influential book *The War against the Jews 1933–1945* (1975), had already pointed out the parallels with the Holocaust, and Schormann took the case further by paraphrasing her title for his own study – *Der Krieg gegen die Hexen*, and sharing terminologies like '*Ausrottung*' ('extermination') and '*Endlösung*' ('final solution') with Dawidowicz's book.[23] As Behringer points out, though, Ferdinand seemed to interest himself in the persecution late in his career, and only after the persecution was well on its way and so the elegant comparison was only vaguely supported by facts.[24]

In *The Rise of a Persecuting Society*, R. I. Moore wrote regarding the rise of the persecution of lepers, Jews and heretics in the Middle Ages that:

> When rulers begin to assert themselves, and to create a recognisable apparatus of state, the earliest developments always include the appearance of a hierarchy of specialized agencies for the enforcement of order ... and law itself becomes coercive, imposing from above a pattern of guilt or authority, rather than a mediatory, seeking agreement or compromise.[25]

Yet the rise of the state and central government not only brought with it stronger central control and oppression, but also produced new ways for the populace to influence matters that interested them through the extension of the judicial system, the parliamentary estate or Diet systems of governance. It could be said that early modern state formation would not have been possible without the co-operation of the people and therefore a real and continuous negotiation with them. The coercive power of Machiavellian princes was more an ideal than a reality.

The theory that the establishment of new central powers will lead to the development of persecution is teleological to an extent that may have served the older sociology of Weberian ideal types, but not perhaps modern historians. Yet it would be erroneous and somewhat unfair to read Trevor-Roper's theory as a mere claim that Protestants persecuted Catholics and Catholics persecuted Protestants as witches – of which there is no evidence – or that they did it simply or cynically for power. The theory claims that the power struggle between the creeds and the new great powers in Europe created both a social and religious fear of ideological deviants and the machinery of legitimate violence to deal with the fear. The link there, too, is 'far from direct', explains William Monter, although religious fervour may have made people more aware of and therefore more likely to demand punishment for sins of all kinds.[26]

The Trevor-Roperian view of confessional conflicts, as well as the Nazi comparison, has attracted few supporters among recent scholars, though both remain influential in general works on history and in the popular media. State formation as a factor behind the trials in some areas and their moderation in other places, however, has attracted several scholars in both a way that emphasised the civilisation process and a way that emphasised the development of judicial systems as part of the power struggle in general. Wolfgang Behringer, who is otherwise very critical of Trevor-Roper, points out that the great German persecutions were used by the princes to gain popular support for the centralising of legal government. One reason why some German territories did not experience as drastic hunts as others, he suggests, was that they did not embark on programmes of state formation or confessionalisation.[27] The reverse point is included in the argument: There has to be a popular interest involved in the methods with which popular support is to be gained.

elite guilt

The conspiracy and state-building theories regarding the witch trials were concerned primarily with elite society. The question of the responsibility of general populaces for persecutions is a difficult one, whether in the past or the present. It is easier to apportion guilt to ruling elites. Sill, Trevor-Roper's focus on elites was understandable; the 'history from below' approach to witchcraft only began to emerge in the early 1970s, most notably in the work of Keith Thomas, and there had been few scholarly local and regional studies. Regarding the latter, Alan Macfarlane's (1970) study of the county of Essex, H. C. Erik Midelfort's (1972) work on south-western Germany, and William Monter's (1976) research on witchcraft in eastern France and Switzerland, would do much to reveal the inadequacies of Trevor-Roper's thesis.[28] Monter revealed, for example, that Trevor-Roper's characterisation of Calvin's Geneva as a centre of witch persecution and terror was far from accurate.

The top-down view was long dominant, partly also because of the way Trevor-Roper and his predecessors had defined witchcraft and 'witch craze' as ideological, dramatic and destructive miscarriages of justice: if it was 'of the people', it was defined as something else – sorcery, or the 'scattered folklore of peasant superstitions' – but not the type of witchcraft which had European significance.[29] During the 1970s, research by Keith Thomas and Alan Macfarlane discovered the importance of the interpersonal relationships in village-level accusations, and Norman Cohn and Richard Kieckhefer almost simultaneously discovered that the thesis of the guilt of the Inquisition, on which Trevor-Roper had placed considerable emphasis, was based on forgeries. Nevertheless, some historians pointed out that the elite diabolical views – or in Cohn's words 'the other notion of a witch' – and the machinery of legitimate violence were necessary for the panics, too. Only if the accused could be made or tortured to tell stories according to demonological theories, of meeting accomplices in Sabbats, could the trials accumulate so that villages would be emptied. The witch hunt as an organised enterprise needed elite leadership, and even though the populace may have been willing to co-operate, for a long time it was not thought to be able to produce a persecution by itself.[30]

Since the 1970s numerous scholars have found that witchcraft trials were initiated by or needed the co-operation of whole communities and not just its elite members.[31] In the case of Valpuri, too, we see a record of an elite-led interrogation after an accusation made by a supposed member of a local elite, the local pastor. Only after that did the record refer to harm suffered by peasants. But how far did that reflect the initiatives in the trial? Before the High Court, Valpuri's case had been tried in the lower courts in the parish of Huittinen. Then, too, the pastor, supported by his maid, made the original accusation. During the trial, however, many villagers volunteered information on the harm Valpuri caused: making people ill, threatening to bring forth wolves if not given enough alms and even causing the death of a child. In all of the testimonies, even those made by the pastor, Valpuri's *maleficium* was a personal threat to the accusers, rather than abstract or targeted against the Crown and religion.[32] Were the neighbours here responding to pressure from the elites or were they using the state judicial machinery for their own purposes? Scholars have also noted that whereas state formation may have created new and tighter judicial machineries in Europe, the grip of central governments also served to moderate and control the excesses of local governments, as well as extra-legal popular justice.

irrational crazes nevertheless?

Irrationality is inherent in the term 'craze', as well as other epithets used to describe the witch trials, such as 'panics', 'psychosis', 'hysteria' or '*Massenwahns*' ('mass delusions') as Kurt Baschwitz described them. Those employing such terms also made the most of extreme examples of torture or seemingly 'absurd' aspects of the trials to highlight the sense of irrationality. For some the use of such terminology can been seen as a way of coping with the tragedy of persecution, but it is also a device for writers and scholars to distance themselves and their readers from the persecutors as well as to condemn the persecutions as a whole.[33]

Historians of the 1960s and 1970s made clear their disbelief in demons and witches flying to Sabbats and causing harm by supernatural means, and in the other features of the stereotype of conspirators. Cohn repeatedly acquitted the victims of the persecutions he investigated of the terrible crimes that formed the stereotypes of persecution.[34] Similar implications appear in the use of the word 'victim'. Those who thought of a persecution of innocents saw the victims of the witch crazes and witch hunts in the accused and convicted, whereas many recent scholars have been interested in the victims of witchcraft, that is those who claimed damage from the witches and consequently feature in the trials as the accusers and witnesses. The interpretations saw the persecutions as labouring to find the accused guilty even against the better judgement of individual judges or other members of the courts. This has also produced a popular rhetoric of victimisation, of innocence employed by individuals who have felt themselves persecuted or threatened – or have wanted to appear so, from feminist and environmental activists to people publicly accused of various vices from political corruption to child abuse. Appropriating the 'witch hunt' serves to highlight innocence. The use of such a comparison often comes in for heavy criticism for being ill-founded, some exaggerating and some belittling the suffering caused to and by different parties. Nevertheless, they persist in the popular understanding and use of the history of witch trials.

The emphasis on irrationality, as well as the emphasis on the elites, was both an adoption from the earlier literature and a by-product of Trevor-Roper's definition of what the important manifestation of the witch trials was: 'the crazes'. If it did not include torture, the Sabbat and an escalation into a 'panic', if the accused were treated with legal caution, punished leniently or freed, it was not important. The effect of Cohn's original interest in the mechanisms of persecution was similar. This is a notion that still influences a great deal of the present scholarship on witchcraft. Scholars whose approaches and interests are vastly different from Trevor-Roper's still share his focus and see the major episodes as the most important kind of witchcraft and witch trials.[35]

By this definition, persecution became an inescapable machine of destruction, which secured its success and produced the evidence for its necessity by torture – every victim would be forced to tell a story to provide further evidence of the terrors of witchcraft. Many students have followed Trevor-Roper and Rossell Hope Robbins in illustrating the mechanisms of persecution by invoking the heart-breaking letter of the tortured burgomaster Johannes Junius (1628), who, when facing execution, tells his daughter how his confessions and the names of accomplices were extracted from him.[36] We could also refer again to Valpuri's trial in Turku. If the record began with leading questions, they turned continuously more menacing as the interrogation went on. After a few more questions the record has a paragraph stating that:

> she was seriously told to confess whether or not she had been questioned many times both in the lower courts and in this High Court for her witchcraft and even been thrown into water by the executioner as a test, as the lower court records show. Finally she admitted this.
>
> Question: Why did she not sink even though her hands and feet were tied? Answer: I kept my mouth shut so I could not sink.
>
> Question: She had well been whipped in the pillory in Tyrvää and then been banished? She could not deny this.

At some point she was again admonished to tell the truth, 'because she had nothing before her but death for all the bad things she had done'. Her reply sounds quite as tragic and almost as articulate as Johannes Junius':

> My death is yours to decide, good lords, many better persons than I have lost their heads. But one thing she new for sure, that before a drop of her blood falls to the ground, God will take her soul.[37]

Valpuri could not be made to confess, and in the end she was told to prepare for her death, for she would get what she deserved. The guards were told to take her to Turku Castle to be detained there – a harsh response not only considering the prison conditions but also the fact that imprisonment was often reserved for those who had a pending capital sentence. The serious warnings to tell the truth might be an indication that torture may have been applied or at least threatened, although it was strictly illegal. The allusions to the former use of violence may have seemed like a veiled threat of its new application. The final command to prepare for her death, despite her professed innocence, could be read as proof of the court's intention to convict her.

It is cases like this on which the interpretation of the courts' eagerness to convict is based. In many European courts this seems to have been a standard procedure. Lyndal Roper, for example, reads the trial records of the Augsburg magistrates as a discourse forced by the threat of torture.[38] There were cases where the careful examination of whether or not someone had done or tried to do some magical harm turned into 'crazes' that convicted on the slightest suspicion. Salem, perhaps, had such a character, although Miller's vocabulary of evil in his portrayal of the judges is a simplistic exaggeration.[39] The witch hunt instigated by Matthew Hopkins in England, for example, and the Swedish Blåkulla trials in Dalarna, Göta, Åland and Ostrobothnia, as well as many other local epidemics, share these features.[40] Much of the best-preserved and most easily accessible material is, of course, that which

attracted most contemporary attention either in pamphlet literature or in the courts themselves. Those were usually the most sensational cases. This even applies to Valpuri's trial record: it has survived because it was published by a late eighteenth-century newspaper for the entertainment and self-gratification of 'enlightened' readers, a task which may have been best served by choosing the most 'irrational cases' for publication.[41]

The emphasis on the most extraordinary and extensive trials, designated as 'crazes', has attenuated since Trevor-Roper's thesis and the initial comparisons with the Holocaust. It is apparent now that most of the trials in Europe were of the milder, endemic type, with none of the excesses that Trevor-Roper felt important. These trial processes were moderate and meticulous according to all legal provisions. This reassessment of the severity and nature of the witch trials over the last three decades is illustrated by the rise and fall of the mythology of the 9 million supposed witch executions, as described by Wolfgang Behringer.[42] It was a number arrived at during the late eighteenth century through a few arbitrary calculations based on an assumption that all places and all decades of Christian history had experienced witch hunts of equal severity. The estimated number had been reduced in scholarly circles by Trevor-Roper's time, although he presented only the numbers involved in specific 'crazes', avoiding any grand total. Since then the number has fallen even further. The current estimates of British and continental trials range from 100,000 to 200,000, and the estimates of death sentences from 40,000 to 50,000. Where cases of minor witchcraft and 'superstition' – often confused and mixed in the courts with witchcraft and *maleficium* – have also been studied, death sentence rates are considerably lower; for example, in Finland only 10 per cent of the lower court sentences were death penalties. The remaining convictions were fines and lesser forms of corporal punishment (birching was common) but half of the accused were actually acquitted. In Sweden and Finland lower court sentences were automatically sent to the High Court for review, and the High Court seems to have mitigated a majority of the death penalties, mostly to fines or lesser corporal punishments.[43] This was the case with Valpuri, too. Despite her gloomy farewell at the end of the hearing, the High Court seems to have shown clemency towards Valpuri. She was still alive ten years later, though she was executed for some other crime in 1665.[44]

Considering the popular influence of Trevor-Roper's essay, his focus on the atypical mass trials and intellectual origins of the witch persecution, and neglect of the popular inspiration for the majority of witch trials, generated a distorted understanding of the early modern experience. Trevor-Roper responded irritably to such criticism: 'those critics who have censured me for not entering more sympathetically into the mental processes of the peasantry are barking up the wrong tree'.[45] He argued that the 'witch craze' was a separate phenomenon from 'mere witch-beliefs', and had much more important historical significance.

moralism and rationality

From their inception, histories of witch persecution have included a strong moral undercurrent. For Enlightenment writers, witch persecution was an example of the evils of religious bigotry and 'superstition'. Today there are also pressures for historians not only to research and understand but also to moralise and condemn persecutions. As far as post-war historians of witchcraft were concerned, this need was heightened by the comparisons with recent persecutions. Consequently, as Trevor-Roper observed, there had been a great deal of reluctance among historians to take up such 'disgusting subject[s], below the dignity of history'.[46] Even those who came after Trevor-Roper were aware of the dangers: explaining could be misinterpreted as excusing.[47]

The terms 'witch craze' and 'witch hunt' were used by post-war generations of scholars until the 1980s, as Joseph Klaits very bluntly put the matter in his introduction to *Servants of Satan,* to 'connote an unfair judicial proceeding of the McCarthyite type, undertaken for cynical purposes'.[48] Moreover the finger pointed mostly at the Churches. For modern historians, accustomed to a modern western separation of religion and secular politics, the early modern Churches' role in the witch persecution was an appalling display of violence and venality. Rossell Hope Robbins asserted, for example, that the witch persecution was 'the official policy of the churches, Roman Catholic and Protestant'. 'What makes witchcraft so repellent, and morally lower than fascism', he suggested provocatively, was that 'the clergy led the persecutions and condoned them in the name of Christianity'.[49] The Inquisition had for long had the worst reputation in this respect, in part due to several centuries of successful Protestant propagandist history.

Historians should have known better, of course. The modern western separation of religion and politics did not exist at the time of the witch trials. Religion and politics were not mutually exclusive but inseparable. Accordingly, the rationalist idea, dismissing the rhetoric of faith and belief as mere recourse to religion for political purposes, has attracted much criticism since the 1980s.[50] The Trevor-Roperian interpretations, however, should not be singled out for particular criticism in this respect. He, like his successors, was careful to point out that the persecuting elites were motivated by real religious zeal, sincere in their own faith and convinced of the existence of an opposing, dangerous, satanic culture.[51] They were not cynically blackening their perceived enemy, since as these elites believed they were on the side of right, it followed from their cosmology that the enemy must be evil. There are echoes of this dichotomy in historians' conceptions of human nature. While earlier generations of historians had seen humans as progressing from a stage of iniquity and ignorance, the twentieth-century post-war generation expressed doubts regarding societal progress, but not necessarily regarding fundamental human nature: failing miserably in their efforts to be good, humans were still not essentially evil.

Corrupt leadership has long been a strong moral theme in the histories and dramatisation of the witch craze, and it became even more so when the witch craze was compared to the twentieth-century persecutions. The dramatic appeal of this can be seen even in Miller's approach. *The Crucible* points an accusatorial finger at the local leaders in Salem and elite judges and their responsibility for the 'evil' that broke out.[52] Trevor-Roper seems to have felt the blame he was placing on his own kind when he claimed that the clergy, as the educated ideological leaders and the generators of new ideas, were responsible for the spread of the persecution as well as for its decline. For many of the established scholars of the 1960s, it was easier to identify with the elites than with the masses of the past.[53] Trevor-Roper took a great deal of trouble to state that the demonological theories of witchcraft, like all stereotypes of fear, were inextricably linked with the learned demonologists' native cosmology. Almost apologetically, he pointed out that a 'majority' of the early modern demonologists were 'philosophers in a wider field' who

> wrote upon demonology not necessarily because they had a special interest in it, but because they had to do so. Men who sought to express a consistent philosophy of nature could not exclude what was a necessary and logical, if unedifying extension of it.[54]

Not everyone was as merciful. The history of the 'witch craze' became a point of identification for many users of history and most of them identified with the victims of persecution rather than the persecutors. One of the most obvious examples was the radical feminists' stance on the witch trials. Witches were made a point of identification, as described in a later chapter in this book, an example of those who had fought and suffered the grip of patriarchy and misogyny throughout history. Radical feminists evocatively used the vocabulary of persecution. Writers like Mary Daly and Andrea Dworkin named the witch craze a 'Women's Holocaust' or a 'gynocide'. It is no coincidence that Daly, too, identified the male historians of her time with the persecutors, accusing them even of continuing the gynocide as their interpretations of the witch craze seemed to hide the persecution of women. What Trevor-Roper and Cohn had interpreted as an elite fear of conspiracy against established society, Daly interpreted as a conspiracy of established male society against women. Both identifications were to crumble, as detailed research from the 1970s onwards made the witch trials look both more moderate and more complicated in nature, and thus less stereotypical.[55]

Historians have largely accepted their task as explaining the phenomena of the past instead of (or sometimes in addition to) passing judgements on them. There has certainly been reluctance in the last 15 years to equate witch persecution with Nazi genocide, at least in academic circles. There is no doubt that chronological proximity to persecutions affects the way they are analysed. Geoffrey Scarre, having written on both witchcraft as a historian and the Holocaust as a philosopher, offers the explanation that whereas the religious zeal that fed the witch trials is not present in our society (despite some newly awakened antagonism between some forms of Christianity and some forms of Islam), racism, which fed the Holocaust, persists.[56] The moral risk in the comparison, however, may be not in explaining – for it is not excusing – but in exaggerating or belittling the suffering of some and the guilt of others. There are also factual risks. Most of the explanations we now find credible for the witch trials have withdrawn from comparisons with any modern persecutions.

notes

1. See Barbara Schier, 'Hexenwahn und Hexenverfolgung. Rezeption und politische Zurichtung eines kulturwissenschaftlichen Themas im Dritten Reich', *Bayerisches Jahrbuch für Volkskunde* (1990) 45–6; Wolfgang Behringer, 'Witchcraft Studies in Austria, Germany and Switzerland', in Jonathan Barry, Marianne Hester and Gareth Roberts (eds), *Witchcraft in Early Modern Europe: Studies in Culture and Belief* (Cambridge: Cambridge University Press, 1996), p. 69.
2. Trevor-Roper's 'The European Witch-Craze of the 16th and 17th centuries' was first published in 1967 in a collection of his essays called *Religion, Reformation and Social Change: The European Witch-Craze*, and republished on its own 'in slightly revised form' in 1969 by Pelican. The Pelican edition is used here because it is used more frequently by students of witchcraft and it has an illuminating foreword.
3. Norman Cohn, *Europe's Inner Demons: An Enquiry Inspired by the Great Witch-Hunt* (London: Chatto & Windus, 1975).
4. Geoffrey Scarre, 'Moral Responsibility and the Holocaust', in Eve Garrard and Geoffrey Scarre (eds), *Moral Philosophy and the Holocaust* (Aldershot: Ashgate, 2003), pp. 111–12.
5. Cohn, *Europe's Inner Demons*, pp. ix and xi (quotation). See also, for example, Hans von Hentig, *Die Strafe I: Frühformen und kulturgeschichtliche Zusammenhänge* (Berlin: Springer, 1954), pp. 193ff.
6. Cohn repeated similar features throughout *Europe's Inner Demons*; for example, pp. 1, 3–4, 6, 7, 17–18.
7. Cohn, *Europe's Inner Demons*, p. xiii.
8. Trevor-Roper, *The European Witch-Craze*, pp. 52–3, 93. See also R. I. Moore, *The Formation of a Persecuting Society: Power and Deviance in Western Europe, 950–1250* (Oxford: Blackwell, 1987), p. 90.
9. Kurt Baschwitz, *Hexen und Hexenprozesse: Die Geschichte eines Massenwahns und seiner Bekämpfung* (München: Rutten & Loening Verlag, 1963).
10. Here is a distinct reference to the supposedly torture-led witch hunts, where confession was the only way out. In Salem it really was a way out, for it was only those who confessed and turned to accusing others who were saved from the gallows. Arthur Miller, 'Introduction', in *Arthur Miller's Collected Plays* (New York: Viking Press, 1957), pp. 39–47, quotation p. 40; Bernard Rosenthal, *The Salem Story: Reading the Witch Trials of 1692* (Cambridge: Cambridge University Press, 1993), pp. 204ff.
11. Robin Briggs, *Witches and Neighbours: The Social and Cultural Context of European Witchcraft* (London: HarperCollins, 1996), pp. 40–4.

12. Stuart Clark, 'Inversion, Misrule and the Meaning of Witchcraft', *Past and Present* 87 (1980) 98–127; Clark, *Thinking With Demons: The Idea of Witchcraft in Early Modern Europe* (Oxford: Oxford University Press, 1999), pp. 3–147.

13. Jan R. Veenstra, *Magic and Divination at the Courts of Burgundy and France: Text and Context of Laurens Pignon's Contre les devineurs (1411)* (Leiden and New York: Brill, 1997), p. 166.

14. Walter Stephens, *Demon Lovers: Witchcraft, Sex and the Crisis of Belief* (Chicago: University of Chicago Press, 2002), pp. 207ff.

15. *Åbo Tidningar*, No. 20, 18 May 1795. My translation. The alternation between the first- and third-person singular in the answers is from the newspaper original.

16. Citing catechism articles and churchgoing or praying as evidence in Finnish witch trials was rare but did occur.

17. *Åbo Tidningar*, 18 May 1795.

18. See Carlo Ginzburg, *Ecstasies: Deciphering the Witches' Sabbath* (London: Hutchinson Radius, 1990).

19. Per Anders Östling, *Blåkulla, Magi och Trolldomsprocesser. En Folkloristisk Studie av folkliga trosföreställningar och av Trollsomsprocesserna inom Svea Hovrätts jurisdiktion 1597–1720* (Uppsala: Uppsala Universitet, 2002), and Jari Eilola, *Rajapinnoilla. Sallitun ja kielletyn määritteleminen 1600-luvun jälkipuoliskon noituus- ja taikuustapauksissa* (Helsinki: Finnish Literature Society, 2003), p. 152, see even the Sabbat stories as popular creations. On other popular antisocial and invertive traits see, for example, Annabel Gregory, 'Witchcraft, Politics and "Good Neighbourhood" in Early Seventeenth-Century Rye', *Past and Present* 133 (1991) 30–66, and Raisa Maria Toivo, 'Marking (Dis)Order: Witchcraft and the Symbolics of Hierarchy in Late Seventeenth- and Early Eighteenth-Century Finland', in Owen Davies and Willem de Blécourt (eds), *Beyond the Witch Trials: Witchcraft and Magic in Enlightenment Europe* (Manchester: Manchester University Press, 2004).

20. Hugh Trevor-Roper, *The Crisis of the Seventeenth Century: Religion, The Reformation and Social Change* (New York: Harper & Row, 1968). The book was first published the year before as *Religion, the Reformation and Social Change*.

21. Max Weber, *Economy and Society* (Berkeley: University of California Press, 1978), pp. 901–4. The English version of *Economy and Society* first appeared in 1968 but most British and US academic historians and sociologists were familiar with Weber's ideas, both from the earlier translations of his various piecemeal works and from the German editions of the works.

22. See also Hentig, *Die Strafe I*, pp. 90ff.

23. Gerhard Schormann, *Der Krieg gegen die Hexen: Das Ausrottungsprogramm des Kurfürsten von Köln* (Göttingen: Vandenhoeck und Ruprecht, 1991); Lucy Dawidowicz, *The War against the Jews, 1933–1945* (London: Weidenfeld and Nicolson, 1975) – published in Germany as *Der Krieg gegen die Juden 1933–45* (Wiesbaden: Fourier Verlag, 1979).

24. Wolfgang Behringer, *Witches and Witch-Hunts: A Global History* (Cambridge: Polity Press, 2004), pp. 115–17. See also Thomas Becker, 'Hexenverfolgung in Kurköln', *AHVN* 195 (1992) 202–14.

25. Moore, *The Formation of a Persecuting Society*, p. 109. On the parallel with witches and earlier persecutions see pp. 35, 79.

26. The critique has been presented in, for example, Geoffrey Scarre, *Witchcraft and Magic in Sixteenth and Seventeenth-Century Europe* (Basingstoke: Macmillan, 1987), pp. 39–40. On later research see William Monter, 'Witch Trials in Continental Europe', in Bengt Ankarloo and Stuart Clark (eds), *Witchcraft and Magic in Europe: The Period of the Witch Trials* (London: Athlone, 2002), pp. 10–12 (quotation).

27. Behringer, *Witches and Witch-Hunts*, pp. 113, 117–19 (quotation).
28. Keith Thomas, *Religion and the Decline of Magic: Studies in Popular Beliefs in Sixteenth and Seventeenth-Century Europe* (London: Weidenfeld and Nicolson, 1971); Alan Macfarlane, *Witchcraft in Tudor and Stuart England: A Regional and Comparative Study* (London: Routledge & Kegan Paul, 1970); H. C. Erik Midelfort, *Witch Hunting in South-Western Germany, 1562–1684: The Social and Intellectual Foundations* (Stanford: Stanford University Press, 1972); E. William Monter, *Witchcraft in France and Switzerland: The Borderlands during the Reformation* (Ithaca: Cornell University Press, 1976).
29. Trevor-Roper, *The European Witch-Craze*, p. 12; Rossel Hope Robbins, *The Encyclopedia of Witchcraft and Demonology* (New York: Crown Publishers, 1959), p. 9.
30. Cohn, *Europe's Inner Demons*, pp. xii, 126ff., 252; Richard Kieckhefer, *European Witch Trials: Their Foundations in Popular and Learned Culture 1300–1500* (London: Routledge, 1976), pp. 16ff., 86–7, 105; Briggs, *Witches and Neighbours*, p. 404; Thomas, *Religion and the Decline of Magic*; Macfarlane, *Witchcraft in Tudor and Stuart England*; Joseph Klaits, *Servants of Satan: The Age of the Witch Hunts* (Bloomington: Indiana University Press, 1985), p. 5.
31. A summary of the still ongoing discussion of the *makt-stat* in Scandinavia. See, for example, Antti Kujala, *The Crown, the Nobility and the Peasants 1630–1713. Tax, Rent and Relations of Power* (Helsinki: Finnish Literature Society, 2003), pp. 16–23.
32. *Åbo Tidningar*, 18 May 1795. On the lower court records see Marko Nenonen, *Valpuri Kyni – noitanainen* (undated), <www.chronicon.com/noita/valpurikyni.html>.
33. In this sense, considering the purpose of showing the persecution fantasies and stereotypes in their absurdity to the presumably western reader, the way Cohn began his book by describing the attitudes against early Christians is excellent. On excesses and obscurities see Klaits, *Servants of Satan*, p. 149. Antero Heikkinen, *Paholaisen liittolaiset. Noita- ja magiakäsityksiä ja -oikeudenkäyntejä Suomessa 1600-luvun jälkipuoliskolla* (Helsinki: Finnish Literature Society, 1969), p. 230, includes a similar and often used story of a witches' salve that turned out to be the prison guard's boot grease. For similar stories during the McCarthy era see Eric Hobsbawn, *The Age of Extremes. A Short Twentieth Century 1914–1991* (London: Abacus, [1994] 1995), p. 234.
34. Cohn, *Europe's Inner Demons*.
35. Again, Trevor-Roper, *The European Witch-Craze*, pp. 9, 12–13. On current scholarship see, for example, Lyndal Roper, *The Witch Craze: Terror and Fantasy in Baroque Germany* (New Haven and London: Yale University Press, 2004), pp. 4–12. For a different approach see Allison Rowlands, *Witchcraft narratives in Germany: Rothenburg 1561–1652* (Manchester: Manchester University Press, 2003). For expressions presuming that witch trials were miscarriages of justice, where the judges laboured to convict instead of finding the truth see, for example, Schormann, *Der Krieg gegen die Hexen*, pp. 18–19; James Sharpe, *Instruments of Darkness: Witchcraft in England 1550–1750* (London: Hamish Hamilton, 1996) p. 7. Even when Robin Briggs, in *Witches and Neighbors* (pp. 399–400), argues that the majority of witch trials might not have been the kind of the crazes, his wording 'the persecution was a relative failure' makes one think that something else was aimed at.
36. For example, Trevor-Roper, *The European Witch-Craze*, p. 84; Jeffrey Russell, *A History of Witchcraft: Sorcerers, Heretics and Pagans* (London: Thames and Hudson, 1980); Klaits, *Servants of Satan*, pp. 128–31; also, in books of rather different emphasis, see Walter Stephens, *Demon Lovers*, p. 6, and Marko Nenonen and Timo Kervinen, *Synnin palkka on kuolema* (Helsingissä: Otava, 1994), pp. 267–70.

37. The wording was not Valpuri's own, but the scribe had translated it. Still, the content of the reply seems articulate for the uneducated peasant/beggar woman that Valpuri was. Sometimes the way in which Valpuri's trial mirrors later fiction makes one wonder if the editors of *Åbo Tidningar* took some liberties with the printing of the High Court record. The actual events and charges, as well as Valpuri's grounds for defence can be confirmed, however, from the extant lower court records.

38. Lyndal Roper, *Oedipus and the Devil: Witchcraft, Sexuality and Religion in Early Modern Europe* (London: Routledge, 1994).

39. Arthur Miller, Introduction, pp. 42–3.

40. See Malcolm Gaskill, *Witchfinder: A Seventeenth-Century English Tragedy* (London: John Murray, 2005); Bengt Ankarloo, *Trolldomsprocesserna i Sverige. Rättshistorisk Bibliotek* (Stockholm: Nordiska Bokhandeln, 1971), pp. 67–76, 231–62; Per Sörlin, *Trolldoms- och vidskepelseprocesserna i Göta Hovrätt 1635–1754* (Stockholm: Almqvist & Wiksell International, 1993), pp. 47–65; Bengt Ankarloo, 'Witch trials in Northern Europe 1450–1700', in Bengt Ankarloo, Stuart Clark and William Monter (eds), *Athlone History of Witchcraft and Magic in Europe: Witchcraft and Magic in the Period of the Witch Trials* (London: Athlone, 2002), pp. 78–80.

41. On similar points concerning the material see Marion Gibson, *Reading Witchcraft: Stories of Early English Witches* (London: Routledge, 1999); Marko Nenonen, 'Historiankirjoittajien Paholainen, Noitavainojen uusi kuva', in Sari Katajala-Peltomaa and Raisa Maria Toivo (eds), *Paholainen, moituus ja magia – kristinuskon kääntöpuoli. Pahuuden kuvasto vanhassa maailmassa* (Helsinki: Finnish Literature Society, 2004). This also points out that research on the witch trials has overemphasised the continental Sabbat trials. Nenonen also sees other and less edifying reasons than the mere fact that some source materials have survived better and some are easier to use than others.

42. Wolfgang Behringer, 'Neun Millionen Hexen. Entstehung, Tradition und Kritik eines populären Mythos', *GWU* 49 (1998) 664–85; Behringer, *Witches and Witch-Hunts*, p. 157.

43. The current European numbers can be read in any textbook. See, for example, Brian Levack, *The Witch-Hunt in Early Modern Europe* (London: Longman, 1987), pp. 19–21; Briggs, *Witches and Neighbours*, p. 8. The figure of 200,000 trials is an estimate by Nenonen, based on the fact that minor witch trials have in many areas not been counted in the statistics. See Marko Nenonen, *Noituus, taikuus ja noitavainot Ala-Satakunnan, Pohjois-Pohjanmaan ja Viipurin Karjalan maaseudulla 1620–1700* (Finnish Literature Society, 1992), pp. 376–7.

44. The original records of Valpuri's High Court interrogation or decisions were lost in the fire of Turku in 1828.

45. Trevor-Roper, *The European Witch-Craze*, p. 9.

46. Trevor Roper, *The European Witch-Craze*, p. 7.

47. Cohn, *Europe's Inner Demons*.

48. Klaits, *Servants of Satan*, p. 3.

49. Robbins, *Encyclopedia of Witchcraft and Demonology*, p. 17.

50. This critique has been levelled by, for example, Diane Purkiss, *The Witch in History: Early Modern and Twentieth-Century Representations* (London: Routledge, 1996), at students of English witchcraft, as well as many scholars after her. As an example of the difficulties, however, see the trouble Sharpe takes in stating his (and his readers' presumed) disbelief in witchcraft: Sharpe, *Instruments of Darkness*, p. 7. These statements of personal disbelief have also been necessary because there are

people today who believe in witchcraft as a satanic pact or as a remnant of an ancient fertility cult.

51. Cohn, *Europe's Inner Demons*, pp. 254–5. Cohn's descriptions of some of the earlier persecutions, and especially those concerning the Knights Templar, are different in this account; Klaits, *Servants of Satan*, p. 61.

52. Arthur Miller, 'The Crucible', in *Arthur Miller's Collected Plays*, pp. 225–9, 234, 239, 242–3.

53. Trevor-Roper, *The European Witch-Craze*, p. 100.

54. Trevor-Roper, *The European Witch-Craze*, pp. 106–7 (quotation).

55. Mary Daly, *Gyn/Ecology. The Metaethics of Radical Feminism* (Boston: Beacon Press, 1978). On feminist readings of witches see Purkiss, *The Witch in History*.

56. Geoffrey Scarre, 'Moral Responsibility and the Holocaust', pp. 112–13. He also claims, perhaps quite rightly, that we tend to feel more personal sympathy for those to whom we feel close than for those who just appear as numbers in a chart – and that we generally know more about the lives of the victims of the Holocaust than those of the witch trials.

7
culture wars: state, religion and popular culture in europe, 1400–1800

marko nenonen

To understand the witch trials it is, of course, necessary to understand broader developments in European society at the time, and, in this respect, general historical accounts of early modern history proffer a fascinating paradox. The period from the mid fifteenth to the eighteenth century was a period of considerable development in science, technology and educational schemes. In the long run, living conditions changed for the better even for the commonalty, no matter how selective and sporadic the enhancement. Up until the early twentieth century, some writers, influenced by the ideas of the Enlightenment, were bold enough to label this development as 'progress' and the course of history was seen to be a progressive advancement.[1] Later generations, though, became much more cautious in attaching value judgements to historical occurrence, especially after the two world wars and other atrocities of the twentieth century. However, notwithstanding the fact that the word 'progress' appears little in current academic writing, many still see the improvement of social conditions and the development of western democracy as a progress incomparable with earlier centuries. To a great extent, European historiography can be seen as a grand narrative showing European nations' – or Europeans' – route from the supposed 'Dark Ages' of the medieval period, towards a modern society which we consider, despite its apparent shortcomings, superior to the living conditions of earlier centuries.[2]

On the other hand, the early modern period is seen as a time of continuous wars, religious intolerance and turmoil, unrest and impoverishment, and social polarisation. How could any progress have taken place under those circumstances? Who progressed? Furthermore, the period also gave birth to a modern state machinery, which meant, at least to some extent, professional and accountable administration. The early modern state, however, is judged unfavourably by many modern historians, especially among those who have studied popular culture. The state is mostly seen as a set of repressive tools by which the elite expanded its control over the masses in order to promote the interests of the governing clique and to suppress the lower orders, along with people's everyday religious beliefs.[3]

Moreover, the modern western Churches at that time, whether Catholic or Protestant, are not primarily remembered for advocating literacy, decent manners, humanity, work ethics, and the like. Instead, scholars, particularly from a socialist perspective, have generally presented that even if the Church authorities acted more or less independently from the state, all major religious groupings provided fundamental support for the repressive state policy, whether for their own ideological or religio-political reasons or just to support the powers of the elite – to which very many of the clergy belonged. The Church also provided efficient disciplinary tools, among others humiliating penances, for social control purposes at the parish level.[4]

More paradoxically, all the controversial trends of the early modern period were followed by the era of the Enlightenment with its emphasis on education, science, rational reasoning and improvement of social and political environments. Some advocates of the Enlightenment became committed to ending ignorance, injustice, bigotry and poverty. The development of new science and technology seemed to support their efforts. What was still missing in the Enlightenment 'programme', though, was an edification of the human mind through mass education. In this sense it is not entirely surprising that some nineteenth-century historians saw progress when addressing the differences between the earlier centuries and their own time, which was full of confidence in new designs for social and political promotion. However, even this human edification had its obvious limits. Religious hatred was alive and well in the age of the Enlightenment, and it was the most thriving period of the slave trade, when millions of people, mostly from Africa, were transported to western countries to enhance the riches of their leading families.[5]

This paradoxical portrayal of the early modern period as being one of progress and yet one of retarding intolerance, war and repression brackets the witch trials as historical phenomena. Furthermore, the rise and decline of the trials can be used to illustrate both these developments in European society. One such approach, that fits the paradox, has been to emphasise the consolidation of the modern state and the Churches' means of social control in fighting religious deviance and 'superstition'. The rise of absolutist states led to the repression of traditional popular culture with its magical beliefs and traditions. Also, new policies of state intervention demolished the long-established and extensive autonomy enjoyed by local communities.

the state, the church, and social control

One of the most influential advocates of this social control theory of early modern state development is the French historian Robert Muchembled. In a series of books and articles he argued how 'the persecution of witches is an effect of the acculturation of rural areas by the religious and political elite'.[6] According to Muchembled,

> Representatives of the Church, the king, and the governing orders of society directed all their efforts to the constraint of bodies and the submission of souls and to the imposition of total obedience to the absolute king and to God.[7]

Even if he over emphasises the repressive nature of the state and religious machinery, his viewpoint is to some extent shared by most scholars. The role of the state has come to the forefront in witch-trial studies during the last few decades, reflecting a broader trend in modern European historiography. Some scholars, such as Boris F. Poršnev (Porchnev), a Marxist historian in the former Soviet Union, have suggested that the transformation of the state was motivated by the ruling elites' fear of rebellions. Other historians have not usually gone so far, although there is not much against this 'Poršnevian thesis' in the literature either, Poršnev's works having been largely neglected by American and British historians.[8] One could summarise the idea of many scholars by stating that the ruling elite saw the guidance of the commonalty necessary in order to develop good administration, to fight crime, to suppress revolts, to promote an efficient economy – loyal workforce included – and to increase taxes. It was not a question of ignorance of the 'primitive mind' that brought about hatred and ignited persecution. It was the question of order and discipline. However, as Brian Levack has pointed out, and as will be discussed later, there are considerable flaws in any overarching explanation of the witch trials in terms of 'state-building'.[9]

Generally, the emergence of the modern state – 'the movement of centralizing power' as Muchembled put it[10] – means almost everything which one nowadays considers self-evident components and functions of states. However, these days the repressive powers of the state are not as apparent for most Europeans as they were for the subordinated in the early modern period. A new state apparatus meant that the accountability of the administration grew gradually and professional office holders became the norm. Also, the hierarchies and powers of governance were classified, and dozens of new institutions were established. In addition, better governance necessitated accurate information, which prompted investigations into economic and social conditions in order to provide the basis for reforms, such as the development of the Poor Law system in England.

Another principal factor in deciphering the social and intellectual environment of the witch hunts is the assertion that before the sixteenth century, most people hardly understood the teachings of Christianity with any depth. Jean Delumeau stated in his *Catholicism between Luther and Voltaire* (1971) that the European peasantry were first truly Christianised during the period of the Reformation and Counter-Reformation in the sixteenth century. According to Delumeau the two Reformations were 'two processes which apparently competed but in actual fact converged, by which the masses were Christianized and religion spiritualized'.[11] For the Protestant Reformed Churches, in particular, it became of paramount importance that the common people understood not only the tenets of Christian faith but also implemented them in every aspect of their daily lives. Not surprisingly, the ecclesiastical authorities, both Protestant and Catholic, launched a concerted effort to eradicate beliefs and practices which they did not consider consistent with their respective doctrines, as a means of reinforcing their own grip on their respective laities. Witchcraft and magic were obvious targets. So the clergy, with the support of secular authorities, undertook numerous efforts to guide and rebuke the people, who now belonged not only to their families and local communities, but were also subordinated to the new centralised state and the Church. People became subjects in two senses of the word: they were subjects of the state and, on the other hand, they were subjects in the meaning that they were personally responsible for their conduct. Paradoxically, subordination to the modern state meant stronger emphasis on personal responsibility and individualism. The very same aspect featured in the transformation of religion during the period of Reformation, with the Protestant emphasis on the personal nature of one's relationship with God. Stuart Clark, contextualising Delumeau's thesis, thought it betrayed 'a degree of cultural condescension', but nevertheless 'allows us to locate the history of early modern witchcraft at the very heart of the reforming process'.[12]

witch hunts as a means of acculturation

Even though the state authorities across western Europe did not act in accordance with a conscious plan or follow a co-ordinated policy, the repression of local autonomy and traditional popular culture resulted from the large-scale efforts by which the authorities pushed forward their reformist policies. Confrontation was inevitable and, without necessarily subscribing to the Poršnevian thesis, it is likely that confrontation itself partly led the state machinery to sharpen its punitive instruments. With regard to this confrontation between elite and popular cultures, Robert Muchembled does not use the term 'class struggle' but he speaks of 'cultural revolution'. His interpretation certainly presents it as a violent conflict between those who ruled and those who were under rule. According to him, this provides a framework for understanding the witch hunts. It was a conflict between people with divergent socio-cultural engagements, the reformers being on the side of the state and the Church.

Like Muchembled, Christina Larner, in her influential work on Scottish witchcraft, emphasised the elite's repressive policy in manufacturing witch hunts. In her *Enemies of God* (1981) she saw witch hunts as 'activity fostered by the ruling class'. Furthermore, she also built upon Jean Delumeau's idea that the Reformation and Counter-Reformation were the first periods of true Christianisation of the rural commonalty. The idea of a godly state on earth required secular authorities 'to impose the will of God upon the people'.[13] Yet, although they are often mentioned in the same context, Larner and Muchembled display different emphases. Larner highlighted the differences between popular magic and the learned notions of diabolical witchcraft, whereas Muchembled did not place much emphasis on diabolical witchcraft theories in his early work, partly because the area of France his research focused on did not produce many sensational diabolical cases. In Scotland, on the other hand, diabolical witch hunts were numerous and this fact drew Larner's attention to the interaction between popular and elite culture.

Larner's study demonstrates that at the early stage of witchcraft suspicions, maleficent magic was the only concern for most people who accused others for bringing harm and misery through witchcraft. The diabolical interpretation was introduced in the course of legal proceedings, thereby turning *maleficium* accusations into satanic witchcraft. In seeing how traditional magical beliefs were reinterpreted in courts, Larner followed Norman Cohn's approach (1975) and Richard Kieckhefer's (1976) dichotomy between popular and learned culture.[14] Cohn demonstrates the gradual evolution of diabolical witchcraft theories and stereotypes. Larner showed how the elite imposed these upon popular conceptions of witchcraft and magic.

Muchembled's particular contribution to witch-hunt historiography, however, is his application of the theory of acculturation. He uses the term 'acculturation' to describe and explain the process through which the elite could convert popular beliefs and traditions into more appropriate forms for the centralised governance and authoritative religion to suppress. As Stuart Clark characterised the process, 'Witches were the deviants of Christianization, with "witchcraft" acting as a catch-all term of cultural censure and conquest.'[15] But what does the idea of acculturation add to what sounds like a straightforward issue of social control? As the acculturation theory constitutes an important framework in Muchembled's work, it is remarkable that he does not present anything about his starting points regarding acculturation theories.[16] He seems to understand acculturation simply as a means of social control:

> In order to accomplish the submission of souls and bodies, in order to achieve the acculturation of rural areas, magical beliefs and practices had to be suppressed. Whether the magistrates were aware of it or not, the stake and pyre enabled a dynamic and learned culture to reject and weaken

a popular culture that was very ancient and nearly immobile, but that nevertheless met all changes with enormously strong resistance.[17]

Much of popular tradition was bound up with seasonal festivities in communities and family celebrations.[18] According to Muchembled, from the sixteenth century onwards, the authorities increasingly took control over traditional feast days, either by suppressing them or by transforming and taming them. Muchembled explains that more light is also shed upon the role of women in witchcraft accusations through examining popular culture in general, rather than concentration solely on witch trials. He remarks how women were 'the receptacle of popular culture'.[19] Regardless of the contempt which women most likely encountered, to a great extent women were important holders of traditions within families and local communities. This made women more vulnerable when traditions came under attack, and their attackers could use the label of witch to help suppress them. Be this as it may, the gender issue has before and afterwards much coloured the witch-hunt accounts.[20]

The timing of the state intervention into the rural communities is interesting, if one subscribes to Muchembled's position. He suggests that in France the process of acculturation reached its peak in the mid seventeenth century, although many actions against popular festivities, for example, had been taken earlier. According to this timetable, then, the witch hunts commenced and peaked earlier than the decisive period of state intervention at the local level. This actually weakens his case in postulating that repressive state policies ignited witch hunts. On the other hand, the Church seems to have been first to promote the new course. As Muchembled writes:

Between 1550–1700 ... the Church managed to impose, at least in appearance, religious individualism and obedience to power ... During the sixteenth century, efforts to repress popular culture had already begun in cities and towns.[21]

It is discomforting for Muchembled, who colourfully draws the picture of ruthless repression, that the masses did not offer stronger resistance and no more revolts emerged. Instead, uprisings as well as witch hunts in France died down during the reign of Louis XIV (1643–1715).[22] There were seemingly many common people, who Muchembled calls the 'peasant elite', who instead of revolting saw obvious advantages in adapting themselves to the new standards of decent life.

In considering why the state succeeded, Muchembled finds an explanation which is very much a main constituent of his acculturation thesis. He borrows Michel Foucault's 'excellent definition' of a 'political technology of the body', a concept idea which he finds utterly fascinating. 'The political technology of the body' was the means 'used by the monarchy and its henchmen during the two last centuries of the Ancien Régime'. The body becomes 'a useful force only if it is both a productive body and a subjected body', he cites from Foucault's *Discipline and Punish* (1975). 'Productive' refers to the idea that a body is not only the source of new generation (of the subordinated) but also a source of workforce creating new riches (for the ruling elite). This condition 'had long been realized in France', states Muchembled. The 'political technology of the body' is something 'designed to obtain the most perfect obedience possible on the part of the subjects'. This would be fulfilled by 'sexual repression, by training in bodily control on all occasions, by judiciary mutilation and by torture, all of which marked the social limits beyond which each individual could no longer make use of his own body as he saw it'.[23] Essential for the constraint of individuals was the humiliation and torture of the body of the individual.

With this 'political technology of the body' comes also a 'vast movement' of sexual repression in the sixteenth and seventeenth centuries.[24] The thrust of this view is that before the new policies people were freer to follow their instincts and impulses, and sexual behaviour in general was less restrained. It is suggested that the early modern period brought about a change in this respect and the sexual etiquette, in conformity with the new teachings of the state and the Church, put more stress on intimacy and restraint.[25] According to Joseph Klaits, influenced by Muchembled's work, 'the quantum leap in witch trials during this era was one outlet for the deep stresses produced at all social levels as the godly reform of sex took hold'.[26]

Norman Cohn has demonstrated that erotic orgies with bizarre practices, ritual murder and cannibalistic feasts had long before the witch hunts 'belonged to one particular, traditional stereotype: the stereotype of the conspiratorial organization or secret society engaged in a ruthless drive for political power'.[27] Witch hunts did not create the combination of persecutions and sexual behaviour. Furthermore, almost certainly the stories of odd feasts related to deviant and non-conformist groups, like the Early Christians, heretics and witches, were untrue, says Cohn. Yet sexuality has been one major theme in witch-hunt historiography, and the idea of repression of sexuality constitutes its own theories of the origin of witch hunts, the more so as most of the accused were women.[28]

Muchembled argued that as well as a stronger emphasis on the control of female sexuality the new ideology also fostered new models of family life. The role of father and his paternalistic rule was much emphasised.[29] Fathers, like kings, should guide their subordinates toward the right norms and values, and both should rule according to the divine instruction. Yet it was clear that the divine order gave much more privileges to the heads of the hierarchy than to those ones who were supposed to be subservient to their masters. Muchembled gives rather substantial space to the new family model. It is understandable, since the theme was not usually introduced in this context. We will see below that there might have been other reasons as well for this emphasis.

the political context of social control theories

The basis of acculturation theory, in its various forms, implies that when two (or more) different cultures confront, the less developed or lower, more archaic or otherwise weaker culture will be changed or wither under the pressure from the more elite, advanced or otherwise stronger culture. Anthropological theories, mostly dated from the late nineteenth century, classified two main categories of the acculturation process, depending on to what extent mere force is used to obtain the dominance of one culture over the other and how deeply the culture in question is affected.[30] First, cultural elements can be modified by borrowing influences from different cultures in the course of interchange, such as through trade and travel. This kind of free borrowing and modification is often seen as incorporation, which does not necessarily embrace force. Second, as distinct of free borrowing and interchange, a change can be put into force by military or political means. Assimilation means a complete or practically complete domination of one culture over the other or others. Also, as a third possible classification it is sometimes seen that a new synthesis may emerge through cultural fusion, as various elements give birth to a new culture. Acculturation need not require or imply force. The ruling ideology and elite's lifestyle can offer natural incentives for others to strive for a new lifestyle which is considered superior to the old.

Acculturation theories arose from the problems of understanding the confrontation of western culture with other cultures. From the very early days of European expansion the non-European cultures were deemed inferior or even primitive by western observers. To a great extent, the Europeans' feeling of superiority might have derived from seeing Christianity as the only true religion. In the early modern period ruling classes in Europe were fully aware of their culture's alleged supremacy in trade and warfare. Social sciences and humanities reflected the superior position of European imperial powers and ruling elites. Following this, the advocates of cultural 'Enlightenment' directed their efforts to education, social reforms, and to the promotion of science and technology. Europeans – or more precisely western Europeans – were seen as bearing the torch bringing light to the human race. The first modern anthropological studies were written in the glow of this nineteenth-century conception of superiority and patronage. The presumption was that under their rule other cultures would follow the European models of modernisation. As we now know, it did not go quite this way.

After the propagation of acculturation theory by anthropologists in the late nineteenth century, the approach certainly influenced anthropology between the two world wars but was then sidetracked. It was in France in the 1960s and 1970s that the thesis experienced a renaissance, this time in historical studies of tensions within European societies, most notable in the work of Pierre Chaunu, Jean Delumeau and Muchembled.[31] Acculturation theory was also adopted in a few historical analyses on the confrontation between western civilisation and other civilisations. Muchembled, for example, mentions Inca civilisations and Peru's history a few times when discussing the similarities of animist beliefs everywhere.[32] However, he did not refer to Nathan Wachtel's 1971 study on the crisis and social change provoked by the Spanish conquest of Peru, even though both wrote in French. Wachtel was among the very few who fully developed the acculturation approach rather than pay lip-service to it. Indeed, Peter Burke has drawn attention to Wachtel's novel theory on acculturation, which he judges sympathetically even though 'the term "acculturation" was originally coined by anthropologists ... and is no longer taken seriously by most of them'. Wachtel introduces the terms 'acculturation' and 'destructuration'. By acculturation he means a cultural interchange in a situation where one society is dominant and the other subordinate. By destructuration, which according to Burke was originally introduced by Italian sociologist Vittorio Lanternari, Wachtel means that the links which made the different parts of the traditional social system as a whole, disappeared and disintegrated in the acculturation process, even though many features of previous traditions and customs survived after the conquest.[33]

One important and influential theorist must also be mentioned in this context, even though he might not have directly influenced acculturation theories. Italian communist Antonio Gramsci's (1891–1937) idea of 'dominant cultures' and 'cultural hegemony' became important reference points for many 1960s' cultural critics in Europe.[34] For Gramsci, it was a question of cultural dominance executed by one or the other social class. According to Burke, Gramsci's influence can be seen in Wachtel's theory, perhaps through Lanternari, though sociology researchers did not focus on the power relations between civilisations. Instead, the focus – like Gramsci's – was shifted to Europeans' own societies. New theories of the most delicate matter were introduced: power relations between the ruling classes and their subjects in Europe.[35] Indeed, I would suggest that acculturation theory related to witch hunts developed from this sociological rather than anthropological standpoint, and so did many other approaches emphasising the means of social control imposed by the elite.

One can find numerous clues in Muchembled's writings on witchcraft to the influence of the modern political scene on his interpretations of early modern society. Bob Scribner convincingly suggested that the notion of acculturation in Muchembled's and other French scholars' work 'had a rather curious origin in political reactions to the French decolonisation of Algeria, whence it slipped into fashionable historical discourse in the early 1970s'. As regards some themes in Muchembled's work, such as popular culture and 'mass' culture, fathers' authoritarian rule, class struggle and the repressive policies of the ruling classes, one might even suggest that Muchembled is speaking within the terms of the revolutionary year of 1968, characterised by massive student demonstrations against the establishment. 'No historian, in my opinion, is a simple observer of the past', he said, and compared the early modern period with 'the great cultural upheaval' that followed the two world wars in the twentieth century, 'a crisis in established values, a crisis of faith, a crisis of family, the problem of power'.[36]

One cannot help noticing the similarities which Muchembled creates by comparing the early modern period with the time of radical social criticism in the 1960s and 1970s. The authoritarianism of patriarchal values and ideal family model at the beginning of the early modern period is put side by side with the younger generations' rebellion against the establishment, often directly against their fathers (and their values) in the 1960s and 1970s. Other analogies can be found as well. Muchembled sees it as paradoxical that despite the harsh repression, peasants did not rebel more forcefully. This, again, parallels the alleged contrariness of the 1960s and 1970s when, despite capitalist exploitation, the working classes did not generally support the student revolutionaries. Muchembled himself implicitly refers to this analogy.[37] In some aspects, it would seem that Christina Larner shared Muchembled's inspiration from the radical political scene of the 1960s, even though there were no obvious mutual influences as they worked separately. For Larner the confrontation of the elite culture and popular culture was a matter of a power struggle. Class struggle between 'the powerful in relation to the powerless', to use Larner's expression, determines the power relations of the people, as it did in much of the political debate in the 1960s and 1970s. However, both Muchembled and Larner seemed to try to avoid the most extremist interpretations of class struggle by noting, as Larner did, that 'the system as described amounts to something more complex than a simple dichotomous model of class exploitation'.[38]

Their sharing of the reformist inspiration of the 1960s and 1970s is worthy of note, since otherwise their most important academic mentors appear to point in another direction – with the exception of Jean Delumeau, to whom both writers referred. However, Larner, who expressed herself to have been greatly influenced by Delumeau, quite interestingly turns his ideas upside down. According to her, he was wrong about the reasons why harsh means of social control, including witch hunts, took place. It was not because Christianity was imperfectly realised but because of its success – the completeness of Christianity's triumph over the political, and its fusion with worldly issues. She explained that from the Reformation to the Industrial Revolution 'Christianity served as the world's first political ideology'. By 'political ideology' she meant a 'total world view which serves to mobilise political action or to legitimise governments'.[39]

divergent models: repression theories and reformist theories

Originally, acculturation theories were rather conservative, as they laid much stress on the point that the acculturation of the inferior cultures caused by the dominant powers were – according to them – to a great extent something that people willingly aspired to. This may have reflected the westerners' self-confidence in the colonial era, which scholars of later generations have found harder to share.[40] Writers of later generations, on the other hand, lay much more emphasis on the repressive modes of social control. They admit, though, that large-scale rebellions did not occur even if, as Larner put it, 'objective' reasons for revolts existed. This is explained by the triumph of the central government's repressive policy. However, it is not evident as to what the repression of popular culture means. It is likely that many actions taken by the state authorities and the clergy were repressive in the sense that ordinary people were guided to the supposed better behaviour in their worldly course. In spite of this, even the transformation of popular culture did not originate only from the elite's policies. Peter Burke has emphasised that popular culture itself comprised elements in constant change. He, too, sees that the rule of the elite was important in the transformation of popular culture, especially after the mid seventeenth century, when the state machinery and the Church had firmly established their new administrative tools, and in western Europe at least the witch trials were in decline. Yet when addressing the changes within the sphere of popular culture he uses the word 'transform' or 'reform', instead of 'repression'.[41] This represents his view of change, which for him is rather constructive, unlike many who see it only as a consequence of depressing social discipline.

In certain respects the 'witch hunts' do not represent much of the general character of the era, especially, as it is rather an ambiguous notion. There were trials for 'superstitious' beneficial magic, maleficent magic, heresies and diabolical witchcraft. There were mass trials – though fewer than once was depicted[42] – and single trials. Trials for 'superstition' and harmful magic dominated in most if not in all areas, though fascination with diabolical witchcraft and flying witches has dominated the research to the extent that in some treatises any detailed clarification of the accusations in courts of law has been deemed unnecessary. As the calculated number of prosecutions has been revised downward from millions to between 100,000 and 200,000,[43] one is forced to conclude that during the 300 years of witch trials in Europe, there were frequently periods of several years when not a single alleged witch appeared in most European courts. The apparent soundness of repression theories notwithstanding, it is not obvious, therefore, how witch hunts were related to the centralisation of state powers and to the repression of popular culture. It is obvious that the Churches attacked traditional magic, along with other sinful activities. Creating new diabolical witchcraft theories, on the other hand, is a different matter. Why did they emerge and why were they vehemently propagated?

Popular culture as a concept is vague and intricate, although it has strong positive connotations for many in contrast to 'elite culture'. Popular culture as a research subject was much influenced by the Romantic Movement in the late eighteenth and nineteenth centuries.[44] Johan Gottfried von Herder's work

Stimmen der Völker ('Voices of the People') (1778–79) was crucial in this sense and his approach soon found followers in many European countries. However, in many fields the true nature of popular culture has been questioned. As early as 1931, Albert Wesselski, from the then Czechoslovakia, cast doubt on whether, for example, fairy tales could be seen as original popular culture – meaning the culture of 'ordinary' people or uneducated masses. He rejected the idea that a genuine, popular oral origin of fairy tales could be found. He suggested that the origin of most fairy tales, which later generations had 'found' and considered as authentic pieces of creative work of the anonymous masses, were, in fact, works of the writers of their time. Moreover, the writers were not from the 'silent masses', and were instead influenced by learned sources and foreign influences.[45] Richard Kieckhefer makes the same point by saying that even if something is intended 'for widespread dissemination' it does not make it popular culture, although the word 'popular' can be used in this sense as well: 'There is no assurance that the stories narrated in such literature had popular origins, or that they did in fact take hold on the popular imagination.'[46]

The conventional categorisation of 'popular culture' and 'elite culture' derives from, as Peter Burke has suggested, Robert Redfield's classical idea of 'the great tradition' or elite culture of the educated few, and 'the little tradition' of the masses. This dichotomy itself is problematic, particularly because it fails to acknowledge that the elite participated in the 'little tradition' which was, in a sense, their 'second culture'. On the other hand, the rest did not have access to the 'great tradition' which was established and reproduced in schools and universities, unlike the 'little tradition' which survived and developed 'informally'. For the majority, the popular culture was the only culture.[47] Another problem is that the 'little tradition' or the culture of the masses, often seems too homogeneous, which, as Burke emphasises, 'was far from being the case in early modern Europe'. He quotes Antonio Gramsci: 'The people is not a culturally homogeneous unit, but it is culturally stratified in a complex way.'[48] Instead of one popular culture there were 'many popular cultures or many varieties of popular culture', observed Burke. Peasants were not all the same. Moreover, there were 'popular sub-cultures' like those of beggars and thieves. Burke straightforwardly warns that what we call 'popular culture' was often 'the culture of the most visible of the people'; in other words, young adult males.[49]

Yet even if popular culture in the early modern period could be categorised with some firmness, it is not clear that it was the commonalty who firstly and mainly became the target of the authorities' reformist programme. On the contrary, measures were first predominantly taken against the members of the elite and those who were supposed to stand as an example, such as the clergy. This is clearly demonstrated in a critique that Jean Wirth raises against Muchembled's acculturation thesis. Muchembled interpreted the story of the Carmelite monk and preacher Thomas Conecte (d.1434), written by chronicler Enguerrand de Monstrelet, as an example of the 'acculturising' preaching campaign in fifteenth-century Flanders. However, it seems that Thomas Conecte targeted mostly the members of the elite, denouncing 'the eccentric appearance of high society women'. He was not a messenger of the lay rulers, nor that of the clergy who – according to the chronicler – detested him, although Muchembled omits this aspect of the story. Jean Wirth states that instead of being unpopular, Conecte 'acted rather as a tribune of the people who ... made many speeches in praise of the common people'. Indeed, anticlericalism may have been a popular theme among the people. Moreover, Wirth continues that the ecclesiastical elite more generally were not 'acculturising' the people but defending themselves against those 'who questioned their function and practices'.[50] Jean Wirth has a point. The Reformation did not start as a calculated policy of the reformers, although it soon led to radical consequences and new practices. Yet nothing of the sort was in Martin Luther's mind at the beginning. Some turning points of the reforms were evidently pushed forward by the common people, influenced by popular anticlericalism. On the other hand, not only the commonality was disciplined; new education schemes with new demands and qualifications of learning were directed at the functionaries of the state and the Church in many European countries. Also in other fields of public life dozens of new edicts set new conduct requirements upon the members of the elite.

All this, again, leads us back to the question of the true historical trends of the period of the witch trials. Much research is still needed to understand the themes of the era. Taking into account the trouble which historians have in understanding the true course of early modern European history, it is no wonder that they still struggle also in placing witch trials into the proper historical context. Although some critical remarks have been raised here against the social control approach in witch-hunt historiography, one does not have to decide between the all too simple repression approach and the obsessively progressive model of the rise and fall of the witch trials.

notes

1. For a classic example, see J. B. Bury, *The Idea of Progress: An Inquiry into its Origin and Growth* (London: Macmillan, 1920). With a more relative point of view, see G. G. Iggers, 'The Idea of Progress in Recent Philosophies of History', *Journal of Modern History* 30 (1958) 215–26; Iggers, 'The Idea of Progress: A Critical Reassessment', *American Historical Review* 71 (1965) 1–17.

2. Many writers in the era of the Enlightenment produced 'Enlightened' or 'cosmopolitan' narratives of European development. However, their 'mindless and overweening' confidence in progress has been overemphasised, if not caricaturised by later writers. According to Johnson Kent Wright, the belief in historical progress was by no means representative of all writers of the era. Wright states that 'legends about Enlightenment faith in historical "progress" persist against all evidence'; Wright, *Historical Thought in the Era of the Enlightenment*, in L. Kramer and S.

Maza (eds), *A Companion to Western Historical Thought* (Oxford: Blackwell, 2002), pp. 123–4, 132–41, citation p. 140. It seems, though, that Roy Porter, in his *The Enlightenment: Britain and the Creation of the Modern World* (London: Penguin Press, 2000), pp. 424–45, sees progress as a justified concept of the Enlightenment – at least in Britain, where both the proper notion of the Enlightenment and the idea of progress seem to have had more practical essence than on the Continent. The more or less explicit idea of progress was prevalent to many influential writers of national histories in the nineteenth century, among them Jules Michelet, Thomas B. Macaulay and George Bancroft. See T. N. Baker, 'National History in the Age of Michelet, Macaulay, and Bancroft', in Kramer and Maza, *A Companion to Western Historical Thought*, pp. 185–201. Present historical writing is by no means free of the idea of a teleological development of the nations or Europe; see Baker, 'National History', pp. 200–1.

3. See Diarmaid MacCulloch's sarcastic comment on the 'modern liberal academic historians' who have taken an interest in writing the history 'from below', but who feel little sympathy for and no sense of kinship with contemporary popular culture; MacCulloch, *Reformation: Europe's House Divided* (London: Allen Lane, 2003), p. 592. For a consistent overview of the period, see Euan Cameron (ed.), *Early Modern Europe* (Oxford: Oxford University Press, 1999).

4. See R. Po-Chia Hsia, *Social Discipline in the Reformation: Central Europe 1550–1750* (London: Routledge, [1989] 1992), p. 122, *passim*; MacCulloch, *Reformation*, pp. 591–600, 633–4, 672–3.

5. For a modern-day classic on the slave trade, see S. M. Elkin, *Slavery: A Problem of American Institutional and Intellectual Life* (Chicago: University of Chicago Press, [1959] 1968). For a different approach see R. W. Fogel and S. L. Engerman, *Time on the Cross: The Economics of American Negro Slavery* (London: Wildwood House, 1974).

6. Robert Muchembled, 'Satanic Myths and Cultural Reality', in Bengt Ankarloo and Gustav Henningsen (eds), *Early Modern European Witchcraft: Centres and Peripheries* (Oxford: Clarendon Press, [1990] 1993), p. 153. See also Muchembled, 'The Witches of the Cambrésis: The Acculturation of the Rural World in the Sixteenth and Seventeenth Centuries', in J. Obelkevich (ed.), *Religion and the People, 800–1700* (Chapel Hill: University of North Carolina Press, 1979); Muchembled, *La roi et la sorcière* (Paris: Desclée, 1993); Muchembled, *La sorcière au village* (Paris: Gallimard, 1979); Muchembled, *Sorcières: justice et société aux 16e and 17e siècles* (Paris: Imago, 1987); Muchembled, *Les derniers bûchers* (Paris: Ramsay, 1981).

7. Robert Muchembled, *Popular Culture and Elite Culture in France 1400–1750* (Baton Rouge: Louisiana State University Press, 1985), p. 269. Originally published in French in 1978.

8. On the 'Poršnevian thesis', see W. Schulze, *Bäuerlicher Widerstand und feudale Herrschaft in der frühen Neuzeit* (Stuttgart: Frommann-Holzboog, 1980), pp. 27–37, 141–2; Günter Vogler, 'Bäuerlicher Klassenkamp als Konzept der Forschung', in W. Schulze (ed.), *Aufstände, Revolten, Prozesse. Beiträge zu bäuerlicher Widerstandsbewegungen im frühneuzeitlichen Europa* (Stuttgart: Klett-Cotta, 1983), pp. 23–40. Some of Poršnev's writings are attainable in German: 'Das Wesen des Feudalstaates' and 'Formen und Wege des bäuerlichen Kampfes gegen die feudale Ausbeutung' are published in *Sowjetwissenschaft Gesellschaftswissenschaftliche Abteilung* 2 (1952) 248–77; 3 (1952) 402–59.

9. Brian P. Levack, *The Witch-Hunt in Early Modern Europe* (London: Longman, 1987), pp. 89–90; Levack, 'State-Building and Witch Hunting in Early Modern Europe', in

J. Barry, M. Hester and G. Roberts (eds), *Witchcraft in Early Modern Europe: Studies in Culture and Belief* (Cambridge: Cambridge University Press, 1996), pp. 96–115.

10. Muchembled, *Popular Culture*, pp. 1, 183ff.

11. J. Delumeau, *Catholicism between Luther and Voltaire: A New View of the Counter-Reformation* (London: Burns & Oats, 1977), pp. 159–61, 175–202, 225. Originally published in French in 1971. See also Delumeau, 'Les Réformateurs et la superstition', *Actes du colloque sur l'Amiral Coligny et son temps* (Paris: Société de L'Histoire du Protestantisme Francais, 1974), pp. 451–87.

12. Stuart Clark, 'Protestant Demonology: Sin, Superstition, and Society (c. 1520–c. 1630), in Ankarloo and Henningsen, *Early Modern European Witchcraft*, pp. 46–7.

13. C. Larner, *Enemies of God: The Witch-Hunt in Scotland* (London: Chatto & Windus, 1981), pp. 1, 5, 25, 157ff.

14. N. Cohn, *Europe's Inner Demons: An Enquiry Inspired by the Great Witch-Hunt* (New York: Basic Books, 1975); R. Kieckhefer, *European Witch Trials: Their Foundations in Popular and Learned Culture, 1300–1500* (London: Routledge & Kegan Paul, 1976).

15. Stuart Clark, *Thinking with Demons: The Idea of Witchcraft in Early Modern Europe* (Oxford: Oxford University Press, 1997), p. 510. See more generally chapter 34, 'Acculturation by Text'.

16. Neither is there more on this issue in his other works. See, for example, his 'Lay Judges and the Acculturation of the Masses (France and the Southern Low Countries, Sixteenth to Eighteenth Centuries)', in Kaspar von Greyerz (ed.), *Religion and Society in Early Modern Europe, 1500–1800* (London: George Allen & Unwin, 1984).

17. Muchembled, *Popular Culture*, p. 236; see also p. 316.

18. On this, see also E. Muir, *Ritual in Early Modern Europe* (Cambridge: Cambridge University Press, 1997).

19. Muchembled, *Popular Culture*, p. 66.

20. For one of the most recent studies on this issue, see Éva Pócs, 'Why Witches Are Women', *Acta Ethnographica Hungarica* 48 (2003) 367–83.

21. Muchembled, *Popular Culture*, pp. 211, 212.

22. Muchembled, *Popular Culture*, pp. 188, 269, 271.

23. Muchembled, *Popular Culture*, p. 188. See M. Foucault, *Discipline and Punish: The Birth of the Prison* (London: Penguin, 1977), pp. 3ff. Originally published in French in 1975.

24. Muchembled, *La sorcière*, p. 177.

25. Muchembled considers this a matter of fact; Muchembled, *Popular Culture*, pp. 189–96, 233. See also J. Klaits, *Servants of Satan: The Age of the Witch Hunts* (Bloomington: Indiana University Press, 1985), pp. 48–85.

26. Klaits, *Servants of Satan*, p. 84.

27. Cohn, *Europe's Inner Demons*, pp. 7, 11, also *passim*.

28. However, ignoring other than diabolical witch trials has possibly led to the overrating of the role of women in witch trials. In many areas, trials other than diabolical panics were perhaps less often related to women than Sabbath beliefs. See Marko Nenonen, *Noituus, taikuus ja noitavainot* (Helsinki: SHS, 1992), pp. 31, 313, 437 (summary); Nenonen, 'Historiankirjoittajien paholainen. Noitavainojen uusi kuva', in S. Katajala-Peltomaa and R. Maria Toivo (eds), *Paholainen, noituus ja magia* (Helsinki: SKS, 2004), pp. 250–3, 281 (English Summary: 'The Devil in the Historian's Mind: Witch-Hunt Stereotypes in European Historiography', pp. 317–20).

29. Muchembled, *Popular Culture*, pp. 224–9.

30. On the old schools of acculturation theories, see M. J. Herskovits, *Acculturation: The Study of Cultural Contact* (New York: J. J. Augustin, 1938), pp. 2ff., also pp. 33ff., where the author reviews some influential works in the field. On modern approaches see N. Wachtel, 'L'acculturation', in J. le Goff and P. Nora (eds), *Faire de l'histoire. Nouveaux problèmes* (Paris: Gallimard, 1974), pp. 124–46; on terminology see pp. 129–34.

31. J. Wirth, 'Against the Acculturation Thesis', in von Greyerz, *Religion and Society in Early Modern Europe*, p. 67. See also Stuart Clark, 'French Historians and Early Modern Popular Culture', *Past and Present* 100 (1983) 62–99.

32. Muchembled, *Popular Culture*, pp. 29–30, 50, 54, 58, 61, 65.

33. Peter Burke, *Sociology and History* (London: George Allen & Unwin, 1980), p. 101. His later book *History and Social Theory* (Cambridge: Polity Press, 1992) covers much the same issues; on acculturation, see pp. 125, 155–7; N. Wachtel, *La vision des vaincus. Les Indiens du Pérou devant la Conquête espagnole 1530–1570* (Paris: Gallimard, 1971), pp. 24–6, 32, also *passim*; Wachtel, 'L'acculturation'. Burke has suggested that the term 'acculturation' should be replaced with the idea of 'negotiation'. According to him, acculturation inadequately describes the mental bargaining process through which individuals consider their position and the possible benefits of the reforms; P. Burke, 'A Question of Acculturation?', in *Scienze, credenze occulte, livelli di cultura* (Firenze: Olschki, 1982), p. 204; Burke, *History and Social Theory*, p. 87. It seems to me, however, that it is only a question of a word and, therefore, does not lead us much further.

34. On Gramsci's theory, see R. Heeger, *Ideologie und Macht. Eine Analyse von Antonio Gramscis Quaderni* (Uppsala: Almqvist & Wiksell, 1975), esp. pp. 69ff. and pp. 117ff.

35. Acculturation as an explanatory factor has often been used when encountering minority groups in European societies. 'The others' include, for example, the Lapps in Northern Europe (see E. Asp, *The Finnicization of the Lapps: A Case of Acculturation* (Turku: Turun yliopiston julkaisuja, 1966) and rural population under the pressurised circumstances caused by the conflict between the declining rural culture and expanding urban culture (see Peter A. Munch, *A Study of Cultural Change. Rural-Urban conflicts in Norway*, Studia Norvegica 9 (Oslo: Aschehoug & Co., 1956)).
    Regarding the weight of anthropological inspiration, emphasised much in many witch-hunt histories, it is noteworthy that for example Max Marwick's compilation *Witchcraft and Sorcery* (Harmondsworth: Penguin, 1970) is influenced not only by anthropology but also sociology.

36. Muchembled, *Popular Culture*, pp. 2, 3; R. W. Scribner, 'Historical Anthropology of Early Modern Europe', in R. Po-Chia Hsia and R. W. Scribner (eds), *Problems in the Historical Anthropology of Early Modern Europe* (Wiesbaden: Harrassowitz Verlag, 1997), pp. 13–14. See also Wirth, 'Against the Acculturation Thesis', pp. 66–78. See also Pierre Bordieu's *Algeria 1960* (Cambridge: Cambridge University Press, [1963] 1979), which considerably influenced the French intellectual climate in the 1960s and later.

37. Muchembled, *Popular Culture*, pp. 2–3, 5, 319–20.

38. Larner, *Enemies of God*, p. 50. The themes of the 1960s and 1970s can also be seen in Larner's other works. See Larner, *The Thinking Peasant: Popular and Educated Belief in Pre-Industrial Culture* (Glasgow: Pressgang, 1982), reprinted in Larner, *Witchcraft and Religion: The Politics of Popular Belief* (Oxford: Blackwell, 1984).

39. Larner, *The Thinking Peasant*, pp. 35, 37–8.

40. See Wachtel, *La vision des vaincus*, pp. 24–5, 212.

41. P. Burke, *Popular Culture in Early Modern Europe* (revised reprint, Aldershot: Ashgate, [1978]1994), pp. 58–64, 113–15, 207ff.
42. See, for example, R. Briggs, *Witches and Neighbours: The Social and Cultural Context of European Witchcraft* (London: Penguin Books, [1996] 1998), p. 402.
43. See, H. E. Næss, *Trolldomsprosessene i Norge på 1500–1600-tallet* (Båstad: Universitet-sforlaget, 1982), pp. 17–18, 372–3; Levack, *Witch-Hunt*, pp. 19–21; R. Golden, 'Satan in Europe: The Geography of Witch Hunts', in M. Wolfe (ed.), *Changing Identities in Early Modern France* (Durham: Duke University Press, 1997), pp. 220–1; W. Behringer, 'Neun Millionen Hexen. Entstehung, Tradition und Kritik eines populären Mythos', *Geschichte in Wissenschaft und Unterricht* 49 (1998), pp. 664ff.
44. Burke, *Popular Culture*, pp. xiv–xxiv, 3ff.
45. A. Wesselski, *Versuch einer Theorie des Märchens*, Prager Deutsche Studien 45 (Reichenberg i. B.: Franz Kraus, 1931), pp. 10ff., 144ff. Wesselski criticised 'the Finnish school' of ethnography which tried to trace the genuine origin and then chronologically and geographically follow its diffusion and transform into more sophisticated – or possible more vulgarised – versions (pp. 144, 157). See also C. Zika, 'Appropriating Folklore in Sixteenth-Century Witchcraft Literature', in Hsia and Scribner, *Problems in the Historical Anthropology of Early Modern Europe*, pp. 175–218. Zika, from a slightly different point of view, supports Wesselski's idea, though without mentioning him.
46. Kieckhefer, *European Witch Trials*, p. 27.
47. Burke, *Popular Culture*, pp. 23–5, 28–9.
48. Burke, *Popular Culture*, p. 29.
49. Burke, *Popular Culture*, pp. 29–36, 46–7.
50. Wirth, 'Against the Acculturation Thesis', pp. 68–71.

8

the return of the sabbat: mental archaeologies, conjectural histories or political mythologies?

willem de blécourt

By the 1980s witchcraft historians approached the complex mixture that made up a witch trial from both a macro-sociological and a micro-anthropo-logical point of view. The anthropological approach, however, was mainly a social anthropology, more concentrated on the people involved in a trial than on the concepts they applied. This was particularly unfortunate in the case of the witches' Sabbat since, as Norman Cohn had rightly remarked, without the imagery of Sabbats and the flights to them there would not have been any mass trial.[1] Moreover, the leading proponents of an anthropological perspective relied on texts from England and New England, which implied that had they wanted to look into Sabbat concepts, there was very little to research. Margaret Murray's theory and its reception had caused witchcraft scholars in general to turn their back on witches' assemblies. This trend was strengthened by Richard Kieckhefer, who subsumed the Sabbat under 'diabolism', since he could not find any trace of it in the early witnesses' depositions. As far as he could see stories 'about women who went on mysterious nightly rides ... with some mysterious goddess' stood apart from bewitchments. There was no evidence (at least not before 1500) that the two ever mingled on a village level.[2] The prosecutors had superimposed the Sabbat on witchcraft accusations when interrogating a witch and it was derived from heretical sects such as the Waldensians and the Cathars. If the notion of the witches' flight (the 'mysterious nightly ride', also called transvection) was derived from popular sources, then it was still grafted onto witchcraft by the same prosecutors in order to justify connections between witches in different localities. This whole picture was painted in terms of a popular/elite dichotomy and the Sabbat in all its detail was considered as belonging to the elite category. Together with the focus on social history and the wish of most witchcraft historians to write a history of people previously neglected, this ensured that the Sabbat was also left out in places where it could have been useful, for instance as a decisive criterion in the categorisation of witch trials. Instead, historians applied numbers.[3]

It was only at the end of the 1980s that researchers regained their interest in witches' assemblies. This included the German historian of everyday life, Richard van Dülmen, the Hungarians Gábor Klaniczay and Evá Pócs, the Danish historian and folklorist Gustav Henningsen, and especially the Italian historian Carlo Ginzburg. In November 1992, French witchcraft historians organised a conference on the Sabbat theme where Ginzburg and Pócs were given a prominent place, their contributions opening the subsequent conference volume.[4] The Paris conference was one in a series of international witchcraft conferences, of which those in Stockholm in 1984 and Budapest in 1988 are also of special interest to the subject of this chapter.[5] While van Dülmen concentrated on different types of witches' meetings and their historical development, Ginzburg, Pócs and Henningsen paid more attention to their presumed popular roots. They argued that behind the image used by demonologists it was possible to discern fairies (Henningsen and Pócs) and eventually shamanism (Ginzburg and later also Pócs). Klaniczay commented on the latter and placed (primarily Hungarian) Sabbat imagery into its historical context. A few years later the German historian Wolfgang Behringer also interpreted the stories of the horse-herder and cunning man Chonrad Stoeckhlin as shamanistic.[6] In this chapter we will unravel their arguments, above all by juxtaposing them.[7] Geographically the focus will be on Italy and Hungary.

Carlo Ginzburg can be characterised as the *enfant terrible* of the historians' craft: he is always trying out new intellectual challenges while leaving it to others to pick up the pieces. He exhibits a vast interest that exceeds his early modern specialism and makes frequent forays into art history. He also combines a keen eye for detail with sweeping conclusions. If a debate catches his interest, it is one conducted with friends rather than with witchcraft historians. He first presented himself in the field of witchcraft history with his thesis *I Benandanti*, submitted in Pisa in 1964 and published in Turin in 1966.[8] This book was initially taken by most as a rare confirmation of the Murray thesis as it ostensibly provided documentation of a fertility cult in connection with witchcraft. Only when it was translated into English as *The Night Battles* in 1984, after the success of the author's *The Cheese and the Worms*, did it become obvious that it had 'nothing to do with the "old religion"'.[9] Instead, it purported to describe a process of acculturation in which the so-called *benandanti* gradually internalised the Inquisitors' interpretation of what they told them about their nightly experiences. Ginzburg's last book on the Sabbat, *Storia notturna* (1989), subtitled *Deciphering the Witches' Sabbath*, was translated unusually quickly and published two years later with the title *Ecstasies*; consequently the English version bears traces of haste. In between he published several articles in which he kept witchcraft historians on their toes; these articles either contained a brief summary of a part of the coming book, such as 'The Witches' Sabbat: Popular Cult or Inquisitorial Stereotype?'[10] and the slightly more extensive 'Deciphering the Sabbath',[11] or they delved into related issues, like 'Germanic Mythology', 'Freud, the Wolf-Man, and the Werewolves'[12] and 'Charivari, Youth Associations and the Wild Hunt'.[13] Afterwards the author turned his back on the subject.[14]

Storia notturna ('Stories in the Dark') had a mixed reception. There was great praise, with one reviewer stating that 'after *Storia notturna*, it is difficult to imagine that the study of witchcraft will remain as tied as it has been in the past to an analysis of the phenomenon only in those periods of persecution and therefore of documentation'.[15] Among historians the reaction of Klaus Graf was perhaps more typical: while he admires Ginzburg's style and presentation and is slightly jealous of the reviews in the (in this case German) press, he is highly critical of the Italian's arguments.[16] But what is difficult to imagine is precisely what happened, at least among the experts and specialists.[17] Why is this? One of Ginzburg's complaints about the reception of *Storia notturna* is that reviewers only considered parts of it. 'Reviews appeared', he said in an interview, 'in which only the first part was discussed – which is of course highly unusual – because the authors did not feel capable of saying something about the rest.'[18] This seems to be the general fate of this book. References to it are selective, whether concerned with witches or fairies, for example.[19] This fragmented response is hardly astonishing when set against the wide array of topics Ginzburg discusses in *Storia*, such as fourteenth-century trials of heretics and lepers in the south of France, seventeenth-century German animal disguise, the prehistoric movements of the Scyths, the distribution of the fairy-tale Cinderella, or arctic shamanism, while in passing giving a new meaning to the concept of the 'Oedipus complex' – and this is only a selection. On top of this, although presented as a search for the popular roots of the Sabbat, the book is not about the Sabbat as such. This all suggests that it is not so much the topics themselves that make most historians cautious, but the way they are linked.

Storia notturna is a very rich and intricate book, with a sophisticated composition and informed by a profound scholarship. As we will see in the course of this chapter, the book is also alarming. Any attempt at a summary remains inadequate, but for the purpose of this chapter the following has to suffice – other bits and pieces will surface later on. The first part of the book contains a reconstruction of the series of fourteenth-century trials leading to a new 'witch cult'. Ginzburg then traces two clusters. One is medieval and concerns women travelling in the train of a supernatural female leader or participating in one of her assemblies. This female figure (Kieckhefer's 'mysterious goddess') was known by the name of Diana or Herodias. Locally she was called Habundia, Richella, Satia, Oriente, or simply the 'mistress of the games'. The second cluster entails groups of men battling for agricultural fertility at particular times of the year. They fought in spirit or in some ritual animal disguise and the winners could be assured of a prosperous harvest. Here the examples generally derive from the early modern period and from eastern Europe and they include Friulian (from the Friuli region of Italy) *benandanti*, Hungarian *táltos*, Romanian *calusari* and Baltic werewolves. According to Ginzburg, these two clusters form the first layer discernable underneath the Sabbat. Writing as if his material represents an archaeological dig, he next discerns the contours of something even older. In the course of his book this becomes more and more concrete. In his own (translated) words: 'Behind the women ... linked to the "good" goddesses of the night, hides a cult of an ecstatic nature'; 'The nightly flights to devilish meetings echoed an age-old theme: the ecstatic journey of the living to the world of the dead.'[20] Several chapters later, Ginzburg writes: 'The ecstasies of the followers of the goddess irrevocably call to mind the ecstasies of the – male and female – shamans from Siberia and Lapland', concluding at that point: 'The popular nucleus of the Sabbat – the magical flight and the metamorphosis – seems to hark back to a very old Eurasian substratum.'[21] The second cluster is dissected in a similar vein. 'All the routes we have negotiated to determine and explain the roots of the witches' Sabbat converge at one point: the journey to the world of the dead.'[22] These far-reaching conclusions are based on an analysis that is vaguely structural, profoundly phenomenological, only morphological in name and hardly historical: it is selective instead of serial and devoid of contexts.

The two strands meet in the late sixteenth- and early seventeenth-century Friulian *benandanti*: 'Friuli ... should be considered as a kind of border area, where the two generally separated versions of the ecstatic cult overlapped and converged.'[23] Thus the *benandanti* rather than Sabbats constitute the central point of *Storia*; in fact, fertility battles are hardly to be found in the records of European witch trials.[24] As Ginzburg formulates it: the 'theme disappears',[25] although this presumes it had been present at some point of time. *Storia notturna* harbours a dimly concealed north Italian and even Friulian bias. The two complexes are only defined and related to each other because of their appearance in Ginzburg's original thesis. There we also find the first launch of the idea of a shamanistic deep layer. Then, however, it is no more than a vague suspicion, based on the reading of German authors. 'I have not dealt with the question of the relationship which undoubtedly must exist between *benandanti* and shamans', he writes in the foreword.[26] But in the run-up to *Storia notturna* this impression turned into an undeniable conclusion without additional evidence or considerations. We can thus consider *I Benandanti* as the matrix of *Storia notturna*: the latter provides a vindication and justification of the thesis Ginzburg wrote in his early twenties. In *I Benandanti* we also already meet with a rudimentary form of what can be seen as one of Ginzburg's main methodological devices: the stringing together of different narrative elements or motifs. For instance, only in one case – of a woman who was already exceptional in several other aspects – is there mention of an abbess who had to be revered at a meeting; but for Ginzburg this provides enough grip to link her to 'that polymorphous feminine divinity' and in the same sentence to the Wild Hunt.[27] In *Storia* this method is perfected and supported by references to Ludwig Wittgenstein and Rodney Needham. However, the practice of establishing series of 'isomorphisms' across societies surpasses any structural analysis;[28] in a sense, founding his method on a particular theory of knowledge is little more than adding an extra, intellectual layer and is comparable to what early modern demonologists did to popular expressions of magic and witchcraft. It also gets the author into serious difficulties in the later stages of the book. How can one put history back into a linking exercise based on superficial resemblances?

I Benandanti contains an elegant and well-considered presentation of surprising material which is still unique in a European witchcraft context. At this early stage of his career it already shows Ginzburg's mastery of bibliography and his engagement in a cultural approach to magic that was in some ways far ahead of its time. It also put people on the historical stage who, around 1600, lived in a remote corner of Europe, Friuli, and who called themselves *benandanti* (literally 'well-walkers'). When interrogated by the Udine branch of the Roman Inquisition, they told stories of how their spirits left their bodies at night to fight evil people with fennel stalks. By analysing the dossiers of about 40 of them, Ginzburg argues that within two generations the *benandanti* came to concur with the Inquisitors and admitted that they had actually visited a witches' Sabbat, thus turning from good into bad (but with the prospect of becoming good Christians again). As the book is also a blueprint for *Storia*, however, the lamentable implication is that the weaknesses it also contains are repeated. This especially pertains to Ginzburg's failure to recognise the *benandanti* as healing specialists (as opposed to ascribed maleficious witches), to the way he conflated the specialisms of men and women, and to his identification of the *benandanti*'s opponents as the wandering dead.[29] In the transition to *Storia*, which covers thousands of years, the acculturation process which Ginzburg situates in the decades around 1600 and which constitutes the main argument in *I Benandanti* has become insignificant, although more in a methodological sense than as to its content. Acculturation as the main thesis in *I Benandanti* is anyhow unconvincing, for (apart from a recurrent problem of the Inquisitors in understanding the local dialect) it is only based on the admissions of four *benandanti* and can be attributed to inquisitorial pressure.[30] We still do not know whether the Sabbat interpretation of their nocturnal forays was also echoed outside the interrogation room, although there is material enough, such as later trials and folklore records, to establish this.[31]

On the other hand, the notion of acculturation does reveal Ginzburg's thinking about witchcraft. He attempted to study witchcraft from below, to find the 'popular' behind the 'elite' inquisitorial procedures. However, he still looked through the lens of the Inquisitors and this affected his perspective on the very subject he proclaimed to be writing about: witchcraft. Trials of *benandanti* only render a partial perspective on witchcraft. The Roman Inquisition prosecuted no malificient witches, since the view was that 'denunciations for *maleficium* ... were based on suspicions rather than on observation of a magical procedure'.[32] The *benandanti* were not only protectors of the crops, but also cunning folk: charmers, fortune-tellers and professional unwitchers. They could recognise witches because, as they told their clients, they had met them in nightly battles. Comparison is therefore only relevant with other cunning folk, and these were generally not prosecuted as witches.[33] If we consider the Sabbat as an intrinsic (but not inevitable) element of a witch trial, in which diabolism met with the malificient deeds ascribed to witches, then it may not be appropriate to take the Sabbat image that was grafted onto the stories of the *benandanti* as a point of departure to search for its conceptual folklore roots – especially when we keep in mind that this image was already in circulation for at least a century (or two centuries, if we agree with Ginzburg). The Sabbats may have been more or less similar in both cases, but the underlying 'popular' concepts vary significantly. It would have been different if the Inquisitions into the *benandanti* had taken place when the new witch trials started, but even then they would have been in the wrong area.

The gender division of the Friulian specialists is another complicated issue. In *Storia* Ginzburg confesses that he had overlooked 'the ecstatic speciali-sations that distinguished male and female *benandanti*'.[34] However, gender remained a weak spot in his work; he maintained that the men and the women 'called themselves *benandanti*', and that the women *benandanti* were seers of the dead.[35] But there is very little evidence of female *benandanti*, while there are signs of the assimilation of female specialists into the group of primarily male *benandanti* by Inquisitors and by Ginzburg. The case of Anna la Rossa, as presented in *I Benandanti*, serves as an illustration here. She 'claimed that she could see the dead and converse with them'. Ginzburg comments next: 'it was not stated that Anna la Rossa was a *benandante*, in fact the word was not even mentioned'. Yet several pages later he has overcome this obstacle and calls Anna 'one of the *benandanti* who claimed she could see the dead'; in the next chapter she has become a *benandante* as a matter of fact.[36] The reason for this incorporation is given by quoting another woman healer who, when asked whether she was a *benandante*, replied: 'No sir, I am not one of the *benandanti*, but my deceased husband was; he used to go in procession with the dead.' At this point Ginzburg decides: 'Here then is the explicit confirmation of a link which had been suggested hypothetically: whoever could see the dead, went with them that is, was a *benandante*.' This is a very weak link, the more so because the husband apparently refused to exercise his ability for the public good.[37] It also does not confirm that *women* who were able to communicate with the dead were *benandanti*.[38] 'On the whole,' Ginzburg concludes, 'in Friuli the myth of the processions of the dead occupied a relatively marginal place ... in that complex of beliefs that we associate with the *benandanti*.'[39] On the basis of his own evidence, 'marginal' is an overstatement. Nevertheless, the notion of the contact with the dead had stuck itself in Ginzburg's mind (I shall later try to explain how); in subsequent publications he states that the *benandanti* as well as a host of other figures all 'claimed to be able to travel periodically (in spirit or in animal form) to the world of the dead'.[40] By the time he came to write *Storia notturna* this notion had acquired overwhelming proportions.

Richard van Dülmen defined four stages in the evolution of the Sabbat, depending on the witch's relation with the Devil. In line with other authors, he stated that the full Sabbat image was only completed around 1600, and even afterwards it did not penetrate everywhere.[41] (For Ginzburg there are no more developments after 1430.) While we may debate the details of van Dülmen's thesis on the basis of additional material, he rightly alerted us to change and the differentiation of the Sabbat image. Fragments of demonological interpretations of bewitchments could also easily spread without the Sabbat. Had Ginzburg given any attention to this uneven development, he probably would not have presented the general image of the witches' Sabbat which opens *Storia notturna*, for it represents a construct which only occasionally was turned into reality when someone questioned witches, with or without using torture.[42] If we follow Ginzburg, the meeting was 'usually' held in 'remote places', whereas trial transcripts show that in a lot of cases witches said that they met near (or even in) their own homes.[43] Their means of transport could vary from 'broomstick, ointment, astride on animals, or changed into animals themselves', as mentioned by Ginzburg, to walking – even Montague Summers noticed that witches went on foot to nearby Sabbats.[44] Metamorphosis during transvection was certainly not a widespread element; metamorphosis at the Sabbat was even more rare, and riding on animals is not the same as turning into them. The next element in Ginzburg's sketch refers to the Black Mass, which is often missing in trial accounts. As van Dülmen concluded, a non-satanic dance was the regular feature, at least north of the Alps where most of the witch hunting occurred.[45] Any attempt to present a general picture of the Sabbat must therefore fail, as it ignores regional differences and historical development.[46] The Basque Sabbat imagery, for example, in which children are abducted to a Sabbat, made into a witches and given dressed toads, would hardly have been recognised in Denmark, where the notion of the Sabbat can merely be 'glimpsed' in the way witches danced around a church.[47] The story a witch told her interrogators about the consumption of a child's corpse at the Sabbat was 'a most unusual element in Lorraine confessions',[48] whereas it was commonplace in the early Swiss trials. Ginzburg nevertheless denotes his Sabbat characteristics as 'basic elements'; both demonological and trial sources would confirm the image 'in essence' – a remarkable contrast with his detailed scrutinising of other texts.

Many elements Ginzburg traces back into a prehistorical nebula were never part of a witches' Sabbat and only come into the picture through his method of stringing different descriptions together on the basis of one or two other similarities in their forms. This especially concerns all the connections to the dead, if they are not merely rhetorical. It is not that various elements were not present in some places at some times. The bone miracle (the ritual resurrection of bovines) in the late fourteenth century, reportedly performed by Oriente, was even a less isolated incident than Ginzburg shows: around 1,400 several more north Italian instances occurred, and the motif figured in the life of St Germain, without, however, becoming a stable Sabbat ingredient outside Italy.[49] Witches certainly visited wine cellars (but that does not make them revenants). Fertility fights appeared to be largely unrelated to meetings of malicious witches – the adversaries were certainly not the dead but neighbouring people.[50] The Wild Hunt, with its primarily male leaders, did not contribute to the Sabbat stereotype, nor to any local variations.[51] On the other hand, it would not be very difficult to find many more examples that would fit Ginzburg's scheme of a journey to the otherworld, but this would be more appropriate in a study about possible shamanistic elements in European culture than in an uprooting of the Sabbat (the legend of Jack and the Beanstalk figures a world tree and stolen treasure, sadly missing from Sabbat descriptions). What follows is that Ginzburg used a questionable ordering principle for his book *Storia notturna*. The rather disturbing conclusion can only be that neither the male *benandanti* nor the female Friulian fortune-tellers contributed anything to the production of the early Sabbat image. The historical backgrounds of the Friulian magical experts and their geographical connections were primarily their own and did not constitute any Sabbat roots. Or to keep with Ginzburg's metaphor of ever deepening layers: digging in Friuli primarily results in finding Friulian remains, and if they happened to have been imported from the Balkans they had still been used in Friuli. If the Sabbat, as a concept developed and applied by prosecutors in the 1420s, contained any 'popular' element, it may be found in female assemblies, foremost of narrative character, possibly associated to some female 'deity' and to fairy lore. Since the demonologists reported this themselves, such an observation is not particularly new. To postulate a Europe-wide occurrence of fairly-like women is tricky, however, because, as historian and folklorist Hans Peter Broedel recently concluded: 'the available evidence is scattered and contradictory, and suggests a group of more or less related components rather than a single, coherent belief-system'.[52] This again points to the importance of regional differentiation.

Gustav Henningsen started his academic career researching actual witchcraft in Denmark in the 1960s, but when he wanted to undertake comparable work in Spain he found this had already been done by anthropologist Líson Tolosana. The historian Caro Baroja pointed him to the then hardly researched archives of the supreme court of the Spanish Inquisition in Madrid.[53] Eventually this led to the discovery of a series of documents from Sicily, where from 1587 people had started to talk about the *donna di fuora*, the 'ladies from outside'. Here then was again evidence of a pre-inquisitorial female assembly and, unlike Friuli (at least Ginzburg's interpretation of it), the Inquisitors did not succeed in demonising the stories beyond their own culture.[54] Henningsen noticed that the Sicilian 'white Sabbat' stood in contrast to the demonological one: 'the beautiful fairies have turned into horrible demons, the splendid food into a rotten, stinking mess; the sweet music has become hateful caterwauling, the joyful dance exhausting capering, and the pleasurable love – painful rape'. The only problem was to find places in Europe where fairy traditions were clearly superseded by a satanic Sabbat, and his attempts only yielded examples 'from regions in the European periphery that never experienced witch-hunts', including southern Spain and Morocco.[55]

For Ginzburg the Sicilian records were disturbing (not to mention that they had been discovered by a Dane in Spanish archives).[56] He first suggested that the 'white Sabbat' had perhaps been created out of a black equivalent.[57] In *Storia* he terms Sicily an 'anomaly' because it did not suit the ethnic map he was redrawing which would explain similarities between 'Sabbat' elements in, for instance, Scotland, Friuli and Romania. As he had concluded in the previous chapter: 'The ecstatic cult of the night goddess seems to be absent in that part of the German world that has not known the influence of the Celts.'[58] The only 'Celtic' traces in Sicily he could detect were the Breton romances, and it was unlikely that they had resulted in lasting traditions on a popular level. This, in one of Ginzburg's unparalleled mental leaps, provided the lever to direct the reader to the pre-Grecian Mediterranean goddess, and – more historically reliable – to the Sicilian cult of the *matrones*. For his part Henningsen fully recognises the stories about mysterious meetings as legends told by the magical experts, the 'local opinion leaders'. As legends are known to cross language barriers, to 'migrate', this leaves open the option of derivations, including from Breton material.

The Sicilian argument is a classic example of a further phase in the archaeology of culture. This not only treats cultural entities – such as a piece of a story, an expression, a custom – as material artefacts to be sifted out of historical texts but also presumes that their distribution can be interpreted in a similar way. If, for instance, the bone miracle is found in both Nordic stories (Thor) and in north Italian witch trials, then either there will have been a way in which (because of seniority) the first influenced the second, or one has to presume that both go back to a common ancestor. In order to allow for the latter, a second presumption has to be admitted, namely that the piece of culture is tied to a particular people. Thus, as Hungarian folklorist Éva Pócs argues in her book *Fairies and Witches*, when there are 'common features of the Slavic and Central European German (or even Scandinavian) traditions', they may possibly be regarded as 'common Indo-European remains'. Or more specifically:

On the basis of striking similarities between Scottish-Irish and the SE and Central European fairy beliefs or the common features of the *Bonnes Dames* and the *Abundia* and *Satia* traditions (related to the Gallic *Matronae* cult), the question of *Celtic* heritage may also arise.[59]

If this way of reasoning bears a remarkable resemblance to some lines in *Storia notturna*, this is not due to copying (both publications appeared in the same year)[60] but to a common reliance on nineteenth- and early twentieth-century folklorists. Pócs, moreover, takes one step further by using the concept of a 'relict area': this presumes that a peripheral region that has not been touched by a certain development from a particular centre may have preserved traditions that were superseded and thereby lost in the affected areas. As she formulates it: 'Witchcraft never became a central belief in the Orthodox, eastern territories of the Balkans, and that is why elements of archaic belief systems preceding the appearance of witchcraft could remain.'[61] This would have solved Henningsen's problem, and indeed Pócs writes elsewhere about 'white Sabbats' turning black: 'We have no reason to doubt that such processes did take place in those areas of Central and Western Europe where a "pre-witchcraft" fairy mythology similar to the Sicilian-Balkan one did actually exist.'[62] What is important here is not the availability or lack of evidence, but the way of thinking which replaces 'origin' by geographical marginality. In the 1970s this became outdated and was rightly replaced with a more dynamic, relativistic and contextualised concept of culture, in which changes of form, content and meaning are acknowledged.

Ginzburg knew this very well; if he had not figured it out himself, he was told so on several occasions by German ethnologists. But for reasons about which we can only speculate he chose to ignore the critique of diffusionism in the second part of *Storia* and to juggle around with it in the third. After having established a Celtic layer and shamanistic features, he combines the two by suggesting historical contact between Siberians and Celts with the Scyths as mediators. He then proceeds to explain that the reason why shamanism could strike a chord was that it appealed to notions already present and to a basic human system of polarities, such as life and death, male and female, body and spirit. To illustrate the latter, Ginzburg analyses the myths clustered around mono-sandalism and the tale of Cinderella (who also lost only one shoe), mainly by again applying his stringing tricks.[63] One way to see this is that his particular methods and determination to alter existing witchcraft historiography led Ginzburg into a cul-de-sac from which he could only escape by an intellectual somersault. We can also take a step back and deduct that Ginzburg, in a very entertaining way, shows how silly diffusionism can be if one carries it to its extreme consequences.

The demise of diffusionism in the sense of a broad theory based on the ancient movements of people does not, of course, imply that the notion of diffusion as distribution is inadequate itself. When we take Hungary as an example, it is obvious that, as a place on the periphery, it experienced its main witch-trial periods much later than the 'heartland of the witchcraze'. This was made abundantly clear by Gábor Klaniczay, who between 1984 and 1994 published a series of (partly overlapping) articles about the Hungarian witch trials.[64] In these he drew on the ongoing results of a Hungarian research team, which he co-founded with Éva Pócs, the aim of which was to make available as much of the Hungarian witch-trial material as possible.[65] One of the themes Klaniczay discusses is foreign influence on Hungarian witch trials, particularly from German states. In northern Hungary and Transylvania, for example, the influence of German Lutheranism can be seen in the dispropor-tionate number of healers in the first, late sixteenth-century trials. At the end of the seventeenth century the prosecution of male shepherds and vagrants spilled over into western Hungary from Catholic Austria. German-speaking soldiers played a main role in instigating trials in the eighteenth-century Hungarian plains. What is more, from 1690 onwards the main legal codex justifying prosecutions was the Austrian edition of Carpzov's *Practica Rerum Criminalium*.[66] This alone shows that various trial series had a distinct social and cultural background; the accusations among the seventeenth-century Transylvanian aristocracy are another illustration of this. The Hungarian witch trials were 'a huge cultural melting pot, collecting and combining magical notions and mythologies of various functions and different regions and peoples'.[67]

The documents of these trials would also be extremely suitable for Sabbat research. As Klaniczay phrases it: 'The lack of innate demonological traditions promises an uncorrupted popular concept of witchcraft; the existence of shamanistic beliefs in the region makes it legitimate to look for local archaic versions of the sabbath.'[68] Here we can already discern the influence of Ginzburg (for the *absence* of shamanism would also support such a search with a similar force of argument). Nevertheless, Klaniczay harboured some doubts, pointing to the very 'small minority' of trials with shamanistic traits. In one of his last articles on the subject[69] he stressed from the start that, under torture, suspected witches very quickly learned what was expected of them, namely to produce believable descriptions of Sabbats.[70] They also told their interrogators what they knew, and in this way any account of group meetings, be they feasts, marches, battles or charivaris, could end up in confessions.[71] In all probability this applies to some extent to witness' depositions, too, since witnesses were still in a subordinate position. One of the remarkable features of the Hungarian witchcraft material is that it contains testimonies about Sabbats, not just by suspects (as elsewhere in Europe) but also by witnesses who had been the target of witches. Theirs were versions of an abduction story, relating how they had been overwhelmed by witches in the middle of the night, changed into a horse and ridden to a meeting. This may be a rare instance of the existence of Sabbat notions among the populace, unrelated to the elite demonologies, but we have to keep in mind that it concerns late eighteenth-century trials, among them those in Hódmezővásárhely, where foreign soldiers played a role as catalyser and susceptible children provided crucial evidence. Before we object that the witnesses could still have tapped into age-old local traditions, we also need to take into account that the principal elements of this particular story can be found all over western Europe and that the main subtype concentrates on the uncovering and shaming of the nightly visitor, rather than on a ride to a witches' meeting. This all makes it more likely that the story was imported in the Hungarian plains and subsequently told as a personal experience – Klaniczay cites a deposition by the Hungarian servant of an Austrian officer in which it is told how witches rode him to the Sabbat, and by a nine-year-old girl who added her own details to a similar story.[72]

This Hungarian example shows the importance of unravelling the dynamics of that storytelling event in which witchcraft testimonies were expressed, with all its extremely uneven power relations. After all, as Klaniczay reminds us, 'with their fantastical sabbat stories the accusers tried to kill the accused. These are stories aimed to convince the judge and to threaten and stigmatize the witch.'[73] To shift the focus from this event in an attempt to reconstruct an origin that can at best only be conjectural carries with it the danger of denying history itself. How then to approach the shamanism which would reveal itself in the Sabbats of Hungarian witches? Here things become puzzling once more. As we have seen, Ginzburg's notion of the shamanistic tendencies of the *benandanti* and the followers of Diana remains no more than a vague suspicion centred around the broad concepts of trance, or ecstasy, which is how he translated flight and shapeshifting (with the bone miracle thrown in for good measure). If there is anything clear about shamans, however, then it is that they certainly did not play in teams.[74]

Yet by discussing eastern European wizards who conformed to the Friulian *benandanti*, both Klaniczay and Pócs take their concept of shamanism from Ginzburg, although other Hungarian scholars preferred a type of shamanism that did not involve group combat.[75] Klaniczay, however, is careful enough to remark that the Hungarian *táltos* only started their 'struggles for fertility' at the end of the eighteenth century.[76] He also observes that the 'shamanistic' fragments he found were merely unconnected traces and that the wizards' profile was 'distinct' from Asian shamanism. At the end of his paper 'Shamanistic Elements in Central European Witchcraft' he moves away from ecstatic fights and concludes:

> There was, indeed, a place in the witchcraft beliefs where shamanic figures could and generally did fit in: in the role of the diviner, the cunning folk, the wizard, the professional magician, who engaged in magical counter-aggression against the presumed witches.[77]

In the early 1980s there was still very little research available on cunning folk; now there is at least enough to state that if they performed any rituals, those differed significantly from shamanistic trance. Moreover, Siberian shamans did not fight or even detect witches. This leaves precious little room for a European variant of shamanism, and in Budapest Henningsen proposed dropping the term altogether when speaking about the predecessors of witchcraft.[78] Indeed, one of the reviewers of *Storia* asked whether it was 'really necessary to go back that far in the past to discover the origins of witchcraft'. In his view the *benandanti* were already profoundly Christianised and any reference to their 'Neolithic ancestors' (meaning shamans) would only diverge from the Counter-Reformation struggle the Friulians were caught in.[79] Nevertheless, in his turn German witchcraft historian Wolfgang Behringer took Klaniczay's suggestion on cunning folk to support his own argument as to why the activities of the late sixteenth-century south Bavarian, Chonrad Stoeckhlin, could legitimately be called 'shamanistic'.[80]

In spite of this, Behringer's case study, which ranges through the Swiss, Austrian and Italian Alps, is a measured response, even refutation of the main arguments in *Storia notturna*. Ginzburg had already paid attention to Stoeckhlin in *I Benandanti* – he had found him mentioned in a publication of 1931 written by a local historian – and indicated that the story of the horse-herder healer would be a valuable subject for analysis.[81] At that time Ginzburg thought that the evidence in the case formed the 'most exact parallel' to the *benandanti*, but Behringer only agrees with this if 'Ginzburg's exaggerated claims to have discovered a mythic essence that supposedly survived over centuries or even across millennia without losing any of its vital force' were tuned down a bit. These were not Ginzburg's exact words,[82] but Behringer's intention seems clear. 'For our interpretation, the question of origins is almost of no significance', he remarks later; it is an 'exercise in anachronism'.[83] Behringer also provides the most promising building blocks for a solution of the search for a hypothetical 'white Sabbat' out of which the diabolical version would have been summoned – if we accept that the nineteenth- and twentieth-century Swiss legends about the *Nachtvolk* (the regional fairies) were already present at the end of the fourteenth century. In fact, he concludes his book by pointing to the geographical correspondence of these traditions with the first witch trials and more or less suggests that the fairies' meetings were forcefully transformed into Sabbats.[84] Shortly before, however, he had shied away from undermining Ginzburg's shamanistic musings, extending his hand to the Italian instead.[85] 'We can still recognize that the mythic world of many early modern Europeans provided fertile soil … for shamanistic ideas', he writes; 'the surviving fragments of myth, these vestigal mythologems, were powerful enough to generate myths over and over again'.[86] Shamanism was a positive possibility of interpretation, and since the Russian ethnographer Vladimir Basilov had shown that it could prosper under Islam, it could do the same under Christianity. Shamanistic ecstasies can be found all over Europe,

Behringer declares, referring to almost the same (unconvincing) collection of examples as Ginzburg. Behringer suddenly throws aside all the sensitivity he displayed when dealing with local traditions and the ideology of folklorists. He uses a definition of shamanism – ecstasy, contact with the spirit world – that is too generalised (it would include spiritist mediums, for instance) and formulated for the purpose of comparative analysis, which he does not undertake. If Stoeckhlin had actually gone into a trance and not just *told* stories, why were there no witnesses to report on the process of inducing this trance, or on the outward signs of his altered state of consciousness? In other words, why is shamanism so alluring?

In one of the endnotes to the chapter on ritual animal disguise in *Storia notturna*, where he refers to his essay on the charivari, Ginzburg notes that the Tübinger ethnologist Hermann Bausinger had commented on the similarities between his (Ginzburg's) argument and that of the Austrian National Socialist, Otto Höfler. On that occasion Ginzburg had neglected to mention explicitly Höfler's *Kultische Geheimbünde der Germanen* (1934).[87] But Ginzburg was quite familiar with the Austrian's work and the prickly exchange with Bausinger only lay at the surface of a sensitive issue that pervaded Ginzburg's work on the *benandanti* and subsequently on *Storia*. For, although from a Jewish and socialist background, as a young student Ginzburg had been inspired by German-speaking folklorists of the interbellum and not just on the subject of diffusionism. He had found the Latvian werewolf Thies in Höfler's book (he copied the latter's spelling of the werewolf's name and never bothered to carefully read the original trial transcript by von Bruiningk).[88] The Wild Hunt, one of Höfler's main areas of interest, had already been connected with the witches' flight by Jacob Grimm. Ginzburg's idea of shamanism was derived from Will-Erich Peuckert and Lily Weiser-Aall; the similarity between witches flying on broomsticks and shamans had been observed in 1927.[89] (The amount Ginzburg owes to Swiss folklorist Karl Meuli deserves a separate study.) The most that can be said of Peuckert's political affiliations is that he carefully avoided clashes with the Nazis and that his 1951 book *Geheimkulte* is a clear refutation of Nazi theories, including those by Höfler. In his later years Peuckert primarily became a compiler of folklore texts (as well as experimenting with 'witches' ointments'). Nevertheless, in *I Benandanti* Ginzburg called him a 'racist', apparently not as one would expect from a Jewish position towards a Nazi, but as an Italian would denote a German.[90] The 'unmistakably Nazi overtones' in Höfler's book were only noticed later[91] and Ginzburg never seems to have discovered the fascist leanings of the Romanian historian of religions Mircea Eliade, whom he quotes favourably in *Storia*.[92] Among other things, *Storia* can be seen as a recasting of the Germanic background of the *benandanti* and as showing Ginzburg's awareness and subsequent attempt at an exorcism of the fascist influence on his work.[93] Whereas, for instance, in *I Benandanti* Friuli was pictured as a meeting place between 'Germanic and Slavic currents',[94] in *Storia* this was replaced by a 'Celtic substrate', which shows that

Ginzburg was thinking in terms of a Celtic–Germanic opposition, an instance of *interpretatio romana*. And although he referred to the Wild Hunt–Wodan–Yggdrasil–shamanism chain in passing, he carefully avoided elaborating on it. At least Ginzburg addressed the question of the relation between political ideology and historical interpretation. However, the notion of the wandering dead, derived from the Wild Hunt, had become too important to abandon or even to change, and the denunciation of Höfler and other Nazis[95] also meant hiding Ginzburg's considerable indebtedness to them, originated at a time when his political antennae were not fully matured. Underneath the layer of fascism we can discern a strong nationalistic presence.

Ginzburg forcefully inserted his fascination with the dead time and time again into his unriddling of the Sabbat. This is hardly borne out by early modern evidence, but directly connected to the folklorists who initially inspired him and some of whom he later renounced. During the 1930s, folklore, as the branch of knowledge which provided the required expertise to discuss the cultural aspects of witchcraft, was still saturated with nationalism. The dead were primarily one's own dead; folklore offered the means of identification with one's perceived ancestors, and by suspending history this comforting feeling was given an absolute certainty. Ginzburg's lyrical enthusiasm for Jacob Grimm extends to his ideological background.[96] The concept of continuity not only provided the framework for the decontextualised details which plague *Storia notturna*; it is also one of the foremost articulations of nationalist proclivities.

The few theories about shamanism which emerged in the 1920s, on the other hand, were very circumscribed as they expressed an awareness of the influence of arctic customs on Norse traditions. When a connection with witchcraft was made, it centred on the broomstick. As such, shamanism came to constitute a counterweight to the militaristic ideas of the *Männerbunde*, the all-male fraternities or societies. In Hungary, shamanism had been used to express a national identity, separate from communism, which is one of the reasons why Ginzburg's findings were appropriated by other Hungarians to underline eastern European unity. In all these cases the concept of shamanism was not only politically loaded, it was also justified as resulting from geographical contacts, with the Laps, the Ugrians and the Friulians. However, when Ginzburg resumed thinking about shamans in the 1980s, they had shed these contexts, even their arctic origins, and were developing into a spiritual force in western society. In *Storia*, Ginzburg thus tries to show how 'the traces of this Eurasian continuum' were not just restricted to 'northern Europe' – that is to say Nordic mythology.[97] Shamans, their romantic quality enhanced by Eliade in another undertaking to atone for Nazi sympathies, became universal New Age role models who could establish their own links to the spirit world. They were also perceived as masculine (their transvestism was dispensed with), and what is in this context perhaps most important, as mediators between the living and the dead, they were the ultimate historians.

This may explain the lure of shamanism, not just for Ginzburg but for other historians, too. Confronted with the choice between the powerful individual male shaman with his anti-totalitarian connotations and the feminine groups of teasing fairies, most witchcraft researchers instinctively went for the first.

notes

1. Norman Cohn, *Europe's Inner Demons* (London: Chatto & Windus, 1975), p. 252.
2. Richard Kieckhefer, *European Witch Trials: Their Foundations in Popular and Learned Culture, 1300–1500* (Berkeley and Los Angeles: University of California Press, 1976), p. 40. See later, Richard Kieckhefer, 'Avenging the Blood of Children: anxiety over child victims and the origins of the European witch trials', in Alberto Ferreiro (ed.), *The Devil, Heresy and Witchcraft: Essays in the Honor of Jeffrey B. Russell* (Leiden, Boston and Köln: Brill, 1998), pp. 91–118.
3. For instance, H. C. Erik Midelfort, *Witch Hunting in Southwestern Germany 1562–1684: The Social and Intellectual Foundations* (Stanford: Stanford University Press, 1972), p. 9: 'I decided that any hunt resulting in 20 or more executions in one year could qualify as large.'
4. Nicole Jacques-Chaquin and Maxime Préaud (eds), *Le sabbat des sorciers en Europe (xve–xviiie siècles)* (Grenoble: Millon, 1993).
5. Bengt Ankarloo and Gustav Henningsen (eds), *Early Modern European Witchcraft: Centres and Peripheries* (Oxford: Clarendon, 1990); Gábor Klaniczay and Éva Pócs (eds), *Witch Beliefs and Witch-Hunting in Central and Eastern Europe* (Budapest: Akadémiai Kiadó, 1993; *Acta Ethnographia Hungarica* 37 (1992)).
6. Wolfgang Behringer, *Shaman of Oberstdorf: Chonrad Stoeckhlin and the Phantoms of the Night* (Charlottesville: University Press of Virginia, 1998). The original German version appeared in 1994. See also Wolfgang Behringer, 'How Waldensians Became Witches: Heretics and Their Journey to the Other World', in Gábor Klaniczay and Éva Pócs (eds), *Communicating With the Spirits* (Budapest and New York: Central European University Press, 2005), pp. 155–92. This important article only became available after the conclusion of the writing of this chapter.
7. On 7 October 1999 a roundtable discussion with Carlo Ginzburg was organised at the Budapest Institute for Advanced Study. Discussants were Wolfgang Behringer, Gustav Henningsen, Gábor Klaniczay and Éva Pócs. A report will be published in Gábor Klaniczay and Éva Pócs (eds), *Witchcraft Mythologies and Persecutions*.
8. His very first article, dealing with a case from the inquisitorial records of Modena, was originally published by his university in 1961. An English translation appeared after 29 years: Carlo Ginzburg, 'Witchcraft and Popular Piety: Notes on a Modenese trial of 1519', in Ginzburg, *Myths, Emblems, Clues* (London: Hutchinson Radius, 1990), pp. 1–16.
9. The quote is from Cohn, *Europe's Inner Demons*, p. 223, who was one of the few to properly assess *I Benandanti*. For the general and more accurate view after the English edition, see, for example, Brian Levack, *The Witch-Hunt in Early Modern Europe* (London: Longman, 1987), p. 53.
10. In Steven L. Kaplan (ed.), *Understanding Popular Culture. Europe from the Middle Ages to the Nineteenth Century* (Amsterdam, Berlin and New York: Mouton, 1984), pp. 39–51.
11. Ankarloo and Henningsen, *Early Modern European Witchcraft*, pp. 121–37; Ginzburg, 'Présomptions sur le Sabbat', *Annales E.S.C.* 39 (1984) 341–54; Ginzburg,

'Nächtliche Zusammenkünfte. Die lange Geschichte des Hexensabbat', *Freibeuter. Vierteljahrschrift für Kultur und Politik* 25 (1985) 20–36.

12. Both in Carlo Ginzburg, *Myths, Emblems and Clues*. The French translation of 'Germanic Mythology' appreared as: 'Mythologie Germanique et Nazisme. Sur un ancien livre de Georges Dumézil', *Annales E.S.C.* 40 (1985) 695–715.

13. Ginzburg, 'Charivari, associations juvéniles, chasse sauvage', in Jacques Le Goff and Jean-Claude Schmitt (eds), *Le Charivari* (Paris, 1981), pp. 131–40. The German translation appeared in the German edition of *Miti emblemi spie*: 'Charivari, Jugendbünde und Wilde Jagd. Über die Gegenwart der Toten', in Carlo Ginzburg, *Spurensicherungen. Über verborgene Geschichte, Kunst und soziales Gedächtnis* (Berlin: Wagenbach, 1983), pp. 47–60. I have not found an English version of this article.

14. See lately, Tony Molho, 'Carlo Ginzburg: Reflections on the Intellectual Cosmos of a 20th-Century Historian', *History of European Ideas* 30 (2004) 121–48, and Ginzburg's most recent collection of essays: *Wooden Eyes: Nine Reflections on Distance* (London and New York: Verso, 2002).

15. John Martin, 'Journeys to the World of the Dead: The Work of Carlo Ginzburg', *Journal of Social History* 25 (1992) 613–26, 621.

16. Klaus Graf, 'Carlo Ginzburg's "Hexensabbat" – Herausforderung an die Methodendiskussion der Geschichtswissenschaft', *Kea* 5 (1993) 1–16; for an extended version see <www.geschichte.uni-freiburg.de/mertens/graf/ginzburg.htm>. The seemingly critical review of *Storia* by Perry Anderson, 'Witchcraft', *London Review of Books*, 8 November 1990, was primarily a publicity stunt to promote the English translation. Anderson, whom Ginzburg counts among his friends, repeatedly overshot his mark and Ginzburg did not have any qualms replying two months later (10 January 1991), when readers could weigh the arguments for themselves. The review circulated among witchcraft historians as a sign of Ginzburg's fallacy, see Behringer, *Shaman*, p. 193, n.6.

17. Since Ginzburg called Stuart Clark's interpretation of the Sabbat in terms of symbolic reversal 'superficial', Clark chose to honour the book with a single footnote in which he directs future research to 'different associative principles': Stuart Clark, *Thinking with Demons: The Idea of Witchcraft in Early Modern Europe* (Oxford: Oxford University Press, 1997), p. 134, n.1.

18. Hans Peter Ros, 'Over de intrinsieke grenzen van onze kennis. Een interview met Carlo Ginzburg', *Groniek* 28 (1995) 325–35, 329.

19. Levack, *The Witch-Hunt in Early Modern Europe*, pp. 35, 44, 50, 51, only quotes from the second chapter of *Ecstasies*; Diane Purkiss, *Troublesome Things: A History of Fairies and Fairy Stories* (London: Penguin, 2000), pp. 142–8, only refers to the third chapter.

20. Carlo Ginzburg, *Ecstasies: Deciphering the Witches' Sabbath* (London: Hutchinson Radius, 1990), pp. 100, 101. In the following quotations I have substituted my own translation of the text for the sake of clarity.

21. Ginzburg, *Ecstasies*, p. 136.

22. Ginzburg, *Ecstasies*, p. 242.

23. Ginzburg, *Ecstasies*, p. 159.

24. These battles were a distinct Slavic phenomenon and may have been more than just dream experiences; see Cathie Carmichael, 'The Fertility of Lake Cerknica', *Social History* 19 (1994) 305–17.

25. Ginzburg, *Ecstasies*, p. 172.

26. Carlo Ginzburg, *The Night Battles: Witchcraft and Agrarian Cults in the Sixteenth and Seventeenth Centuries* (London: Routledge & Kegan Paul, 1983), p. xxi (preface to the Italian edition).

27. Ginzburg, *The Night Battles*, p. 100. The link with the Wild Hunt had already been established earlier in the book, see p. 40. See also *Ecstasies*, pp. 101–2.

28. See Carlo Severi, 'Le Chamanisme et le dame du Bon Jeu', *L'Homme* 32, 121 (1992) 166–77; also Graf, 'Carlo Ginzburg's "Hexensabbat"', 8–11, who accuses Ginzburg of having built his method on an 'essentialist fiction'.

29. I have voiced my reservations of *I Benandanti* previously in German. See Willem de Blécourt, 'Spuren einer Volkskultur oder Dämonisierung? Kritische Bemerkungen zur Ginzburgs "Die Benandanti"', *Kea. Zeitschrift für Kulturwissenschaften* 5 (1993) 17–29.

30. Ginzburg argues that when *benandanti* identified witches, they also acquired a bad reputation on a village level and finally became to be seen as witches themselves (*The Night Battles*, pp. 78, 94, 96, 118, 121). There is no reason, however, to suppose that a certain increase of the ambivalence of unwitchment experts coincided with the presumed success of the Inquisitors' persuading them to be joined with the Devil.

31. See Franco Nardon, *Benandanti e inquisitori nel friuli del seicento* (Trieste: Università di Trieste, 1999).

32. Mary Rose O'Neil, 'Missing Footprints: Maleficium in Modena', in Klaniczay and Pócs, *Witch Beliefs*, pp. 123–43, 124.

33. Willem de Blécourt, 'Witch Doctors, Soothsayers and Priests: On Cunning Folk in European Historiography and Tradition', *Social History* 19 (1994) 285–303; Owen Davies, *Cunning Folk: Popular Magic in English History* (London: Hambledon and London, 2003), esp. chapter 7.

34. Ginzburg, *Ecstasies*, p. 26, n.42.

35. Ginzburg, *Ecstasies*, pp. 13, 159.

36. Ginzburg, *The Night Battles*, pp. 33, 35, 41, 72.

37. Ginzburg, *The Night Battles*, pp. 38–9.

38. There is only one other case where the two overlap, that of Florida Basili, but she later denied that she was a *benandante*. See Ginzburg, *The Night Battles*, pp. 62–5.

39. Ginzburg, *The Night Battles*, p. 68.

40. Carlo Ginzburg, 'Freud, the Wolf-Man, and the Werewolves', in Ginzburg, *Ecstasies*, p. 147.

41. Richard van Dülmen, 'Imaginationen des Teuflischen. Nächtlichen Zusammenkünfte, Hexentänze, Teufelssabbate', in Richard van Dülmen (ed.), *Hexenwelten. Magie und Imagination vom 16.–20. Jahrhundert* (Frankfurt: Fisher, 1987), pp. 94–130; van Dülmen, 'Die Dienerin des Bösen. Zum Hexenbild in der frühen Neuzeit', *Zeitschrift für Historische Forschung* 18 (1991) 385–98.

42. See Robert Rowland, '"Fantasticall and Devilishe Persons": European Witch-Beliefs in Comparative Perspective', in Ankarloo and Henningsen, *Early Modern European Witchcraft*, pp. 161–90, who also writes about the 'underlying uniformity' of the witches' confessions, although his general sketch differs from Ginzburg. Rowland, however, calls the concern with origins 'misplaced' (p. 166, n.8).

43. The question of place may have to be partially recast in terms of the numinous: see Melinda Égetö, 'Witches' Sabbath on the Vineyard Hills', in Éva Pócs (ed.), *Démonológia és boszorkányság Európában* ['Demonology and Witchcraft in Europe'] (Budapest: L'Harmatton–PTE Néprajz Tanszék, 2001), pp. 365–7 (English summary).

44. Montague Summers, *The History of Witchcraft and Demonology* (London: Kegan Paul, 1926), p. 118.

45. See also Gabór Klaniczay, 'Hungary: The Accusations and the Universe of Popular Magic', in Ankarloo and Henningsen, *Early Modern European Witchcraft*, pp. 219–55, esp. p. 252.

46. See also Elisabeth Biesel, 'Les descriptions du sabbat dans les confessions des inculpés lorrains et trévirois', in Jacques-Chaquin and Préaud, *Le sabbat des sorciers*, pp. 183–97; Biesel, 'Die Pfeiffer seint alle uff den baumen gesessen. Der Hexensabbat in der Vorstellungswelt einer ländlichen Bevölkerung', in Gunther Franz and Franz Irsigler (eds), *Methoden und Konzepte der historischen Hexenforschung* (Trier: Spee, 1998), pp. 289–302.

47. Gustav Henningsen, *The Witches' Advocate: Basque Witchcraft and the Spanish Inquisition (1609–1614)* (Reno: University of Nevada Press, 1980); Jens Christian V. Jensen, 'Denmark: The Sociology of Accusations', in Ankarloo and Henningsen, *Early Modern European Witchcraft*, pp. 339–65, 362.

48. Robin Briggs, *Witches and Neighbours: The Social and Cultural Context of European Witchcraft* (London: HarperCollins, 1996), p. 39.

49. Maurizio Bertolotti, 'The Ox's Bones and the Ox's Hide: A Popular Myth, Part Hagiography and Part Witchcraft', in Edward Muir and Guido Rugierro (eds), *Microhistory and the Lost Peoples of Europe* (Baltimore and London: Johns Hopkins University Press, 1991), pp. 42–70. Behringer, *Shaman*, pp. 59–60, 138, cites trials in South Tirol and Voralberg; see also chapter 8.

50. Gábor Klaniczay, 'Shamanistic Elements in Central European Witchcraft', in Klaniczay, *The Uses of Supernatural Power: The Transformation of Popular Religion in Medieval and Early-Modern Europe* (Princeton: Princeton University Press, 1990), pp. 129–50, esp. p. 136.

51. See Behringer, *Shaman*, p. 80: the Wild Hunt 'appears to have little to do with the fairy-like appearance of the night people'; see also pp. 36–8, 45–6, 136, n.28.

52. Hans Peter Broedel, *The Malleus Malificarum and the Construction of Witchcraft: Theology and Popular Belief* (Manchester: Manchester University Press, 2003), p. 102.

53. Henningsen, *The Witches' Advocate*, p. xxvii.

54. See, for the end of the eighteenth century, Giovanna Fiume, 'The Old Vinegar Lady, or the Judicial Modernization of the Crime of Witchcraft', in Edward Muir and Guido Ruggiero (eds), *History from Crime* (Baltimore: Johns Hopkins University Press, 1994), pp. 65–87.

55. Gustav Henningsen, '"The Ladies from Outside": An Archaic Pattern of the Witches' Sabbath', in Ankarloo and Henningsen, *Early Modern European Witchcraft*, pp. 191–215, esp. pp. 207–14.

56. He had heard Henningsen's presentation in Stockholm (published in Swedish in 1987) and was later given the then unpublished English version (see *Ecstasies*, p. 140, n.1) with a misdating of the Stockholm conference.

57. Henningsen, '"The Ladies from Outside"', appendix, p. 207.

58. Ginzburg, *Ecstasies*, p. 103.

59. Éva Pócs, *Fairies and Witches at the Boundaries of South-Eastern and Central Europe* (Helsinki: Suomalainen Tiedeakatemia, 1989), p. 13.

60. 'Tündér és boszorkány', the Hungarian version of *Fairies and Witches*, was published in *Ethnographia* in 1986, but we can be sure that Ginzburg does not read Hungarian (*Ecstasies*, p. 177, n.40). Ginzburg's publications before 1986 do not contain this particular line of thought.

61. Pócs, *Fairies and Witches*, p. 9.

62. Éva Pócs, 'The Popular Foundations of the Witches' Sabbath and the Devil's Pact in Central and Southeastern Europe', in Klaniczay and Pócs, *Witch Beliefs*, pp. 305–70, 327. When this paper was presented at the Budapest conference, Henningsen's article (originally in Danish) had not been published in English yet. But he had

circulated an English version which was, moreover, translated and published in Hungarian in 1985 in the journal *Világosság* ('Light'). See note 56 above.

63. Given the trade contacts between northern Italy and East Asia it is hardly surprising that a version of the Italian story of Cinderella found its way to the other end of the Eurasian continent.

64. Klaniczay's first essay in this series, 'Shamanistic Elements in Central European Witchcraft', was originally published in Hungarian in *Ethnographia* in 1983; an English version appeared in 1984. See also Klaniczay, *The Uses of Supernatural Power*, pp. 129–50.

65. Gábor Klaniczay, 'Hungary: The Accusations and the Universe of Popular Magic', in Ankarloo and Henningsen (eds), *Early Modern European Witchcraft*, pp. 219–55, esp. pp. 241–3. See also Éva Pócs, 'Preface: Fifteen Years of a Research Team', in Pócs, *Demonológia*, pp. 337–46.

66. Klaniczay, 'Hungary', pp. 229, 226–7.

67. Gábor Klaniczay, 'Witch-Hunting in Hungary: Social or Cultural Tensions', in Klaniczay, *The Uses of Supernatural Power*, pp. 151–67, esp. pp. 158–60, 165. This was also Klaniczay's contribution to the Budapest conference: Klaniczay and Pócs, *Witch Beliefs*, pp. 67–91.

68. Klaniczay, 'Hungary', p. 244.

69. In an obligatory summary of eastern European witchcraft history, Klaniczay repeats the findings of Pócs (as expressed at the Budapest conference) when dealing with the popular background of the Sabbat. See Gábor Klaniczay, 'Bûchers tardifs en Europe centrale et orientale', in Robert Muchembled (ed.), *Magie et sorcellerie en Europe du Moyen Age à nos jours* (Paris: Colon, 1994), pp. 215–31.

70. Gábor Klaniczay, 'Le sabbat raconté par les témoins des procès de sorcellerie en Hongrie', in Jacques-Chaquin and Préaud, *Le sabbat des sorciers*, pp. 227–46, esp. p. 227.

71. See also Pócs, 'The Popular Foundations', p. 334. Van Dülmen and Muchembled argued that in Germany and northern France Sabbat accounts referred to popular feasts: van Dülmen, 'Imaginationen des Teuflischen', p.118; Robert Muchembled, 'Satanic Myths and Cultural Reality', in Ankarloo and Henningsen, *Early Modern European Witchcraft*, pp. 139–60, esp. pp. 149–50.

72. Klaniczay, 'Le sabbat', pp. 237–9. See also Pócs, *Between the Living and the Dead: A Perspective on Witches and Seers in the Early Modern Age* (Budapest: Central European University, 1999), pp. 79–80.

73. Klaniczay, 'Le sabbat', p. 230.

74. Ronald Hutton, *Shaman: Siberian Spirituality and the Western Imagination* (London: Hambledon & London, 2001).

75. Klaniczay, 'Shamanistic Elements'; Éva Pócs, 'Hungarian *Táltos* and his European Parallels', in Mihály Hoppál and Juha Pentikäinen (eds), *Uralic Mythology and Folklore* (Budapest and Helsinki, 1989), pp. 251–76. See also Pócs, *Fairies and Witches*, p. 81, n.232.

76. Klaniczay, 'Shamanistic Elements', pp. 141–2. Ginzburg, *Ecstasies*, p. 132, derived his remark about the fertility activities of the *táltos* from these later instances, although he was familiar with Klaniczay's article, which he read in the 1984 English version.

77. Klaniczay, 'Shamanistic Elements', p. 149.

78. Gustav Henningsen, 'The White Sabbath and other Archaic patterns of Witchcraft', *Acta Ethnoraphica Hungarica* 37 (1991–92), pp. 301–2.

79. Jean-Michel Salman, review of *Le sabbat des sorcières*, *Annales* 50 (1995) 183–7, esp. 186.
80. Behringer, *Shaman*, pp. 88, 140.
81. Ginzburg, *The Night Battles*, pp. 52–3, 191, n.59 (the 1931 publication is available at the Warburg Institute); see also Behringer, *Shaman*, p. 137; Wolfgang Behringer, *Hexenverfolgung in Bayern. Volksmagie, Glaubenseifer und Staatsräson in der Frühen Neuzeit* (München: Oldenbourg, 1987), pp. 131, n.31, 189–91.
82. Behringer, *Shaman*, p. 138. See also Ginzburg, *The Night Battles*, p. 52: 'the closest resemblance'. The 'mythic essence' stems from an unacknowledged quote of Ginzburg's preface to the Italian edition of *Myths, Emblems, Clues*, which is translated in English as 'a mythical nucleus which retained its vitality fully intact for centuries – perhaps for millennia' (pp. xii–xiii).
83. Behringer, *Shaman*, p. 146. The last phrase fails in the German original, but is actually an improvement.
84. Pócs started her research among similar lines (*Fairies and Witches*; relicts in: *Between the Living and the Dead*, pp. 49–50, 88–91), but abandoned them in favour of Ginzburg's shamanism.
85. Pócs noticed later that 'the Stoeckhlin fairy magicians' confirmed Ginzburg's findings: Pócs, *Between the Living and the Dead*, p. 95.
86. Behringer, *Shaman*, pp. 143, 145.
87. Ginzburg, *Ecstasies*, p. 201, n.47.
88. See Willem de Blécourt, 'A Journey to Hell: Reconsidering the Livonian "Werewolf"', *Magic, Ritual and Witchcraft* (forthcoming). Pócs also relies on Höfler in this, see her *Fairies and Witches*, p. 82.
89. Ginzburg cites an entry in the *Handwörterbuch für Theologie und Religionwissenschaft*, Vol. 3, cols 307–8, which clearly represents an intermediate phase and will itself have been derived from the *Handwörterbuch des deutschen Aberglaubens*. See Lily Weiser-Aall, 'Hexe', *HdA* III, cols 1827–1920, esp. cols 1849–51: 'Hexenritt'; see also Artur Haberlandt, 'Besenritt' in *HdA* I, cols 1147–50.
90. See Ginzburg, *The Night Battles*, p. 174, n.8 (to the preface of the Italian edition), where Ginzburg mentions Peuckert's 'racist antithesis between the virile Germans … and the effeminate Mediterraneans'. At p. 192, n.69, Peuckert is attributed with 'racist presuppositions'.
91. Carlo Ginzburg, 'Deciphering the Sabbath', in Ginzburg, *Ecstasies*, p. 135. See also the balanced rebuttal of Höfler's work in *Ecstasies*, p. 173, n.2. See also Behringer, *Shaman*, p. 77, who exposes Höfler more directly.
92. See Mircea Eliade, 'Some European Secret Cults', in Helmut Birkhan (ed.), *Festgabe für Otto Höfler zum 75. Geburtstag* (Wien and Stuttgart: Braumüller, 1976), pp. 190–204, in which the Romanian *calusari*, and by implication the *benandanti* and the Latvian werewolf Thies, are presented as the remnants of a *Männerbund*.
93. In this sense Ginzburg's discussion of the politics of George Dumézil in 'Germanic Mythology' (see note 12 above) acquires an extra, biographical meaning.
94. Ginzburg, *The Night Battles*, p. 58.
95. Such as the Dutch Germanist Jan de Vries; see *Ecstasies*, pp. 114–15, n.45.
96. In *Storia* one of Grimm's insights is described as born 'in the flash of a fading bolt of lightning' (*Ecstasies*, p. 139). In contrast, Behringer blames Grimm for distorting the whole picture of the 'Nachtvolk', which led to decennia of misconception.
97. Ginzburg, *Ecstasies*, p. 220.

9

crime and the law

brian p. levack

The essential steps in the process of witch hunting – the accusation, arrest, trial and execution of witches – all involved the operation of the law. Ever since the scholarly study of witchcraft began with the publication of Wilhelm Soldan's monumental work in 1843, the main sources for the history of witchcraft have been trial records.[1] Witchcraft historians therefore have always had to deal with the law in their work, even when their interest in witchcraft might lie in other dimensions of the subject, such as the content of witch beliefs, the gender of witches or the religious zeal of witch hunters. Somewhat surprisingly, however, very little witchcraft scholarship has focused exclusively on the law.[2] The subject has usually been dealt with indirectly, as part of a broader narrative or as an explanation of the prerequisites for prosecution.

Historical scholarship regarding witchcraft and the law can be divided into four main historiographical traditions, each of which corresponds to a different function that the law serves in society. The first and oldest of these traditions is primarily concerned with the law as an agent of repression or coercion. Historians working in this tradition, which remains dominant in the historical literature on witchcraft today, focus on the use of the legal machinery and legal procedures that made possible the conviction and execution of tens of thousands of witches. The second historiographical tradition, which has come into prominence only in the past 50 years, focuses on the way in which lawyers brought witch hunts to an end. Historians who take this approach are concerned mainly with the historic role of the law as the guardian of due process and individual right. The third historiographical tradition, which also has developed only in recent years, studies witchcraft prosecutions in the larger context of the history of crime. This tradition, which intersects with social and cultural history, views the law primarily as the means by which society establishes social and behavioural norms and controls deviance. The fourth and most recent historiographical approach to witchcraft and the law deals with the connection between state formation and witch hunting. Historians who take this approach are concerned mainly with the way in which the law served as a tool of state authority, the means by which rulers could enhance their own power and bring their subjects more firmly under their rule.

inquisition, torture and witch hunting: the law as an agent of repression

Late nineteenth- and early twentieth-century historians of witchcraft constructed a paradigm of the relationship between the witchcraft and the law that endured until the later decades of the twentieth century. These historians belonged to what has been described as the liberal-rational school of witchcraft scholarship. Like members of any historiographical school, the individual historians differed in their choices of subject matter, emphasis, and interpretation. Nevertheless, they all shared the conviction that witchcraft was an irrational delusion, the product of confessions adduced under torture, the content of which was determined in large measure by the witch beliefs of university-educated clergy who staffed the courts of the papal Inquisition. These historians displayed little interest in the popular dimension of witchcraft or in the social conflicts that underlay most prosecutions. What mattered most to them were the ways in which judges and Inquisitors coerced innocent people to confess to activities they had never engaged in. The law, conceived as both a body of legislation and a system of criminal justice, was viewed by them primarily as a system of repression and an instrument of religious persecution. It was the means by which members of an educated and ruling elite responded to their fears of religious, political and moral subversion.

The most influential historians who belonged to this group were Joseph Hansen, Henry Charles Lea and George Lincoln Burr. All three of these scholars made enduring contributions to witchcraft scholarship. Hansen's massive study of witchcraft prosecutions from 1258 until 1526 and his edition of the sources for the history of witchcraft during that period set the standard for this genre.[3] The fact that Hansen, the archivist at Cologne, was a medievalist had a great bearing on his approach, since the role of the Inquisition in trying witches was greatest during that early period of witch hunting. Hansen's most enduring contribution to witchcraft studies was to show how European witch beliefs developed gradually over a period of three centuries, resulting in the formation of the cumulative or composite notion of witchcraft by the end of the fifteenth century. His extensive coverage of the *Malleus Maleficarum* (1486) helped to establish that work as the most famous demonological treatise of all time, a reputation it still retains today. Hansen's focus on early witchcraft treatises also helps to explain his interest in the law, since the *Malleus*, like many demonological treatises, was intended as a manual for Inquisitors and utilised testimony from trials that the author, Heinrich Kramer, had adjudicated.

Lea (1825–1909) was also a medievalist, although he was more concerned than Hansen with the development of legal procedures, most notably the introduction of inquisitorial procedure and torture in the ecclesiastical and secular courts between the thirteenth and the late fifteenth centuries. Lea placed these legal changes within a broad geographical and chronological framework in a four-part book, *Superstition and Force*, which studied the history of the duel, the oath, the ordeal, and torture.[4] Lea also wrote a three-volume history of the papal Inquisition that retains some value even today.[5] His most significant contribution to witchcraft studies, however, was his three-volume collection of notes for a projected history of witchcraft that he was planning at the time of his death. His abstracts of demonological works and local studies by nineteenth- and early twentieth-century historians strengthened the emphasis that Lea placed on the crucial role that legal developments played in the rise and fall of witchcraft prosecutions.[6]

The third of the early liberal rationalists was George Lincoln Burr (1857–1938), a disciple of Lea who in 1891 published an account of the trial of Dietrich Flade, the lay councillor and judge of the Prince-Archbishop of Trier who, after trying to reduce the severity of the witch panic in that electorate, was himself tried and executed for witchcraft in 1589. Burr paid tribute to Flade as man of common sense 'in that most drearily doctrinaire of ages'.[7] Burr made available to an English-speaking audience some of the most important documents in the history of witchcraft, including sections from Friedrich Spee's *Cautio Criminalis* (1631) and the report of the torture and confession of Johannes Junius, the burgomaster of Bamberg who was tortured and executed in the massive witch hunt that took place in that city in 1628.[8] Burr also edited a collection of some of the most important texts in the history of witchcraft in colonial New England.[9]

The liberal-rationalist interpretation of witchcraft came under attack in the late twentieth century because of its assumptions that all witch prosecutions came from above and that all witch beliefs were irrational.[10] The recognition that most witchcraft accusations originated among villagers and that demonological ideas were imposed at a later stage of the judicial process demanded a new attention to the voluntary testimony of the witches' neighbours. What is more, historians recognised that the stereotype of the witch, while certainly shaped by judges and demonologists, also incorporated elements drawn from popular culture. This new awareness did not, however, alter the view of the legal process that characterised the work of the liberal-rationalist historians. As might be expected from studies that focused on the legal machinery that had been used to prosecute witches, the law continued to be viewed mainly as an instrument of repression.

Three medieval historians writing in the 1970s reflected this historiographical shift in writing about witchcraft and the law. Like Hansen and Lea, all three were concerned primarily with the development of the cumulative concept of witchcraft in the late Middle Ages. In his study of witchcraft trials between 1300 and 1500, Richard Kieckhefer discovered that all confessions to the diabolical aspects of witchcraft (the pact with the Devil and the Sabbath) came from 'contaminated' documents, that is, from the confessions adduced under torture, whereas the depositions of the witches' neighbours prior to the interrogation of the accused revealed a concern with sorcery or *maleficium*. The inquisitorial process therefore led to the 'imposition' of learned notions of witchcraft on the 'popular tradition' of sorcery.[11]

In a broad-ranging study of the origins of the stereotype of the witch, the religious historian Norman Cohn also demonstrated the importance of legal procedures in the development of witch beliefs.[12] Cohn's main contribution to the historiography of witchcraft and the law was to describe in its broad outlines the replacement of an accusatory system of criminal procedure with an inquisitorial system between the thirteenth and fifteenth centuries. Cohn argued that this transformation, which involved the adoption of torture as an inquisitorial tool, was essential to the successful prosecution of witchcraft on a wide scale. Unfortunately, Cohn ended his study at the 'threshold' of the great witch hunt and did not therefore study the different systems of criminal procedure that individual European jurisdictions used during the most intense period of witch hunting in the sixteenth and seventeenth centuries. Local, regional and national variations in criminal procedure had a significant bearing on the differential rates of convictions and executions of accused witches at the height of the trials.

The third medievalist who worked within the historiographical tradition of Hansen, Lea and Burr was Edward Peters, who served as the curator of the Lea collection at the University of Pennsylvania Library. Peters published a new edition of Lea's nineteenth-century work on torture, wrote his own general history of torture from ancient times to the present, and produced a detailed study of the relationship between the learned magician, the witch and the law in the late Middle Ages.[13] In studying the condemnation of magic by the Inquisition (a process partially inspired by the clerical invective against these heretics) Peters discussed a number of legal treatises that were later put to use against witches, most notably the inquisitorial manual produced by Nicolas Eymeric in 1376. As a medieval historian, however, Peters did not venture far beyond the Middle Ages, when the prosecution of witchcraft had only just begun, and when witch trials took place only in a limited geographical area.

A number of recent local, regional and national studies of witchcraft prosecutions during the height of the trials have depicted a complex legal landscape in which courts employed a wide variety of criminal procedures. These studies have shown that it is no longer possible to draw a simple contrast between an accusatorial system with a strict prohibition of torture in England and an inquisitorial system of criminal justice, based on Roman law, on the Continent.[14] Dagmar Unverhau and others have studied the initiation of cases by accusation in certain jurisdictions in Germany, while a number of witchcraft historians have emphasised the contrast between the strict rules regarding inquisitorial procedure laid down in the imperial law code known as the *Carolina* (1532) and the way in which those rules were disregarded in specific German territories.[15] John Langbein and C. R. Unsworth have both shown how difficult it is to label the English system of criminal procedure simply as accusatorial,[16] while I have argued that the differences between the administration of law in England and Scotland are central to understanding the differentials in the execution rates of the two countries in witchcraft cases.[17] Bengt Ankarloo has linked the late seventeenth-century legal revolution in Sweden, which involved the belated adoption of some of the features of inquisitorial procedure, to the intensification of witch hunting in that kingdom at that late date.[18]

Similar qualifications have emerged in the historical work on the connection between torture and witch hunting. Studies of the application of torture in specific jurisdictions have led to the creation of a much more nuanced picture of how torture was administered and how it influenced the intensity of prosecutions. Even within the German territories, where the most severe tortures were administered, the pattern is complex. Unrestricted torture clearly had a bearing on the large witch hunts that occurred in Bamberg, Eichstätt and Würzburg, but Alison Rowlands has shown that judicial caution in Rottenburg ob der Tauber resulted in a fairly low level of convictions and executions.[19] William Monter has demonstrated that adherence to the strict rules regarding the administration of torture in the Jura region prevented the occurrence of large chain-reaction hunts such as those that took place in some of the German territories, while Alfred Soman has studied a pattern of even more remarkable restraint in the appeals of witchcraft convictions to the *parlement* of Paris.[20] Brian Levack has explored a further dimension of this problem by showing how in Scotland, where torture was to be administered only by special warrant by the privy council, just as in England, local authorities frequently tortured suspected witches illegally, usually shortly after arrest.[21]

The administration of torture and the confessions that it often adduced have also attracted the interest of scholars interested in the psychological dimensions of witch hunting. Etienne Delcambre pioneered this approach in his psychological study of the judges as well as the witches in Lorraine,[22] while Lyndal Roper has proposed that witchcraft confessions were the product of collusion between the witch and the torturer.[23] More recently, torture has attracted the interest of cultural historians interested in the human body. Lisa Silverman has investigated how judges approving of the administration of torture in the *parlements* of Old Regime France based their decisions on theories regarding the location of truth in the body, while Elaine Scarry has explored both the effects of bodily pain on suffering and the political dimension of torture.[24] Scarry's work also fits into a large corpus of scholarship on confessions, many of which address the important question in regarding the reliability of 'free' confessions.[25] Some of these contributions say little about witchcraft per se, but they nonetheless have direct relevance to witchcraft studies. Like the others books discussed in this section, moreover, they all view the law in early modern Europe, at least in large part, as an agent of repression or coercion.

the restriction of witch hunting: law as the guardian of due process

The second major historiographical tradition regarding the law and witchcraft presents a much more positive view of the historical role of law in society. Historians working within this tradition have focused their attention on the success that some lawyers and jurists had in reducing the number of witchcraft convictions and executions and eventually in bringing them to an end. Instead of focusing on the claims of sceptics like Johann Weyer, who argued that witches were not legally culpable for their actions,[26] they have studied the actions taken by judges and Inquisitors to protect accused witches from arbitrary prosecution. The function of the law that concerns historians writing in this tradition was its provision for due process and the protection of individual legal right.

The two most influential historical studies written in this tradition are Robert Mandrou's *Magistrats et sorciers en France au XVIIᵉ siècle* and Gustav Henningsen's *The Witches' Advocate*. Mandrou's book is a study of the men who staffed the *parlement* of Paris, the most important of the nine provincial appellate courts in seventeenth-century France.[27] Sitting in the capital and claiming jurisdiction over two-thirds of northern France, it was for all intents and purposes a central judicial institution. Mandrou tells the story of how the Parisian *parlementaires* acquired the right to review all witchcraft convictions by 1624 and how their scepticism regarding the guilt of those accused brought witch hunting in their jurisdiction to almost a complete end. Mandrou's arguments that the *parlementaires* suffered a 'crisis of conscience' that led them to doubt the reality of the crime had little evidence to support it, and Alfred Soman corrected many of Mandrou's errors in using the records of the courts.[28] But Mandrou presented the members of the legal profession, who have almost always been vilified for their role as witch judges, in a much more positive light. His book represented an important legal revisionism in witchcraft studies.

The hero of Henningsen's book is Alonso de Salazar Frías, the Spanish Inquisitor who became involved in a massive witch hunt in the Basque lands in northern Spain between 1609 and 1611.[29] The hunt began with the prosecution of a few witches whose confessions were coerced, but it grew exponentially when hundreds of children freely confessed that adults had transported them to the witches' Sabbath. An Edict of Grace, which allowed the witches to confess with impunity, facilitated these confessions. Salazar, a university-trained lawyer who did not in any way deny the reality of witchcraft, conducted an investigation of the confessions and concluded that none of the meetings of the witches ever took place. In order to prevent such miscarriages of justice from occurring again, Salazar proposed a new set of rules to govern all future witchcraft prosecutions. The enforcement of these rules, which were adopted by the central tribunal of the Spanish Inquisition in 1614, in effect brought an end to executions for witchcraft by that judicial body.

The work of John Tedeschi on the Roman Inquisition – the institution established in 1542 to maintain religious conformity in Italy – belongs to the same historiographical tradition as that of Mandrou and Henningsen. In a series of articles Tedeschi has documented not only the Roman Inquisition's strong tradition of leniency in sentencing witches, but also its insistence upon adherence to strict procedural rules in the conduct of witchcraft trials.[30] The main document that Tedeschi has studied in this regard is Cardinal Desiderio Scaglia's *Instructio pro formandis processibus*.[31] The *Instructio* dealt with all aspects of criminal procedure, establishing strict rules for examining accused witches, calling for restraint in the administration of torture, and recommending particular care in the evaluation of witches' confessions. Like the rules promulgated by the Spanish Inquisition in 1614, the *Instructio* had a lasting negative effect on the intensity of witchcraft prosecutions in Italy.

The scholarship on German witch hunting that belongs to this historiographical tradition has for the most part focused on the decisions and dissertations of the law faculties of the German universities in the late seventeenth and early eighteenth centuries. The German law faculties played a crucial role in witchcraft prosecutions, since the courts of small territories were obliged by the *Carolina* to seek the advice of university jurists regarding how to proceed in such matters. During the period of intense prosecution the jurists for the most part encouraged convictions and executions, but towards the end of the seventeenth century they began to demand procedural restraint and recommend acquittals. The two historians who have done the most work on these consultations are Sönke Lorenz and Gerhard Schormann.[32] Lorenz's discussion of the case of Barbara Labarentin, determined by the law faculty of the University of Halle in 1694, is especially significant, since it led one member of that faculty, Christian Thomasius, to develop his ideas regarding the conduct of witchcraft trials. These ideas found their place in Thomasius' treatise on the crime of magic (1701), which argued that witchcraft prosecutions should end, and his dissertation on torture (1705), which condemned the procedure on legal and religious grounds.[33]

Two historical works published in the past two decades have placed these specific studies of legal scepticism within the broader European context of the decline and end of witchcraft prosecutions. Alfred Soman, who has written extensively on the *parlement* of Paris, has compared the French experience to those of other countries, showing a correlation between the degree of higher central control and the success of governments to bring witchcraft trials to an end.[34] Building to some extent on Soman's model, Brian Levack has emphasised a broader set of judicial developments, including the regulation of local justice, restrictions on the use of judicial torture, and the application of more demanding standards of evidence in witchcraft trials.[35] Winfried Trusen also emphasised the importance of judicial developments in bringing about an end to witch hunting in Germany, while Thomas Robisheaux showed that jurists in the county of Hohenloe in the 1660s and 1670s scrupulously adhered to regular inquisitorial procedure (as set forth in the *Carolina*) in the trying of witches, thereby slowing the momentum of the larger witch hunts.[36] By emphasising the importance of judicial developments in bringing about the end of witch hunting, these authors support an emerging consensus in witchcraft studies that minimises the importance of learned witch beliefs in bringing witch hunting to an end. The men responsible for stopping the trials took action because miscarriages of justice were occurring or because the crime could not be proved in law, not because the crime was impossible to perform.

witchcraft and the history of crime: the law as social control

The third historiographical tradition of witchcraft scholarship dealing with the law is that which treats witchcraft within the broader history of crime. The history of crime is a relatively new sub-field of history that grew out of the new social history that came into fashion in the 1960s and 1970s and has more recently intersected with the new cultural history.[37] Historians of crime have used judicial records to construct a social portrait of the criminal class, to explore the social interactions between the criminals and their victims, to study the power relationships between the accused and the judges, and to discover how different groups viewed transgressions of the social order. The history of crime is almost always based on archival sources: either a large body of judicial records covering trials for a particular crime over a number of years or the records of individual trials that historians subject to a close textual reading.

Many of the regional and local studies of witchcraft prosecutions written in the past four decades might be viewed at least to some extent as contributions to the history of crime. A more distinct set of historical works, however, has specifically investigated the relationship between witchcraft and other crimes. Such an approach has challenged a long-standing tendency to consider witchcraft separately, as a crime that must be explained on its own special terms rather than as one crime among many. That tendency was reinforced by the early modern designation of witchcraft as a *crimen exceptum*, a special crime that allowed judges to modify, suspend or simply ignore the procedural rules courts were obliged to follow in admitting testimony and torturing the accused. The special status of witchcraft as a crime was also encouraged by the assumption that witchcraft, unlike other crimes, was imaginary: that those accused did not actually engage in criminal conduct and therefore could not be discussed in the same criminal context as murderers, thieves or rapists.[38]

Without denying the special or exceptional status of witchcraft, historians during the past two decades have nonetheless begun to treat witchcraft as one crime among many. Christina Larner made a strong case for taking this broad, comprehensive approach on the grounds that it could illuminate a number of changes in the criminal law during the early modern period. These included the 'judicial revolution' or the introduction of inquisitorial procedure, the statutory designation of religious or sexual crimes like blasphemy and adultery as punishable in the secular rather than the ecclesiastical courts, and the criminalisation of women.[39] There is little doubt that these developments in the prosecution of all crimes took place during the period of the witch trials, but the precise role that witchcraft prosecutions played in these broader legal changes can only be determined by detailed archival work. A collection of essays on female criminality in early modern Germany, for example, has revealed that witchcraft was not as central to the criminalisation of women as Larner and others assumed.[40]

Some historians who have placed witchcraft in a broader criminal context have conducted comparative research in the records of a specific locality. Recent studies of this sort include Elisabeth Biesel's essay on 'witchcraft and other crimes' in the city of Toul between 1570 and 1630, William Monter's work on the 'mixed crimes' of sodomy and witchcraft in Aragon, and Malcolm Gaskill's book on witchcraft, coining and murder in early modern England.[41] Gaskill's study possesses considerable methodological interest, since he is concerned more with the cultural rather than the social context of crime, exploring the mentalities of the people who were involved in the prosecution of these crimes. By investigating the 'social meaning' of witchcraft, Gaskill reveals the way in which people from different social environments viewed the transgression of witchcraft.

Studies of witchcraft included in collections of essays on early modern European crime serve similar comparative purposes. James Sharpe's essay on witchcraft, women and the legal process used the prosecution of witches to offer a broader understanding of the role of women in the legal process, both as witnesses and as defendants.[42] An essay by Giovanna Fiume on a case of witchcraft in mid eighteenth-century Palermo showed how the judges reclassified a case of witchcraft as poisoning, thereby demonstrating the influence of Enlightenment thought.[43] An essay in the same collection by Sabina Loriga used the prosecution of more than 70 persons for allegedly using magic to kill a prince from Piedmont in 1715 to document the continuation of elite belief in witchcraft in the country's highest court.[44] Her essay took issue with Mandrou, who argued that the judges of higher courts were contemptuous of the superstitions beliefs of local judges who heard witchcraft cases. Mandrou can be criticised for exaggerating the modernity of his Parisian *parlementaires*, but an investigation of royal murder by the Piedmont senate, even if witchcraft were allegedly involved, should not surprise us, especially when the defendants were accused of using ritual magic to kill a member of the royal family. The case belongs in the same category as the attempts of medieval and early modern sorcerers to kill Pope John XXII and King James VI of Scotland.

Statistical studies of the criminal caseloads of various jurisdictions have also allowed us to assess the level of judicial concern with witchcraft at a particular time and place. Contemporaries may have believed that witchcraft was the most horrific crime one could commit, that its practitioners were legion, and that it presented a grave threat to society, but the actual records of criminal courts, especially the secular courts where most witches were tried, often present a picture of less urgent concern. Only during periods of intense prosecution, such as when large chain-reaction hunts took place, did witchcraft cases occupy a significant percentage of judicial time and effort. Even within the Mediterranean Inquisitions, which tried only spiritual crimes, witchcraft was not the dominant concern, perhaps because strict procedural rules discouraged the number of cases. Monter and Tedeschi's 'Statistical Profile of the Roman Inquisition' provides a useful way of determining the relative importance of witchcraft among the various crimes prosecuted by that tribunal.[45]

Investigations by historians into early modern European definitions of the crime of witchcraft have also raised the question of the relationship between witchcraft and other crimes. Witchcraft, as it was defined in the sixteenth and seventeenth centuries, was a composite crime. In its most elaborate form it combined the crime of maleficent magic with that of diabolism, although different social and professional groups tended to place different degrees of emphasis on one component or the other. More specifically, the crime of witchcraft encompassed a variety of activities that could be prosecuted as crimes in the ecclesiastical or secular courts. Witchcraft could denote heresy, apostasy, blasphemy, maiming, murder, poisoning, theft, destruction of crops, killing of livestock, arson, sodomy, fornication, adultery, infanticide and conspiracy. In 1584 the English sceptic Reginald Scot identified 15 different crimes attributed to witches.[46]

Most historical studies that deal with these definitions and relationships focus either on the late Middle Ages, when the definition of witchcraft was in the process of formation, or in the late seventeenth and early eighteenth centuries when the concept was disintegrating. In the formative peri-
crimes that eventually became identified with witchc
heresy, although separate prosecutions for both crimes co
the period of witch trials. In the period of decline and dis
accused of witchcraft were often prosecuted for committin

specific acts encompassed in the composite notion of witchcraft. The most common of these was the crime of poisoning, which could be interpreted in either natural or supernatural terms. Less common were trials for making pacts with the Devil, such as those prosecuted in Sweden between 1680 and 1789; the trials for sacrilege in late seventeenth and eighteenth-century France, and the trials for magic that took place in Finland when witchcraft trials began to wane.[47]

A further dimension of this study of witchcraft and its relationship to other crimes was the legal situation caused by the repeal of witchcraft statutes or edicts. The two legislative acts that have received the greatest historical attention are the French edict of Louis XIV in 1682 and the British statute of 1736. The French edict, the product of the 'affair of the poisons' that rocked Parisian high society as well as a series of miscarriages of justice in the provincial *parlements*, in effect reclassified witchcraft as fraud while reserving the severest penalties for the crimes of poisoning and blasphemy.[48] One effect of this legislation was the pursuit of 'false sorcerers' by the Paris police in the eighteenth century, an operation studied in detail by Ulriki Krampl.[49] The criminality of these people, who were versed in various occult practices, including fortune telling, alchemy and the sale of talismans and philtres, represented a transformation rather than an elimination of the crime of witchcraft.

The British statute of 1736, which repealed the English witchcraft statute of 1604 and the Scottish act of 1563, has an even more complex historiography. Drafted long after the last English execution for witchcraft (1685) and 14 years after the last Scottish execution, it appears to be a much more comprehensive decriminalisation of witchcraft than the French edict of 1682. It did in fact lead to an end of witchcraft prosecutions as they had been conducted in the past. Yet like the French edict, the statute left the door open for prosecutions of people for 'the pretended use of any kind of witchcraft, enchantment, sorcery or conjuration'. The act therefore punished any pretence to witchcraft, just like the French act of 1682. As Owen Davies has shown in his study of witchcraft in England after 1736, prosecutions for the deceptive, fraudulent practice of any occult art continued until 1951, when legislation finally brought witchcraft prosecutions in any form whatsoever to an end.[50]

In *Witchcraft and its Transformations, c.1650–c.1750*, Ian Bostridge offered a political interpretation of the passage of the British act of 1736. Bostridge argued that demonological theory, which had flourished at the time of the Restoration because it helped to define a model Christian society, lost its force in the early eighteenth century when it became a tool in bitter factional political strife between Whigs and Tories. Ridiculed by the Whigs as part of the ideology of conservative, superstitious High Church Tories, witchcraft theory was linked permanently to a concept of the sacral state that had vanished with the demise of divine right theory at the time of the Glorious Revolution of 1688–89. Opposition to repeal was led by the Earl of Mar, a Scottish peer who viewed the proposed legislation as an attack on religious belief and, by extension, on the identity of the Scottish nation, where belief in witchcraft persisted longer than in England. Even at this late date, the definition of the secular crime of witchcraft in Britain could not be easily detached from its religious foundations.[51]

witch hunting and state formation: the law as a tool of political authority

A fourth historiographical tradition regarding the role of law in witch hunting focuses on the role of witch hunting in the process of state formation. One of the most significant developments in early modern European history was the growth of large, centralised bureaucratic states that exercised effective control over the populations under their rule. This process, which was marked by a general increase in the level of governance, involved the assumption of state or secular control over the Church, the growth of bureaucracy, the suppression of local and regional autonomy, and the cultivation of patterns of obedience and allegiance to sovereign rulers. A number of historians have argued that the prosecution of witches was one of the means by which such states emerged and developed. These historians have emphasised the assumption of state control over witchcraft trials, the definition of the crime of witchcraft in the temporal law, the use of witchcraft prosecutions to legitimise or enhance the authority of secular rulers, and the prosecution of witches as a means of disciplining the population and creating a more homogeneous society that is obedient to the king and God.[52]

It is not surprising that the operation of the law was central to this historical development. The state is essentially a coercive structure that uses the law to enforce obedience to a public, sovereign authority.[53] Historians who have interpreted witchcraft prosecutions in these terms have viewed the law not so much as an agent of repression, a guarantee of due process or even a method of social control, but as a means to enhance or consolidate political power.

Some witchcraft studies have emphasised the deliberate, conscious use of witchcraft prosecutions to this end. Michael Kunze's book on the brutal treatment of the Pappenheimer family in Bavaria at the beginning of the seventeenth century showed how Duke Maximilian, who venerated the principle of the all-powerful state, tried and executed members of the family for witchcraft in order to bolster his own stature as a guardian of public virtue and thus legitimise his rule. Kunze implicates the members of the entire governing class of clerics and jurists in this persecution.[54] Rita Voltmer has argued that the ruling elite of some German states 'exploited' witchcraft prosecutions in order to enhance their power and the authority of the state.[55] Wolfgang Behringer's comprehensive study of witch hunting in Bavaria shows how the growth of state legal power and the development of the concept of reason of state facilitated the effective prosecution of witches.[56] By focusing on divisions within the council regarding the prosecution of witches, Behringer also illustrates the importance of central tribunals in determining the intensity of witch hunting.

A very different approach to the connection between state-building and witch hunting appears in Robert Muchembled's study of popular culture in France. Viewing the entire hierarchy of officials in the provinces, from the king down to the local judges and even the local parish priests, as agents of the state, Muchembled interprets witch hunting as a means of disciplining the local population and eliminating superstitious folk beliefs. According to Muchembled, this process of 'acculturation' served the interests of the post-Tridentine Church as well as the emerging absolutist state.[57] These proclaimed connections between witch hunting and state-building have proved to be problematic, mainly because Muchembled exaggerated the success of both state and Church in penetrating the localities in the seventeenth century. Even if we concentrate on *efforts* at state-building, rather than their success, we still have difficulty viewing the local judges and officials, whom the provincial *parlements* struggled to control throughout the seventeenth century, as agents of state authority. The courts that tried and executed witches for violating the law often did so without the approval of 'the state'. Witchcraft prosecutions, therefore, often served as an indication of the weakness of early modern states rather than their strength.

Among those scholars who have advanced the 'state thesis', Christina Larner has made the operation of the law most central to her argument. In a book on witch hunting in Scotland and in a series of articles, Larner argued that when the Scottish privy council, the main executive arm of royal power, acquired the power to review and approve all local witchcraft trials in 1597, witchcraft became a 'centrally managed crime'.[58] The state could therefore co-ordinate prosecutions, making possible the large 'national' witch hunts that took place in 1628–29, 1643, 1649 and 1661–62. The key to this central government co-ordination and encouragement of witchcraft prosecutions was the practice of the privy council reviewing all petitions for local witchcraft trials and then authorising a commission of local authorities to try witches in the communities where they had been arrested.

There is no question that the Scottish state, which was relatively weak in the late sixteenth century, was striving to acquire more power, especially during the reign of James VI (1567–1625), and that the privy council was the instrument through which royal power was manifested.[59] There is also no question that the great majority of Scottish witchcraft trials that resulted in executions were conducted by commissions of justiciary granted by the council. The problem with Larner's thesis is twofold. First, the privy council, which represented sovereign state power, did not effectively manage the prosecutions, much less co-ordinate them on a national scale. Quite to the contrary, it granted most of the commissions routinely and exercised no effective control over the conduct of the trials once the commissions authorising them were granted. The most severe witch hunts in Scotland were therefore the product of weak central control of local prosecutions rather than of a powerful state machine that was using prosecutions to consolidate its power. Second, the original purpose of the privy council's insistence on approving all local commissions was to restrain witchcraft prosecutions that had spun out of control rather than to encourage them. The privy council did not always exercise such restraint, but it did so periodically, and in the late seventeenth century it took the lead in bringing witch hunting to an end.[60]

The historiography of witch hunting in Africa in the nineteenth and twentieth centuries provides an interesting footnote to the state thesis regarding witchcraft in Europe. Recent historical research has discovered that before the advent of colonial rule, when there were no political structures comparable to European states, African communities intermittently conducted witch-cleansing movements, usually without the sanction of law.[61] In the late nineteenth century, after European countries gained control of African lands and established dependent African states, colonial courts refused to prosecute individuals who were accused of bewitching their neighbours. This legal position, which was influenced by European scepticism regarding the existence of witches, had the ironic effect of leading Africans to claim that the colonial courts were protecting witches.[62] Only in post-colonial Africa did the courts, now under the control of Africans more sympathetic to popular witch beliefs, begin to convict people for practising witchcraft, although they did not administer very severe sentences.[63] Most witch-cleansing movements in the late twentieth and early twenty-first century have operated without the sanction of the law, just as they did before the introduction of colonial rule. In Africa, as in Europe, the state has done more to restrain than to encourage witch hunting.

notes

1. W. G. Soldan, *Geschichte der Hexenprozesse: aus den Quellen dargestellt* (Stuttgart: Cotta, 1843).
2. Two exceptions to this general pattern are P. Hoffer, *The Salem Witchcraft Trials: A Legal History* (Lawrence, Kans.: University Press of Kansas, 1997) and G. Durston, *Witchcraft and Witch Trials: A History of English Witchcraft and its Legal Perspectives, 1542 to 1736* (Chichester: Barry Rose Law Publishers, 2000).
3. J. Hansen, *Zauberwahn, Inquisition und Hexenprozess im Mittelalter* (Munich: Oldenbourg, 1900); *Quellen und Untersuchungen zur Geschichte des Hexenwahns im Mittelalter and der Hexenverfolgung* (Bonn: C. Georgi, 1901).
4. H. C. Lea, *Superstition and Force* (Philadelphia: Henry C. Lea, 1866).
5. H. C. Lea, *A History of the Inquisition of the Middle Ages* (New York: Macmillan Co., 1908–11).
6. II. C. Lea, *Materials toward a History of Witchcraft*, ed. A. C. Howland (Philadelphia: University of Pennsylvania Press, 1939).
7. G. L. Burr, *The Fate of Dietrich Flade (d. 1589)* (New York: G. P. Putman's, 1891).
8. G. L. Burr (ed.), *The Witch Persecutions* (Philadelphia: University of Pennsylvania Press, 1897).

9. G. L. Burr (ed.), *Narratives of the Witchcraft Cases, 1648–1706* (New York: Barnes & Noble, 1914). Another historian in the liberal-rationalist tradition who wrote about witchcraft in colonial America was J. M. Taylor, *The Witchcraft Delusion in Colonial Connecticut* (New York: Grafton Press, 1908).

10. The last significant work on witchcraft written in this tradition is H. R. Trevor-Roper, *The European Witch-Craze of the Sixteenth and Seventeenth Centuries and Other Essays* (Harmondsworth: Penguin, 1969). Trevor-Roper's dismissal of popular culture as 'the rubble of peasant superstition' and his perpetuation of the word 'craze' to describe the irrationality of the great witch hunt place it squarely in the tradition of Hansen, Lea and Burr.

11. R. Kieckhefer, *European Witch Trials: Their Foundations in Popular and Learned Culture 1300–1500* (London: Routledge & Keegan Paul, 1976).

12. N. Cohn, *Europe's Inner Demons: An Enquiry Inspired by the Great Witch-Hunt* (London: Chatto & Windus, 1975).

13. H. C. Lea, *The Ordeal*, ed. E. Peters (Philadelphia: University of Pennsylvania Press, 1973); Edward Peters, *Torture* (Oxford: Blackwell, 1985); Peters, *The Magician, the Witch and the Law* (Philadelphia: University of Pennsylvania Press, 1978).

14. On torture in England during the early modern period see J. Langbein, *Torture and the Law of Proof* (Chicago: University of Chicago Press, 1977).

15. D. Unverhau, 'Akkusationsprozess-Inquisitionsprozess: Indikatoren für die Intensität der Hexenverfolgung in Schleswig-Holstein', in C. Degn, H. Lehmann and D. Unverhau (eds), *Hexenprozesse: Deutsche und skandinavische Beitrage* (Neumünster: K. Wachholtz Verlag, 1983), pp. 59–143. On the *Carolina* and the controversy regarding the nature of inquisitorial procedure see J. Langbein, *Prosecuting Crime in the Renaissance* (Cambridge, Mass.: Harvard University Press, 1974).

16. C. R. Unsworth, 'Witchcraft Beliefs and Criminal Procedure in Early Modern England', in T. G. Watkin (ed.), *Legal Record and Historical Reality* (London: Hambledon, 1989), pp. 71–98; Langbein, *Prosecuting Crime*.

17. B. P. Levack, *The Witch-Hunt in Early Modern Europe* (London: Longman, 1995); Levack, 'State-Building and Witch-Hunting in Early Modern Europe', in J. Barry, M. Hester and G. Roberts (eds), *Witchcraft in Early Modern Europe* (Cambridge: Cambridge University Press, 1996).

18. B. Ankarloo, 'Sweden: The Mass Burnings (1688–1676)', in B. Ankarloo and G. Henningsen (eds), *Early Modern European Witchcraft* (Oxford: Clarendon Press, 1990), pp. 285–317. On changes in Swedish legal practice after 1600 see Per Sörlin, *'Wicked Arts': Witchcraft and Magic Trials in Southern Sweden, 1635–1754* (Leiden: Brill, 1999), pp. 47–50.

19. Alison Rowlands, 'Eine Stadt ohne Hexenwahn: Hexenprozesse, Gerichtspraxis und Herrschaft in frühneuzeitlichen Rothenburg ob der Tauber', in H. Eiden and R. Voltmer (eds), *Hexenprozesse und Gerichtspraxis* (Trier: Spee, 2002), pp. 331–47.

20. W. Monter, *Witchcraft in France and Switzerland* (Ithaca: Cornell University Press, 1976); A. Soman, 'The Parlement of Paris and the Great Witch-Hunt (1565–1640)', *Sixteenth Century Studies* 9 (1978) 30–44.

21. B. P. Levack, 'Judicial Torture in Scotland during the Age of Mackenzie', in H. MacQueen (ed.), *Miscellany IV* (Stair Society, Vol. 49, Edinburgh, 2002), pp. 185–198.

22. Etienne Delcambre, 'La Psychologie des inculpés Lorrains de sorcellerie', *Revue historique de droit français et étranger* 4, 32 (1954) 383–403; Delcambre, 'Les procès de sorcellerie en Lorraine. Psychologie des juges', *Tijdschrift voor Rechtsgesciednenis* 21 (1954) 389–419.

23. L. Roper, *Oedipus and the Devil: Witchcraft, Sexuality and Religion in Early Modern Europe* (London: Routledge, 1994), esp. chapter 9.

24. L. Silverman, *Tortured Subjects: Pain, Truth and the Body in Early Modern France* (Chicago: University of Chicago Press, 2001); E. Scarry, *The Body and Pain* (Oxford: Oxford University Press, 1987).

25. J. O. Rogge, *Why Men Confess* (New York: T. Nelson, 1959); P. Brooks, *Troubling Confessions; Speaking Guilt in Law and Literature* (Chicago: University of Chicago Press, 2000).

26. H. C. E. Midelfort, 'Johann Weyer and the Transformation of the Insanity Defense', in R. Po-Chia Hsia (ed.), *The German People and the Reformation* (Ithaca: Cornell University Press, 1988), pp. 234–61.

27. R. Mandrou, *Magistrats et sorciers en France au XVIIᵉ siècle* (Paris: Plon, 1968).

28. Soman, 'The Parlement of Paris'.

29. G. Henningsen, *The Witches' Advocate: Basque Witchcraft and the Spanish Inquisition, 1609–1611* (Reno: University of Nevada Press, 1980).

30. J. Tedeschi, 'Inquisitorial Law and the Witch', in Ankarloo and Henningsen, *Early Modern European Witchcraft*, pp. 83–118; Tedeschi, *The Prosecution of Heresy* (Binghamton: Medieval and Renaissance Texts and Studies, 1991).

31. On the inclusion of this set of instructions in a treatise by Cesare Carena see R. Martin, *Witchcraft and the Inquisition in Venice, 1550–1650* (Oxford: Blackwell, 1989), pp. 71–3.

32. S. Lorenz, *Aktenversendung und Hexenprozess. Dargestellt am Beispiel der Juristfakultäten Rostock und Greifswald (1570/82–1630)*, 2 vols (Frankfurt am Main: P. Lang, 1982–83); 'Die letzten Hexenprozesse in den Spruchakten der Juristfakultäten: Versuch einer Beschreibung,' in S. Lorenz and D. R. Bauer (eds), *Das Ende der Hexenverfolgung* (Stuttgart: Franz Steiner Verlag, 1995), pp. 227–47; G. Schormann, *Hexenprozesse in Nordwestdeutschland* (Hildesheim: Lax, 1977).

33. C. Thomasius, *De Crimine magiae* (Halle, 1701); *De Tortura ex foris Christianiorum proscribenda* (Halle, 1705).

34. A. Soman, 'Decriminalizing Witchcraft: Does the French Experience Furnish a European Model?', *Criminal Justice History* 10 (1989) 1–22.

35. B. P. Levack, 'The Decline and End of Witchcraft Prosecutions', in B. Ankarloo and S. Clark (eds), *Witchcraft and Magic in Europe: The Eighteenth and Nineteenth Centuries* (London, 1999), pp. 1–93.

36. W. Trusen, 'Rechtliche Grundlagen der Hexenprozesse und ihrer Beendigung', in Lorenz and Bauer, *Das Ende der Hexenverfolgung*; Thomas Robisheaux, 'Zur Rezeption Benedict Carpzovs im 17 Jahrhundert', in Eiden and Voltmer, *Hexenprozesse und Gerichtspraxis*, pp. 527–43.

37. For a guide to some of this work see J. A. Sharpe, 'The History of Crime in Late Medieval and Early Modern Europe: A Review of the Field, *Social History* 7 (982) 169–89; J. Innes and J. Styles, 'The Crime Wave: Recent Writing on Crime and Criminal Justice in Eighteenth-Century England', in A. Wilson (ed.), *Rethinking Social History: English Society 1570–1920 and its Interpretation* (Manchester: Manchester University Press, 1993), pp. 15–26.

38. E. P. Currie, 'Crimes without Criminals: Witchcraft and its Control in Renaissance Europe', *Law and Society Review* 3 (1968) 7–32.

39. Christina Larner, 'Crimen exceptum?: The Crime of Witchcraft in Europe', in Larner, *Witchcraft and Religion: the Politics of Popular Belief* (Oxford: Blackwell, 1984), pp. 35–67.

40. O. Ulbricht (ed.), *Von Huren und Rabenmüttern: weibliche Kriminalität in der Frühen Neuzeit* (Cologne: Böhlau, 1995).

41. E. Biesel, 'Hexerei und andere Verbrechen: Gerictspraxis in der Stadt Toul um 1570–1630', in Eiden and Voltmer, *Hexenprozesse und Gerichtspraxis*, pp. 123–69; W. Monter, *Frontiers of Heresy: The Spanish Inquisition from the Basque Lands to Sicily* (Cambridge: Cambridge University Press, 1990), pp. 255–99; M. Gaskill, *Crime and Mentalities in Early Modern England* (Cambridge: Cambridge University Press, 2000).

42. J. A. Sharpe, 'Women, Witchcraft and the Legal Process', in J. Kermode and G. Walker (eds), *Women, Crime and the Courts in Early Modern England* (Chapel Hill: University of North Carolina Press, 1994), pp. 106–24.

43. G. Fiume, 'The Old Vinegar Lady, or the Judicial Modernization of the Crime of Witchcraft', in E. Muir and G. Ruggiero (eds), *History from Crime* (Baltimore: Johns Hopkins University Press, 1994), pp. 65–87.

44. S. Loriga, 'A Secret to Kill the King: Magic and Protection in Piedmont in the Early Eighteenth Century', in Muir and Ruggiero, *History from Crime*, pp. 88–109.

45. E. W. Monter and J. A. Tedeschi, 'Toward a Statistical Profile of the Italian Inquisitions, Sixteenth to Eighteenth Centuries', in G. Henningsen and J. Tedeschi (eds), *The Inquisition in Early Modern Europe* (Dekalb, Ill: Northern Illinois University Press, 1986), pp. 130–57.

46. Reginald Scot, *The Discoverie of Witchcraft* (London, 1584), pp. 32–9.

47. Soili-Maria Olli, 'The Devil's Pact: A Male Strategy', in O. Davies and W. de Blécourt (eds), *Beyond the Witch Trials: Witchcraft and Magic in Enlightenment Europe* (Manchester: Manchester University Press, 2004); Mandrou, *Magistrats et sorciers*; M. Nenonen, *Noituus, taikuus ja noitavainot: Ala-Satakunnan, Pohjois-Pohjanmaan ja Viipurin Karjalan maaseudulla vuosina 1620–1700* ['Witchcraft, Magic and Witch Trials in Rural Lower Satakunta, Northern Ostrobothnia and Viipuri Carelia, 1620–1700'] (Helsinki: SHS, 1992).

48. Ian Bostridge, *Witchcraft and its Transformations, c.1650–c.1750* (Oxford: Clarendon Press, 1997), pp. 203–31.

49. U. Krampl, 'When Witches Became False: Séducteurs and Crédules Confront the Paris Police in at the Beginning of the Eighteenth Century', in K. A. Edwards (ed.), *Werewolves, Witches and Wandering Spirits: Traditional Belief and Folklore in Early Modern Europe* (Kirksville, Mo: Truman State University Press, 2002), pp. 137–54.

50. O. Davies, *Witchcraft, Magic and Culture, 1736–1951* (Manchester: Manchester University Press, 1999).

51. Bostridge, *Witchcraft and its Transformations*; Ian Bostridge 'Witchcraft Repealed', in Barry, Hester and Roberts, *Witchcraft in Early Modern Europe*, pp. 309–34.

52. On this literature see B. P. Levack, 'State-Building and Witch-Hunting in Early Modern Europe'; R. Walinski-Kiehl, 'Godly States: Confessional Conflict and Witch-Hunting in Early Modern Germany', *Mentalité-Mentalities* 5 (1988) 13–24.

53. The state can be viewed as a political community, a set of arrangements for rule or an abstraction to which loyalty is owed. I define it as a formal, public and autonomous political organisation, staffed by officials who have the legally sanctioned authority to require obedience from the inhabitants of a specific territory over an extended period of time.

54. M. Kunze, *Der Prozess Pappenheimer* (Elsbach: Gremer, 1981); Kunze, *High Road to the Stake* (Chicago: University of Chicago Press, 1982).

55. R. Voltmer, 'Hexenprozesse und Hochgerichte: zur herrschaftlich-politischen Nutzung und Instrumentalisierung von Hexenverfolgungen', in Eiden and Voltmer (eds), *Hexenprozesse und Gerichtspraxis*, pp. 475–525.

56. W. Behringer, *Witchcraft Prosecutions in Bavaria: Popular Magic, Religious Zealotry and Reason of State in Early Modern Europe* (Cambridge: Cambridge University Press, 1997).

57. R. Muchembled, *Popular Culture and Elite Culture in France, 1400–1750* (Baton Rouge: Louisiana State University Press, 1985), esp. chapter 5.

58. C. Larner, *Enemies of God: The Witch-Hunt in Scotland* (Baltimore: Johns Hopkins University Press, 1981); Larner, *Witchcraft and Religion*, esp. chapter 2.

59. J. Goodare, *State and Society in Early Modern Scotland* (Oxford: Oxford University Press, 1999).

60. B. P. Levack, 'The Decline and End of Scottish Witch-Hunting', in J. Goodare (ed.), *The Scottish Witch-Hunt in Context* (Manchester: Manchester University Press, 2002), pp. 166–82.

61. See, for example, S. Ellis, 'Witch-Hunting in Central Madagascar, 1828–1861', *Past & Present* 175 (2002) 90–123; W. Behringer, *Witches and Witch-Hunts: A Global History* (Cambridge: Polity Press, 2004), chapter 6.

62. R. D. Waller, 'Witchcraft and Colonial Law in Kenya', *Past & Present* 180 (2003) 241–75; J. Hund (ed.), *Witchcraft, Violence and the Law in South Africa* (Pretoria: Protea Book House, 2003).

63. C. Fisiy, 'Containing Occult Practices; Witchcraft Trials in Cameroon', *African Studies Review* 41 (1998) 148–51.

10
thinking witchcraft:
language, literature and intellectual history

marion gibson

In the late 1970s and 1980s, a range of new and old ideas began to come together to change the way that some historians read and wrote the history of witchcraft. But these ideas did not originate exclusively within history departments. Rather, historians borrowed them from linguists, philosophers and literary and cultural critics, and in doing so changed the ways in which they could choose to see their own discipline – a process known as the 'linguistic turn'. The history of witchcraft – along with many other histories – became a far more multidisciplinary affair. To the existing contributions from anthropologists were added new ideas from scholars who had been used to writing about Shakespeare, the development of language or the politics of popular culture.

Perhaps the most prominent figure on the history-linguistics boundary in the 1980s was the American literary scholar Stephen Greenblatt.[1] Greenblatt, and others who thought as he did, came to be known as New Historicists. Like the original historicists – literary critics of the 1940s such as E. M. W. Tillyard – New Historicists were interested in the ways in which political and social history impacted on the creation of contemporary literature, especially the literature of the early modern period. But unlike earlier historicists, their work was informed by a range of modern theories that had contributed to the concept of postmodernism. They were not interested as much as earlier historicist scholars in describing the core beliefs with which all educated Elizabethans were supposedly indoctrinated – such as the Chain of Being from God to the lowest aspects of creation – and from this extrapolating a 'world picture' that Shakespeare, Marlowe and other writers had probably shared.[2] Instead, they focused on what could be inferred from the more ephemeral texts of Shakespeare and Marlowe's time – letters, journals, sermons, accounts of topical events. What happened if one juxtaposed these texts with the canonical works of English literature? Suppose a scholar were to juxtapose a private letter giving an account of the New World with *The Tempest*, or an account of Catholic exorcisms with *King Lear*? To put it simply, these scholars hoped to expose debates, processes and concerns – open questions and niggles rather than truths – that were hidden in the less obvious by-ways of culture, rather than trumpeted in every Elizabethan schoolroom.

New Historicism was built on the foundations of post-structuralism with a dash of Marxism and on the work of five twentieth-century thinkers in particular: Ferdinand de Saussure, Jacques Derrida, Roland Barthes, Louis Althusser and Michel Foucault.[3] The first, Saussure (1857–1913), was a Swiss linguist. He argued that language was based on a system of difference: to take a simple example, the word 'cat' was so because it was *not* 'mat'. There was no essential connection between the word 'cat' and the domestic animal that it denoted in English. The signifier (or word) had no natural affinity with what it signified (the idea of the named object). The word's meaning was thus not inherent in the signifier, but made by the speaker and listener who linked signifier and signified to produce a 'sign' that meant something (the referent, the real cat). For many later writers, such as the French philosopher Derrida (1930–2004), Saussure's work opened up a whole range of possibilities previously unnoticed in discussions of culture. Nothing could be taken for granted, for much of what seemed natural was in fact arbitrary. Meanings had to be learned – and so they could be unlearned. The identity of a thing (or person or concept) depended on negatives and absence, upon what it was not, rather than any positive essence. Words thus brought ideas and realities into being, rather than vice versa, and language shaped perception, rather than simply describing what was perceived. Some writers began to discuss what this meant for supposedly natural signifiers like 'patriotism' or 'madness'. One of these writers was the French critic Roland Barthes.

Barthes (1915–1980) was interested in anything that could be seen as a text intended to be read (and thus interpreted) by its observers, which included written and visual representations, performances and public events. In particular, he focused on the way that the least-analysed images and events often contained powerful (and now, it seemed, arbitrary and challengeable) significances, which themselves signified a whole range of other assumptions and codes. Even trivial-seeming objects or occurrences – a wig, a wrestling match – were subtly indoctrinating or 'interpellating' those who saw or read about them. Barthes drew on the theories of the French Marxist philosopher Althusser (1918–1990), who had argued that people were constantly being subjected to a coercive pressure to conform to existing societal norms and structures, which directly or indirectly served 'the state'. Nobody was able to escape from this pressure of 'ideology', because it was reinforced by every learned word or thought. Whatever one believed, even if one rebelled, there was no position that had not been colonised already by the forestalling powers-that-be, and already responded to in ways that disempowered it. But, like Derrida, Barthes believed that if a reader determinedly pulled apart a text, he or she would at least be able to see through its attempt to force upon the reader its message. Interpellation could thus be resisted.

The French historian Michel Foucault (1926–1984) brought together all these ideas in the form that would finally reach the New Historicists and historians of witchcraft. He wrote a series of books on themes that he believed were central to modern society, tracing their origins and deployment as political forces. Foucault's theme was power in its many forms, and especially the power to define and control by naming. Once armed with the ability to perceive the interpellating tricks of language, he decided that it was possible to write a history of the most apparently natural notions about the world and humankind. One had only to study the origins of their definition, the isolating and naming of some states or actions as wrong and others as right. Foucault thus embarked on histories of the definition of sexuality, madness and criminality. Each of these definitions of acceptability appeared to be instinctive, or reasonably defensible as a natural response to the way the world was and is. But each could be shown, said Foucault, to have a traceable origin in the fairly recent past. When was it decided, for example, that the behaviour that twentieth-century western people usually regard as schizophrenic was not attributable to angelic voices and part of a higher wisdom? When did it become madness, or unreason – the Saussurean negative, or opposite, of reason – to be feared and stamped out? Ironically for those who found his work most interesting, Foucault tended to date the inception of most modern forms of control – such as imprisonment or institutionalisation – to the eighteenth century. But for Renaissance scholars like Greenblatt, his ideas resonated in the sixteenth and seventeenth centuries too.

And so, by the mid 1980s literary scholars and those who became interested in their work had a new set of theoretical tools for analysing the world. New Historicists did not actually see themselves as embodying a coherent theoretical position, but rather a series of concerns and 'reading practices'. Some were particularly fascinated by the idea of binary oppositions: the rigid categories of good and evil, masculine and feminine, Protestant and Catholic that structured many Renaissance texts. These oppositions need not now be seen as natural at all – rather, they were created by the way that language works. Like Foucault's 'reason' and 'madness', each pairing contained a privileged term (good, masculine, Protestant) and one that was often imagined as its stigmatised opposite (evil, feminine, Catholic). So in Edmund Spenser's epic poem *The Faerie Queene* (1590) holiness was portrayed as a manly knight learning the ways of Protestant virtue, whilst one of his most dangerous foes was the wicked Catholic female, Duessa. She misled the knight into sin, but eventually she was exposed and defeated and the knight rode on to victory over the Devil. Thus – to summarise crudely – the text had used one half of a binary opposite to create a subversive danger, and then suppressed it using the other half.[4] From this, the early modern reader was supposed to learn what was good, what was evil, and to act accordingly.

In some ways, New Historicism was exactly the same as old historicism in that it ended up producing a series of scholarly orthodoxies about early modern culture.[5] Two of these are demonstrated in the example above: that many early modern texts (and events) depended on the arousal and alleviation of anxieties, and that they both generated and contained their own subversions. Readers might, for example, be expected to react to Duessa with fear and unease, anxiously noting the supposedly seductive and deceitful power of Catholicism. But Duessa's defeat demonstrated the superiority and strength of Protestant belief, and showed that it was a fairly simple matter to detect Catholic lies and errors if one followed the correct path. Thus readers had been given a salutary experience of what they must fear and hate, followed by a reassurance. Duessa was also a witch, and so the text suggested particularly clearly to New Historicist readers that it was creating a subversive threat where actually none existed. Her magic power was a metaphor for the lure of Catholic beliefs and for the power of women as temptresses, and it was described by Spenser with images that were compellingly fearful. But this subversive potency was illusory – Duessa and her colleague, the magician Archimago, simply fled into textual oblivion once challenged by the forces of control and containment. Like 'real' witches, they possessed no effective magic powers against good. They were there as bogeymen, spectres raised only to be driven out by the interpellating power of the text.

This process, the generation and containment of subversion, was of great interest to New Historicists. In particular, they were keen to examine the way in which those who had created a threat, or dwelt fearfully on an existing one, were fascinated by the monster that they described. Images of lurid grotesquerie, the darkest imaginings of the mind, were embodied with unexpectedly joyous creativity in figures like Duessa and Archimago (or Marlowe's eponymous hero Doctor Faustus, or Shakespeare's Sycorax in *The Tempest* or Middleton's Hecate in *The Witch*). These figures were given power to rampage through the ordered decencies of early modern life. No matter how carefully each was forced back into its hellmouth or wilderness or cave, no matter how thoroughly dead was each threat, the monsters still lingered in the imagination. Really, that was their purpose – they were created to titillate and chill, and each expressed a compelling need for its creator and readers. And if witches in fiction fulfilled such a need, what of witches in the texts of demonologists, preachers and natural philosophers?

It was to these that British historian Stuart Clark turned in the late 1970s. He began to think, as he said in his article 'Inversion, Misrule and the Meaning of Witchcraft', that: 'we no longer readily understand the language of early modern witchcraft beliefs'.[6] The phrase chosen – 'the *language* of ... witchcraft beliefs' – was striking, because it did not presuppose that scholars could have any direct understanding of witchcraft and beliefs about it at all. Instead, as for Saussure and his philosophical descendants, language stood between the scholar and the events and beliefs under scrutiny. In order to have any access to these beliefs, readers of texts about witchcraft had to crack the language code, as the New Historicists were trying to do. Thus the assumptions and obsessions of the New Historicists reached witchcraft studies – in advance of many of the seminal New Historicist publications, in fact. Witches seemed to be the ultimate example of a threat created by a society which had an unconscious need of one, manifested through language's emphasis on contrasts and negatives. As the literary scholar Gareth Roberts, who had long been a student of Spenser and of Renaissance witches, noted:

> if we follow Stuart Clark's lead and talk of 'the language of early modern witchcraft beliefs' then one can say that in terms of Saussure's theory of the linguistic sign we can deal with the signifier and the signified, but much research admits to severe problems about witchcraft's referent.[7]

The language of witchcraft was unreadable, because signifier and signified combined to produce a sign that – to the modern mind – meant nothing real at all.

As Clark said, to the reader struggling with such lack of obvious reference,

> demonological classics like the *Malleus Maleficarum* (1486–7) or Jean Bodin's *De la demonomanie des sorciers* (1580) seem to reveal only an arcane wisdom. It is not apparent what criteria of rationality are involved, nor how the exegesis of authorities or use of evidence support the required burden of proof. Since individual steps in the argument are difficult to construe, its overall configuration often remains impenetrable.[8]

Demonological texts tended to be read, therefore, as if there was something wrong with them. To the modern eye – even that of the trained historian – the authorities cited by such writers as Kramer and Sprenger seemed irrelevant and poorly deployed. Even when one could understand the reliance on biblical wisdom and classical authorities, it still seemed astounding that the *Malleus* could adduce as evidence of women's inherent wickedness the story of

> a man whose wife was drowned in a river, who, when he was searching for the body to take it out of the water, walked up the stream. And when he was asked why, since heavy bodies do not rise but fall, he was searching against the current of the river, he answered: 'When that woman was alive she always, both in word and deed, went contrary to my commands; therefore

I am searching in the contrary direction in case even now she is dead she may preserve her contrary disposition.'[9]

As if marking a poor student essay, scholars had been unable to see the relevance of this argument and had responded with the exasperated commentary summarised by Clark. The arguments offered by demonologists not only did not support the weight of the conclusions, which had led to so much death and misery, but they could never have supported it. Why, then, were they being used? What was the purpose of creating a 'hammer of witches' out of such material? It was almost as inexplicable to professors as it was to their students that such writers as Kramer and Sprenger were taken seriously in their own times.

Yet there was also a reluctance to regard demonologists as 'deluded'. This was an 'Enlightenment' position of the eighteenth and nineteenth centuries, with which modern scholars also felt uncomfortable. Such texts could not simply be dismissed as irrational because they came from a world with a different outlook and mental landscape. This would be anachronistic and reductive. But what else could be done with them? One answer was to look for other factors to explain their arguments. A popular choice was the misogyny of many demonological writers, which seemed self-evident to a twentieth-century reader, especially the feminists of the 1970s and thereafter. Perhaps witch hunting was not actually witch hunting at all, but woman hunting and woman-hating. This made such apparently feeble attempts at rationality as the story of the drowned woman more explicable: they were not produced as reasonable evidence, but were instead outpourings of uncontrollable hatred and fear. Men of the early modern period were not really looking for witches to prosecute – they were seeking women to persecute. This argument was advanced with varying degrees of subtlety, and clearly it was necessary to account for the fact that in most witch-prosecuting areas, women were far more likely to be accused than were men. Some writers argued simply that, as Anne Llewellyn Barstow put it,

> witch charges may have been used to get rid of indigent elderly women, past childbearing and too enfeebled to do productive work ... witchcraft accusations often served as a cover for other problems.[10]

Others argued that witchcraft accusation was not woman-hunting as such, or cynically instigated for its functional value. But it might be 'sex-related' because accusers were persecuting certain traits associated with notions of femininity.[11] Christina Larner's firm statement, in making this argument, that 'the prime interest of the authorities at the time was the pursuit of witches as such' came closest to Clark's increasing conviction that witchcraft should be studied in its own right and not as a by-product of misogyny.[12] But her argument was still offering a discrete explanation for such an interest, and was open to challenge and refinement by those who wanted to emphasise that explanation. So, for Marianne Hester, witch hunting was 'sex-specific', 'one means of maintaining and reconstructing male dominance and male power'.[13]

Another, more clearly 'functionalist', approach was to look for economic factors in the lives of accused witches and their accusers.[14] Accused witches had often, it was noticed, begged an item or help from their accusers, and been turned away, before the accuser had then suffered a misfortune for which he or she blamed the accused. Perhaps accusers were persecuting the poor, rather than prosecuting witches as such? Keith Thomas and Alan Macfarlane argued this point, with such a barrage of rich anecdote, scholarship and statistic that it became the orthodox reading after Thomas' *Religion and the Decline of Magic* was published in 1971. Yet this too was an approach suggesting that witchcraft accusations were the result of something other than a straightforward conviction on the accuser's part that a particular woman or man was, quite simply, a witch.

> Witch-beliefs ... upheld the conventions of charity and neighbourliness, but once these conventions had broken down they justified the breach and made it possible for the uncharitable to divert attention from their own guilt by focusing attention on that of the witch. Meanwhile, she would be deterred from knocking at any more unfriendly doors ... [15]

Witchcraft accusation had a use, and in concentrating on discovering this functionality, scholars were not analysing it with the purity of focus that Clark sought. But it was the idea that witchcraft persecution could be viewed solely as part of a longer history of persecution to which he perhaps took most exception. In *Europe's Inner Demons* (1975), Norman Cohn had argued that in the later Middle Ages and Renaissance witches simply took the place of other traditional victims of persecution – lepers, Jews and heretics.[16] As he said in his preface, he believed that the witch hunt was based on 'the urge to purify the world through the annihilation of some category of human beings imagined as agents of corruption and incarnations of evil.'[17] Cohn saw this desire as a universal, unchanging urge, manifested most recently in the twentieth century by the Nazis. But Clark found this reading too little focused on the specific features of witchcraft itself. Like all the other explanations, it seemed to him to evade the real issue, the one to which Christina Larner gestured when she said that maybe 'witch-hunting was actually witch-hunting'.[18]

At least some of the community of scholars working on witchcraft were ready for such a view to be taken forward, and Clark's article 'Inversion, Misrule and the Meaning of Witchcraft', published in *Past & Present* in 1980, began to do so. In the years of debate after *Religion and the Decline of Magic*, there were so many explanations of why people were believed to be witches and prosecuted accordingly that Robin Briggs was able to title an article 'Many Reasons Why: Witchcraft and the Problem of Multiple Explanation'. It was not that no one believed they knew the explanation for the witch hunts – the problem was that there were so many plausible, and sometimes contradictory, explanations that Briggs despaired. It seemed to him, as he reflected in the published proceedings of a conference on the cultural contexts of the European witch hunts held at Exeter University, UK, in 1991, that the only result of 'a golden period for the historical study of witchcraft' had been 'ever-increasing complexity'.[19] All the explanatory arguments had their merits, and some were even implicitly relying on the theory of binary oppositions in which Stuart Clark had become interested. But in Clark's view, such approaches in tracing learned witchcraft beliefs 'to the periodic social need to relocate moral and cultural boundaries by means of accusations of deviance, or ... to the neuroses which are said to accompany the repression of erotic or irreligious impulses in devout minds' were 'bypassing the problem of their meaning by reducing them to epiphenomena', by-products of other trends or factors.[20] Witchcraft beliefs were not, he argued, being studied in their own right. They were instead being regarded as dependent on other, more solidly 'real', phenomena. Witchcraft not 'about' itself, in this reading. Each theory missed the point, because none took the language of the demonologists and accusers at its word, and then investigated what the word, and its relationships with other words, actually meant.

Clark wanted to try to read the works of demonologists in something approaching their own terms. He believed that other philosophical and cultural strands influenced and interwove with their thinking on witches, and just as New Historicists sought to illuminate one area of culture by looking at another, so Clark began to look for the big ideas that held together early modern thinking on subjects as diverse as science and politics. One of the first strands of imagery that he came across was inversion. Witches did everything backwards, as demonology and fiction agreed. So what else was done backwards in the Renaissance? Clark examined festivals of misrule, where scholars ruled over masters for a day, or priests used smelly shoes instead of incense in holiday mockery of the usual rites. What did misrule mean? Probably it had certain social functions, since saturnalian riot and carnival had traditionally been seen as acting as a safety valve for the everyday world of submission and deference, enabling order to be restored afterwards with more general assent. But, in a purer linguistic reading, misrule related exclusively to rule, its binary opposite. It need not have any relevance at all to the actual, physical world because linguistically it made complete sense. The signifier and signified, sign and referent 'misrule' existed because 'rule' did, and 'rule' existed because 'misrule' did, too. Misrule might be socially necessary, but it was also inevitably produced by language structures regardless of any social usefulness.

One of the most obvious examples of misrule and inversion, the witches' Sabbat so often described by demonologists, made much more sense to Clark if viewed in this way. Descriptions of Sabbats and what was said to go on at them were purely and simply a form of structural opposition in a world where it was axiomatic that binary oppositions existed. They made sense as the 'world turned upside down' where this idea had existing cultural currency; and they implied right rule and correct order by their performance. By the conventions of early modern discourse, it was inevitable that Sabbats existed, since church services and godly behaviour did. No one need actually ever see or attend one – indeed, almost certainly such gatherings did not exist – but neither were Sabbats merely a delusion or fantasy. As Clark explained:

> What it made sense for demonologists to say depended partly on traditional metaphysical notions about the logical shape and moral economy of the world, and partly on shared linguistic patterns for describing its most disturbing aspects.

The Sabbat was 'evil for the sake of structural coherence' and 'inversion (both in forms of thought and forms of words) to ensure linguistic felicity'. Clark became so convinced by this view that he felt it became 'difficult to explain, not how men accepted the rationality of the arguments [of demonologists], but how, occasionally, sceptics doubted it'.[21]

This was a controversial argument, and not to everybody's taste. It demanded an easily exaggerated cultural shift of its adherents: from ancient to faddishly postmodern, from solid fact to abstract, from reality to theory and fiction. It, and other similar works of the linguistic turn, opened up a debate about the discipline of history that had been raging quietly for centuries. In the 1570s, Philip Sidney had in his *Defence of Poesy* teased the precursors of modern historians with the assertion that, unlike poets and philosophers, they lacked imagination. They were, firstly, too trustful, in that they overlooked the fictions in their archives. He caricatured 'the historian ... laden with old mouse-eaten records, authorising himself (for the most part) upon other histories, whose greatest authorities are built upon the notable foundation of hearsay'. Historians could only work by ignoring that 'hearsay', Sidney sniped. Then, once falsely convinced that they had established the 'truth', they ploddingly proceeded to recount it baldly. Secondly, therefore, their works consisted of a flatly factual narration, a 'bare "was"' which brought no real enlightenment. In this, historians seemed to Sidney to be uninterested in the true concern of the intellectual, ideas: they were 'tied ... to the particular truth of things, and not to the general reason of things'. Historians, he thought, were anti-theoretical and therefore anti-intellectual in important respects.[22] Four hundred years after Sidney's witty barbs it sometimes seemed that nothing had changed.

The terms of reference may have been vastly different, but the rhetorical positions were much the same, each with its traditional baggage. In a thoroughly postmodern yet oddly familiar way, Stuart Clark suggested that historians like Cohn were hampering themselves by their reluctance to accept the 'manifestly impossible' elements in texts about witchcraft. They ignored these simply by adopting the position that such demonology was 'an intellectual fantasy'. As such, it needed explanation through 'an alternative socio-psychological causation', rather than exploration in its own right.[23] Clinging to the detailed search for causation, implied Clark, Cohn and others missed the point about the self-referentiality of intellectual discourse, the power of fiction and the world of the word. Yet the ideas of the linguistic turn seemed to some to depend upon a willingness to jettison what they saw as the traditional virtues of historical scholarship: the commitment to a search for truth, in particular. Robin Briggs protested against 'the fashionable move ... an appeal to some aspect of post-modernist theory'. He regarded such theories as 'remote' from the details of witch trials and, moreover, as anachronistic 'condescension'. Briggs preferred 'the more traditional assumption that there was a past reality ... which historians are best able to cope with in a relatively artisanal fashion', having, he said, himself 'spent many years working on one of the richest archives in Europe'.[24]

It was fascinating that the disputes between scholars over the nature of history 400 years after Sidney were expressed in exactly the same traditionally insulting stereotypes. Once again, a wilful blindness to fiction was ascribed to traditional historians by admirers of the linguistic turn, along with its old companion, a refusal to explore ideas. In response there was its inverse from those who were not for turning: a proud difference in emphasis between factual 'archives' and fictional 'texts', between particular and general, and finally, an unbridgeable cultural gap between the condescending and remote intellectual and the studious artisan. The re-emergent Sidneyan stereotypes did not do justice to the historians involved, but they did point to the fact that Stuart Clark was an unusual kind of historian. The poet and philosopher too seemed to have some share in his work, which in 'Inversion, Misrule and the Meaning of Witchcraft' moved easily from Jean Bodin's demonology to Ovid's *Medea* and Ben Jonson's masques. As the dates of several of the 'functionalist' works cited above will demonstrate, after 'Inversion, Misrule', many historians continued to write perfectly acceptable traditional histories of witches as women, the poor or those persecuted for the sake of persecution.

Clark's major work, published in 1997, was *Thinking with Demons*, and it was harder to ignore.[25] The title was in itself provocative – the idea of demons was set alongside that of rational thought, as Clark offered a reinterpretation of Cohn's title *Europe's Inner Demons* which suggested that demonology was far from unthinkable, and was not a by-product of neurosis at all. The enormous book – over 800 pages – considered precisely the manifest impossibilities that populated the works of early modern intellectuals. It was not even particularly interested in them as 'demonologists', for one of Clark's arguments was that ideas outside the supposed remit of demonology were in fact profoundly connected with it. The book was thus divided into sections whose titles pointed in other directions: Science, History, Religion, Politics.

But before these was a section on 'Witchcraft and Language', and this was the heart of Clark's project, begun with 'Inversion, Misrule and the Meaning of Witchcraft'. This time, Clark's assertion was stronger – not just that we no longer readily understood the language of witchcraft beliefs, but that 'to make any kind of sense of the witchcraft beliefs of the past we need to begin with language'. Clark asserted that the linguistic turn confronted all historians, whether they liked it or not, but noted that it was hard for many to accept.[26]

Once again, he was concerned with linguistic structures – binaries, contrariety, inversion and parody. Clark used explicitly literary terms, and chose words that taken together were more often associated with literary criticism than traditional historical writing: 'anti-realism', 'idiom', 'difference', 'metaphor' and 'text'.[27] He also threw in a few New Historicist-sounding wordplays: *Thinking with Demons'* chapter 7 is entitled 'Witchcraft and Wit-Craft'. In chapter 8, 'Women and Witchcraft', Clark began by quoting George Puttenham's *Arte of English Poesie*, and including one of Puttenham's verses. He was looking for a 'poetics' of witchcraft.[28] He used Puttenham's theorising on the subject of poetic art to begin an argument that witch hunting was not conducted by woman-haters, but was a result of the natural tendencies of language towards 'deliberately juxtaposed opposites'. Puttenham was discussing rhetorical figures of speech and writing, and for his illustration of the figure *contentio* he wrote a poem about men and women:

> My neighbour hath a wife, not fit to make him thrive,
> But good to kill a quicke man, or make a dead revive.
> So shrewd is she for God, so cunning and so wise,
> To counter with her goodman, and all by contraries.[29]

As Clark points out, this passage echoes many descriptions of witches, with their inherent rebellious contrariety. It is about opposites, and it is expressed in structures and terms that are antithetical.

Using Puttenham's poetic structure of antithesis, Clark proceeded to take up Christina Larner's argument that witchcraft was 'sex-related' not 'sex-specific'. He argued partly from statistical analysis of the demonologists whose texts he had read, and partly from a wider theoretical reading of language itself. Firstly, he said that demonologists – and he had read hundreds of them: Bodin, Crespet, Perkins, Elich, Samson, Benoist, Delrio, Thumm, Hocker, de Lancre, Hemmingsen, Birette, Serclier, Maldonado, Torreblanca, Cooper, Witekind, Holland, and many, many more – were often far less insistent that non-demonological writers in putting forward women's faults. They were not, as a group, especially obsessed with female sexuality either. It was those who were sceptical about witchcraft, such as Johann Weyer in his 1583 *De praestigiis daemonum*, who spoke at length about women's bodily and mental frailties, especially the melancholia thought to be brought on by the menopause. Demonologists were only averagely misogynistic, in fact. How, then, to explain why prosecuted witches were so often women? For Clark, this was – in a sense – once again missing the point. The prosecution of women as women was not, for him, the central issue of the witch hunts. What was important were the reasons women were *thought* to be witches. Their actual femaleness was the epiphenomenon here. The real phenomenon was language, and the way it constructed superior and inferior, good and bad, masculine and feminine. He explained:

> the gendering of witchcraft ... turns out to be reliant on the binary thinking we examined in earlier chapters. We would expect systems of dual symbolic classification, where they obtain, to embrace the categories of gender, and in every instance this seems to be the case ... Whatever its influence over representational systems as a whole, the gender relation is hierarchically weighted so that, once the processes of interchangeability, reinforcement and correlation have had their effect, men are symbolically associated with a range of other positive items and categories, and women with their negative counterparts.[30]

Once men began to think of the women that they knew as 'Woman', it was a short step – linguistically-speaking – to thinking of them as potential witches. Good women had certain attributes, whilst bad, anomalous, women were shrews, dominant wives or female rulers. 'Witchcraft took its place ... as another obvious example of feminine deviance – a deviance that could only take inversionary forms', said Clark.[31] He explained that what he was offering was not 'an account of why witches were "in fact" women', as many witches were in fact men, but 'an account of why (from a particular male cultural perspective) they were *conceived* to be women'. He concluded his chapter by stating that 'the conceptual link between witchcraft and highly anomalous women was provided by the symmetries of inversion'.[32] It was the keynote of *Thinking with Demons* and of Clark's contribution to the study of witchcraft.

Clark's work and that of the 'school' of which he is part has turned attention increasingly to writing and reading about witchcraft. There is a growing number of scholars who have examined the representation of witches, magicians and magical creatures in overtly literary works. Mainstream and minor early modern dramatists such as Shakespeare, Jonson, Middleton, Lyly and Heywood all chose at some point in their career to write about witches, and Diane Purkiss, in her 1996 study *The Witch in History*, examined Elizabethan and Jacobean plays such as Shakespeare's *Henry VI* and *Macbeth*, Middleton's *The Witch* and John Ford, Thomas Dekker and William Rowley's *The Witch of Edmonton*, as well as other kinds of staging such as Jonson's *Masque of Queenes*. As a literary scholar, Purkiss saw her role not as uncovering 'the occulted "truth" about witches' but as assembling evidence 'of a number of different investments in the figure of the witch, trying to tell or retell the rich variety of stories about her and to analyse the way those stories work'.[33] Purkiss presented herself, then, as a reader like Clark. But even more so than Clark, she saw herself as an outsider. She was from a different disciplinary background – English literature – and she proposed to launch two savage assaults on the traditional expositors of the meaning of witchcraft: radical feminists and historians. Both of these groups, she thought, had misread the witch and tried to use their interpretations of stories about witches to further their own dubious ends. American radical feminists had invented, in Purkiss' acid phrase, 'a holocaust of one's own' for themselves, wallowing in historically inaccurate self-pity. Meanwhile, in tenured comfort, British male scholars had been engaging in 'academic self-fashioning', advancing their careers within the established academy by attacking the easier targets of feminist scholarship. Yet, she argued, they had made no attempt to give serious attention themselves to women's legitimate concerns about oppression, and they had not kept abreast of the exciting scholarly developments coming from more radical scholars. The traditional historian, in Purkiss' book, was 'deaf to the methodological questions raised inside and outside history in the past twenty years' and demonstrated a woeful 'reluctance to engage with theory'.[34]

Despite her praise of the work of some of these historians – notably Keith Thomas – as possessing 'richness' and a 'still-amazing breadth of ... knowledge', what remains in the reader's mind from Purkiss' analysis is her invective. Where Clark had dissented from his colleagues' views almost entirely without rancour, Purkiss dismissed the traditional historians with words like 'narcissistic', 'casual' and 'naive'.[35] Yet she and Clark were making an essentially similar point: it was important to read texts about witches *as* texts. Purkiss in particular was inclined to follow scholars like Natalie Zemon Davies, Annabel Gregory and Robert Darnton in seeing 'popular texts as stories', and then taking their insights further.[36] She did not discuss demonology, as Clark did, but looked instead at what witches and accusers said. She applied to their words theories of agency, self-fashioning and authorship drawn from literary studies, and psychoanalytic and feminist theories from the wider philosophical world into which literary scholars often dipped. What if the witch was accorded the role of author in texts said to have spoken by her (and Purkiss was interested primarily in female witches), such as stories of how she became a witch? Purkiss particularly admired the work of Lyndal Roper on German witches, and two other scholars who shared her approach were Deborah Willis and Frances Dolan.[37] Using psychoanalytic theory, Willis studied witches as mothers in pamphlet accounts and in drama in her book *Malevolent Nurture*, whilst Dolan looked at female murderers, infanticide and petty treason as well as witchcraft in her *Dangerous Familiars*. These scholars all shared a feminist agenda, and all used literary techniques and the tools of linguistic analysis. They focused on printed texts rather than archival records, and by doing so they set witchcraft in a wider cultural context of debates about literature and culture just as Clark hoped to do.

Some American scholars also began to apply these readings to their witchcraft history. Bernard Rosenthal's *Salem Story* of 1993 argued that the myths and fictions of witchcraft events needed to be studied alongside actual trial records to gain a true understanding of what witchcraft meant to America's past and present citizens. Rosenthal saw some texts as more accurate than others, and in this sense remained within the traditional framework of empiricist history. But he also recognised clearly that in many ways untrue stories of witches offered more insight into the ideological world of the seventeenth century than did the supposed realities. Hence he chose the subtitle *Reading the Witch Trials of 1692* for his book.[38] Rosenthal's most influential chapter, 'Dark Eve', revisited the events surrounding Tituba, the 'black witch' of Salem, noting, like Chadwick Hansen before him, the different racial identities ascribed to her by different commentators, creative writers and historians.[39] Each carried its own cultural baggage: Tituba appeared as the ignorant 'negro' slave and the the dangerous voodoo priestess, as well as the 'Indian' herbalist. Finally, for Rosenthal, Tituba could be read as an archetypal alien, the 'dark Eve' who had brought the American fall. Tituba was being blamed for transforming America, in a pivotal cultural moment, from an idyllic land of freedom and harmony to a society in crisis, riven by racial, gendered and religious tension. Rosenthal arrived at this conclusion by reading Tituba in some of the terms that Stuart Clark might have chosen: binary opposition, inversion and misrule. He also, like other literary scholars, allowed creative writers such as Arthur Miller to assume the same status as commentators as traditional economic and social historians, such as Paul Boyer and Stephen Nissenbaum.[40] As a result of work of this kind, all scholars are now more likely to be sensitised to witchcraft's place in broader cultural history and cultural poetics.

If witchcraft might be seen as a phenomenon susceptible to literary analysis, it could also be seen in terms of the generic demands made by the specific situations in which stories of witchcraft were likely to be told. When Stuart Clark organised a conference on witchcraft at the University of Wales, Swansea, in 1998, two papers dealt with this aspect of the linguistic turn. Peter Rushton pointed out that historians 'have always relied upon stories from the past for their sources', and that now their production and reception had become an important focus.[41] Accordingly, he examined records of testimony from courtroom situations, where stories of several particular types were likely to be required. He looked in particular at accusatory stories and asked how both the content and persuasive elements within these stories of witchcraft worked to secure the result that the teller needed to win the case. In this he was building on Natalie Zemon Davies' work on tales designed to secure a pardon. Meanwhile, in her paper Marion Gibson discussed the structuring of stories of witchcraft in news pamphlets. Some of these were originally told before magistrates or in Assize courts, whilst others were produced as independent narratives. Gibson argued that stories from different sources were produced for different reasons, and they tended to portray accuser and witch very differently. Stories told by accusers in pre-trial examinations, for example, tended to establish a motive for the witch to attack her or his accuser by explaining how the accuser had behaved badly to the witch and so brought about the attack. But stories created away from a courtroom situation, where a prosecution had often already occurred and the story did not need to provide a plausible motive for the alleged crime, were different. These stories tended to deflect blame from the accuser, either by portraying their relationship with the supposed witch as apparently amicable, or by minimising any injury done to the accused by the accuser. Often these stories were written by or for wealthy victims, who apparently wanted to look virtuous in print. In printed pamphlets produced over a 60-year period there were thus several changes in the genre of stories told about witches, and these seemed to depend on the material conditions that produced each account. Gibson summed up her belief that 'we cannot treat stories of witchcraft as "data" or "evidence" without enquiring whose stories we are reading and in what context they were told and recorded'.[42] Gibson expanded this case in her 1999 study *Reading Witchcraft*, and Clark edited a collection of the conference papers in 2001, commenting that

> in place of the social scientific approaches of previous decades, with their search for explanation – which, in effect, meant explaining witchcraft away – recent research has concentrated on witchcraft as a cultural phenomenon with a reality of its own.[43]

There are, of course, problems to be anticipated in seeing witchcraft as having 'a reality of its own'. If witchcraft accusation was merely a matter of linguistics, when and why did language structures change sufficiently to allow European and American societies to stop investing in witchcraft and demonology? Clark addressed this question in a postscript to *Thinking with Demons*. He suggested that witchcraft did not, in fact, lose its relationship with reality as was conventionally thought. There was no great realisation that it was all nonsense, and must be replaced. But the binary ways of thinking on which witchcraft accusations thrived were increasingly themselves seen to be unstable and partial. An increased cultural relativism meant that intellectuals began to analyse the way that discrimination (especially sectarian and sexual) worked, across politics, science, history and religion. It was not, therefore, ever necessary for early modern people to decide conclusively that witches did not exist, but only for them to note that the ways of describing and proving witches' existence were flawed. After that realisation, 'it faded from thought'.[44] Other questions asked of the linguistic turn generally have concerned its morality. In this case, what is the relationship between linguistic thought-worlds, and the lost realities of the actual men and women who were accused and executed as witches? Is it methodologically meaningful or ethically possible to stand back from the great and terrible events of the past or the present, positioning oneself only as a 'reader'? Clark himself was worried that in trying to understand the world of the demonologists, he might even be seen as 'excusing the inexcusable', defending those whom the Elizabethan sceptic Reginal Scot labelled as 'witchmongers'.[45] His response to this charge initially seemed aloof: it was 'not worth while to adopt any particular moral stance' on such past events as witchcraft prosecutions. But the final sentence of his preface to *Thinking with Demons* suggested a far more subversive 'lesson' of the book. Clark had said that he felt it more worth while to focus his moral energies on issues 'closer to home', and here he sketched the way in which an attention to the processes of language can offer a form of resistance to even the most assured of modern witchmongers: 'the purportedly most essential, objective, and timeless truths have nothing to commend them but the descriptions of those who happen to call them true'.[46]

notes

1. See in particular Stephen Greenblatt, *Renaissance Self-Fashioning* (Chicago: University of Chicago Press, 1980) and Greenblatt, *Shakespearean Negotiations* (Oxford: Clarendon, 1988).
2. See, for example, E. M. W. Tillyard's classic *The Elizabethan World Picture* (London: Chatto & Windus, 1943).
3. Many other influences might also be traced, including the anthropologist Clifford Geertz: readers should consult Terry Eagleton's *Literary Theory: An Introduction* (Oxford: Blackwell, 1983) and Peter Barry, *Beginning Theory: An Introduction to Literary and Cultural Theory* (Manchester: Manchester University Press, 1995) for introductory accounts, or Michael Groden and Martin Kreiswirth, *The Johns Hopkins Guide to Literary Theory and Criticism* (Baltimore and London: Johns Hopkins University Press, 1994) for more detailed discussion and bibliographies.
4. As readers who know *The Faerie Queene* will have noticed, this reading omits the good Protestant female Una, Duessa's other binary opposite. But Duessa is also opposed and paired with the Redcrosse Knight and this opposition (male, female) is the one that is further developed in the texts to be discussed.
5. For a collection giving an overview of New Historicism and its critics see H. Aram Veeser (ed.), *The New Historicism* (London and New York: Routledge, 1989).
6. Stuart Clark, 'Inversion, Misrule and the Meaning of Witchcraft', *Past & Present* 87 (1980) 98–127. A substantial extract is reprinted in Darren Oldridge (ed.), *The Witchcraft Reader* (London and New York: Routledge, 2002) and subsequent references will be to that extract.
7. Gareth Roberts, 'The Descendants of Circe: Witches and Renaissance Fictions', in Jonathan Barry, Marianne Hester and Gareth Roberts (eds), *Witchcraft in Early Modern Europe: Studies in Culture and Belief* (Cambridge: Cambridge University Press, 1996), p. 185.
8. Clark, 'Inversion', p. 149.
9. Heinrich Kramer and Jacob Sprenger, *Malleus Maleficarum*, ed. and trans. Montague Summers (1486; New York: Dover, 1971), p. 45.
10. Anne Llewellyn Barstow, *Witchcraze: Our Legacy of Violence Against Women* (London: Pandora, 1994), pp. 29, 33.

11. Christina Larner, *Witchcraft and Religion*, ed. Alan Macfarlane (Oxford: Blackwell, 1984).

12. Extract from *Witchcraft and Religion* reprinted in Oldridge, *Witchcraft Reader*, as the essay 'Was Witch-Hunting Woman-Hunting', p. 275.

13. Marianne Hester, 'Patriarchal Reconstruction and Witch-Hunting', in Barry, Hester and Roberts (eds), *Witchcraft*, p. 305. Also an extract in Oldridge, *Witchcraft Reader*, pp. 276–88.

14. Hildred Geertz attacked 'functionalist' explanations in her exchange with Keith Thomas, 'An Anthropology of Religion and Magic', *Journal of Interdisciplinary History* 6 (1975) 71–109.

15. Keith Thomas, *Religion and the Decline of Magic* (London: Peregrine, [1971] 1978), pp. 676–7. See also Alan Macfarlane, *Witchcraft in Tudor and Stuart England* (Prospect Heights, Ill.: Waveland Press, [1970] 1991).

16. Norman Cohn, *Europe's Inner Demons: The Demonization of Christians in Medieval Christendom* (London: Pimlico, [1975] 1993).

17. Cohn, *Europe's Inner Demons*, p. xi.

18. Oldridge (ed.), *Witchcraft Reader*, p. 275.

19. Robin Briggs, 'Many Reasons Why: Witchcraft and the Problem of Multiple Explanation', in Barry, Hester and Roberts (eds), *Witchcraft*, p. 49.

20. Clark, 'Inversion', p. 149.

21. Clark, 'Inversion', pp. 157–8.

22. Philip Sidney, *The Defence of Poesy* (1579) in Katherine Duncan-Jones (ed.), *Sir Philip Sidney* (Oxford: Oxford University Press, 1989), pp. 220–4. Sidney argued that the golden fictions of the poet, beautiful but factually untrue, were preferable – hardly Clark's point. But Clark and Sidney share a concern with intellectual spaces – abstract, non-real and theoretical truths – with which many historians are uncomfortable.

23. Clark, 'Inversion', p. 159.

24. Briggs, 'Many Reasons Why', p. 50.

25. Stuart Clark, *Thinking with Demons: The Idea of Witchcraft in Early Modern Europe* (Oxford: Clarendon, 1997).

26. Clark, *Thinking*, pp. 3–4.

27. Clark, *Thinking*, pp. 4, 6–8.

28. Clark, *Thinking*, p. 133.

29. Clark, *Thinking*, p. 106.

30. Clark, *Thinking*, p. 119.

31. Clark, *Thinking*, p. 132.

32. Clark, *Thinking*, p. 133.

33. Diane Purkiss, *The Witch in History: Early-Modern and Twentieth-Century Representations* (London: Routledge, 1996), p. 2.

34. Purkiss, *Witch*, p. 60.

35. Purkiss, *Witch*, pp. 60, 63, 68.

36. Purkiss, *Witch*, p. 74. Natalie Zemon Davies, *Fiction in the Archives: Pardon Tales and their Tellers in Sixteenth-Century France* (Cambridge: Polity Press, 1988); Annabel Gregory, 'Witchcraft, Politics and "Good Neighbourhood" in Seventeenth-Century Rye', *Past & Present* 133 (1991) 31–66; Robert Darnton, *The Great Cat Massacre and Other Episodes in French Cultural History* (Harmondsworth: Penguin, [1984] 1987).

37. Lyndal Roper, *Oedipus and the Devil: Witchcraft, Sexuality and Religion in Early Modern Europe* (London and New York: Routledge, 1994); Deborah Willis, *Malevolent Nurture: Witch-Hunting and Maternal Power* (Ithaca: Cornell University Press, 1995); Frances

Dolan, 'Witchcraft and the Threat of the Familiar', in Dolan, *Dangerous Familiars: Representations of Domestic Crime in England, 1550–1700* (Ithaca: Cornell University Press, 1994).

38. Bernard Rosenthal, *Salem Story: Reading the Witch Trials of 1692* (Cambridge: Cambridge University Press, 1993).

39. Chadwick Hansen, *Witchcraft at Salem* (New York: George Braziller, 1969). However, Hansen believed that actual witchcraft *had* taken place at Salem.

40. Paul Boyer and Stephen Nissenbaum, *The Salem Witchcraft Papers: Verbatim Transcripts of the Legal Documents of the Salem Witchcraft Outbreak of 1692*, 3 vols (New York: Da Capo, 1977); Boyer and Nissenbaum, *Salem Possessed: The Social Origins of Witchcraft* (Cambridge, Mass.: Harvard University Press, 1974).

41. Peter Rushton, 'Texts of Authority: Witchcraft Accusations and the Demonstration of Truth in Early Modern England', in Stuart Clark (ed), *Languages of Witchcraft: Narrative, Ideology and Meaning in Early Modern Culture* (Basingstoke: Macmillan, 2001), p. 21.

42. Marion Gibson, 'Understanding Witchcraft? Accusers' Stories in Print in Early Modern England', in Clark, *Languages of Witchcraft*, p. 53.

43. Marion Gibson, *Reading Witchcraft: Stories of Early Modern Witches* (London and New York: Routledge, 1999). The conference, too, had been called 'Reading Witchcraft' but as Gibson had inadvertently stolen his title, Clark called the collection *Languages of Witchcraft*. Clark, 'Introduction' to *Languages of Witchcraft*, p. 6.

44. Clark, *Thinking*, p. 686.

45. Reginal Scot, *The Discoverie of Witches* (1584), *passim*.

46. Clark, *Thinking*, p. x.

11
gender, mind and body: feminism and psychoanalysis

katharine hodgkin

Gender occupies a paradoxical place in the historiography of witchcraft. The persecution of witches is not about gender alone, but it has a continuing, knotty and intractable relation to gender that comes in and out of focus in the historiography of witchcraft; the fact that the vast majority of those accused of witchcraft in most European countries were women is at once the most and the least visible feature of the persecution of witches. Popular perceptions of witch hunting focus above all on the burning of women, often associated with specific hostile male groups – doctors jealous of midwives, clerics driven mad by celibacy, or religious authorities aiming to obliterate ancient female-centred religions. In the field of academic research, however, gender has often been overlooked, or treated as a side issue. Until quite recently it attracted relatively little attention.

But over the last few decades explanations have multiplied. If once the predominance of women among the accused could be taken for granted (witches are stereotypically female, so naturally more women were accused; or, implicitly or explicitly, women, especially elderly, are likely to be hysterical, menopausal, muddled or malevolent), more recently it has become a much investigated and at times fiercely argued topic. Economic competition, patriarchal control, misogyny and the conceptual structure of demonological thought are among the many explanations put forward for the identification of women as witches; and if none of these has provided a sufficient answer, all have had some influence.

The question of gender has also increasingly become tied to an exploration of mind and body. This link is not inevitable, and is worth pausing on. In part it has to do with a turn to cultural accounts of witchcraft, focusing on language, belief and representation; earlier emphases on economic relations, or the role of the state, for example, have been displaced not only in relation to gender but in witchcraft historiography more generally. However, it is also a consequence of the relations between feminism, history and witchcraft over the last 30 years, which have been complex and at times uneasy. Through feminism, the history of witchcraft was set in dialogue with a political movement, which reclaimed and appropriated the figure of the witch for political ends; and the politics of that movement, however much scholars of witchcraft may have been in sympathy with it, had a problematic impact on the historiography.

Among the most striking features of witchcraft historiography throughout the 1970s and 1980s was the silence of most feminist historians during a period in which the history of women was the focus of intense interest. The history of witchcraft was dominated by men, and, as a rule, men with little interest in gender as an analytical category (not to mention feminism as a politics).[1] The moment when gender starts to be more generally foregrounded by serious researchers is also the moment when historians of witchcraft turn their attention to what was going on in the minds of those involved in the process of identifying a witch. In a sense the possibility of intervening on the subject of gender was reopened by the move to culture and the history of mentalities as governing concepts in understanding witch persecution, and the most substantial explorations of gender in witch persecution over the last decade have been characterised by their openness to forms of analysis that engage with questions of mind and body. Witchcraft is a subject which puts the traditional conceptual frameworks of historiography under pressure, making disciplinary boundaries more permeable.[2] And as far as gender is concerned, two areas of interdisciplinary dialogue are particularly important: the turn to literature, and the turn to psychoanalysis.[3]

This chapter will first fill in this brief narrative, discussing some historical approaches to the question of witchcraft and gender from the 1970s on. It then turns to the shift that took place in the 1990s, when several significant studies appeared, all drawing on psychoanalysis as a way of understanding the part played by gender in witch persecution. The turn to psychoanalysis is of course by no means uncontentious for historians, and I explore some of the problems that arise in relation to psychoanalytic approaches to history as well as the possibilities it opens up. Finally I briefly discuss the recent extension of the question of gender to address the problem of witches who did not conform to the stereotype as well as those who did.

heroines of the sex war: feminism and the witch

The dominant model for thinking about witchcraft in Britain in the 1970s, following the work of Keith Thomas and Alan Macfarlane, focused on neighbourhood dynamics and community responses to economic and religious change. This was a powerful and persuasive account, to the extent that it seemed almost to close off the possibility of others; little substantial work on witchcraft was published in Britain in the 15 years or so following Thomas' *Religion and the Decline of Magic* in 1971.[4] It also seemed simultaneously to account for and to dismiss the predominance of women in witch persecutions. If accusations were generally directed at the poorest and most dependent members of the community, women were vulnerable because of their poverty and dependence on charity; there was no need of more arcane explanations, although both Thomas and Macfarlane note contemporary stereotypes of women as 'weak and vicious', as well as lustful.[5] 'The idea that witch-prosecutions represented a war between the sexes must be discounted', Thomas comments, since women appeared in large numbers as accusers and witnesses in witch trials.[6] And though some researchers into witchcraft in Europe were more concerned with gender, their analyses seldom went beyond a few general remarks about stereotypes. Thus William Monter discusses misogyny and religious stereotypes of women as factors, but his discussion focuses on the disturbed minds of a few – clerical – individuals.[7] In academic witchcraft studies, gender was barely an issue.

Meanwhile, however, outside the academic context, a very different and highly politicised version of witch persecution was taking shape. The feminist movement of the early 1970s was keen to reclaim negative stereotypes: magazines were called *Shrew*, printing presses 'Virago'.[8] The witch was another figure ripe for reclamation, a type of the assertive woman crushed by patriarchy; the judicial execution of thousands (millions, according to many) was an instance of hatred of such women let loose in the world. Mary Daly's *Gyn/Ecology* (1978) celebrated the witch's unquenchable spirit and defiance in the face of oppression, creating a heroine of the women's movement.[9] The publication of Barbara Ehrenreich's and Deirdre English's *Witches, Midwives and Nurses* (1973) added another dimension. The witch, they argued, was a woman performing women's traditional tasks, a midwife and healer who looked after her community, caring for the sick and ensuring fertility through ritual and remedy, and victimised – in their version, by envious men trying to establish medicine as a male profession, but this element was not central to the myth of the midwife-witch that with astonishing rapidity became embedded in many people's understanding of witchcraft.[10] Meanwhile the growing number of people involved with modern witchcraft-based religions (particularly Wicca) represented the witch as an elemental female principle, holding magic powers and mystical significance. The early modern witch became an important precursor for modern movements.[11]

By the early 1980s, then, the witch had become a powerful counter-cultural figure; but that figure could not be taken seriously by academic researchers into witchcraft. Historians pointed out that the version of the witch trials appearing in works like *Gyn/Ecology* could not be reconciled with the historical record, and that the estimated number of executions was fantastically in excess of the real; no correlation could be established between witchcraft accusations and midwifery; the lack of evidence of any kind of organised underground religion at all, never mind Wicca, in early modern records was easy to demonstrate; but to no effect.[12] The myth of 'the burning times', with the witch as proto-feminist and/or quasi-mystical heroine, had for many become an article of faith and a given truth.[13]

For feminist scholars during the 1980s, this situation was challenging. On the one hand, the dominant academic version of witchcraft had no room for serious analysis of gender; on the other hand, the version of witchcraft that did pay attention to gender was happily dismissive of the scholarly reservations of researchers in the field. This may explain the remarkably small number of women academics working on witchcraft during those years. The 1970s and 1980s were an immensely active and fruitful period in women's history, but among all the researchers busy uncovering the hidden histories of women, very few turned to the study of witchcraft. In this particular instance, the question of gender had become very hard to address from within the historical profession. However, during the 1980s, some studies attempted a more serious engagement with the problem.

conceptualising gender: historians and the witch

Two points are repeatedly cited in contesting the idea that gender is an important issue in witch persecution: the fact that a proportion – variable, but almost never insignificant – of those accused of witchcraft were men; and the fact that women's involvement in witch trials was often on the side of the courts, with women playing important parts as accusers and witnesses. Neither of these points, of course, actually means that gender was irrelevant. But they do require a more sophisticated way of thinking about gender, in order to move beyond the argument over whether the witch hunt was driven by the ill-defined and vague notion of 'sex war'. The development of feminist theory throughout the 1980s was in principle providing precisely the necessary conceptual and analytical tools. The emphasis on the construction of gender identities within an entire social context, and the recognition of how unequal power relations shape the identification of both men and women with their assigned gender identities, have a clear bearing on the understanding of gender in witch persecution. It was in the context of such work that feminist scholars began to turn their attention to witchcraft, and to try and arrive at a better model for thinking about gender historically.

The most influential of these writers was Christina Larner, whose detailed study of Scottish witchcraft, *Enemies of God*, appeared in 1981, followed by a posthumous volume of essays in 1984.[14] Larner's central concern was the role of the state in witch persecution. However, she was also interested in the question of how witches were identified at the local level, and in the part played by gender in this process. Witches, she argued, were positioned as deviants and scapegoats, in part because of their age and poverty, but also because they could be perceived as difficult neighbours: they were often regarded as bad-tempered and aggressive, sometimes accused of being scolds as well as witches. Moreover, women who are at odds with the patriarchal order are seen as threatening not only by men, but by more compliant women; it is thus hardly surprising that women were involved in making accusations and testifying against their neighbours. In this sense witches might indeed be seen as women failing to comply with patriarchal norms of female virtue. 'The stereotype witch', notes Larner, 'is an independent adult woman who does not conform to the male idea of proper female behaviour.' But rather than the benevolent midwife-healer of popular myth, the witch was argumentative and cantankerous, a disturber of the social order.[15]

The connection between women and witchcraft, however, Larner insists, is not inevitable; not all witches were women, and in a very influential formulation she argues that witchcraft is sex-related but not sex-specific.[16] On one level, of course, this is a purely descriptive distinction, drawing attention to the difference between an act which can only be performed by one sex, and an act which is predominantly but not exclusively associated with one sex. In the case of a crime, it also distinguishes between crimes for which by definition only one sex can be prosecuted (in this period, rape, or infanticide) and crimes committed predominantly but not exclusively by one sex (murder, or witchcraft). But it also represents a challenge to the model of witchcraft that sees it as a direct attack on women. Witches, Larner declared, were not hunted by the state because they were women; they were hunted because they were witches. The effects of this might indeed be the execution of large numbers of women, but this was not the underlying purpose: 'The purpose of a witch-hunt was the prising out of dangerous persons who were enemies of God, the state and the people.'[17]

For Larner, to see witch hunting purely as woman-hunting narrows the question; the fundamental problem becomes 'why women appeared particularly threatening to patriarchal order at this time, and why they ceased to be so threatening about 1700'.[18] The question of historical change is of course a central problem in understanding witch persecution: why does it begin when it does, and end when it does? There is a general criminalisation of women's behaviour during this period. Infanticide and prositution as well as witchcraft were persecuted more vigorously than before; the early modern legislative process was intervening in women's lives to an extent that it had not previously done. But Larner argues that a focus only on gender cannot account for the larger meanings of witch persecution. Witch persecution, in her account, is undoubtedly framed by patriarchy. Political, juridical, economic and social power in early modern Europe were concentrated in the hands of men; women were more vulnerable to poverty and social ostracism, less able to defend themselves before an aggressive legal system, and generally seen as intellectually and morally inferior. But while an understanding of patriarchal society helps us to make sense of the association of women with witchcraft, it is not patriarchy that *causes* witch persecution; patriarchy is the context rather than the reason.

Other studies of the 1980s and 1990s were more willing to identify both patriarchy in general and a specific attack on women in the early modern period as responsible for the persecution of witches, seeing it as an especially sharp but not fundamentally untypical moment in a transhistorical narrative of misogyny. Of these, Carol Karlsen's study *The Devil in the Shape of a Woman: Witchcraft in Colonial New England* (1987) is the most substantial.[19] Based on archival research, it offers a rich and detailed account of the economic, social and ideological context of New England witch persecution, emphasising the rooted Puritanism of the culture and the difficulties facing the early colonists. In marked contrast to other regional analyses, particularly those of Thomas and Macfarlane, the accused witches Karlsen studied in New England were more likely to be financially independent than impoverished; and she found that tensions around inheritance often lurked behind witchcraft accusations, although there were also various characteristics that made accusations more plausible (such as anger, sexual misconduct and malice).[20] This is an intriguing pattern, but her interpretation of it leads to problems. The name 'witch' comes to be one that may potentially be attached to any threateningly independent woman, and the implication is not only that women's independence is threatening to men, but also that witch persecutions were in some way a coded version of this sense of threat. Karlsen's account at times makes witches sound like proto-feminist heroines of the anti-patriarchal resistance, suffering for their free-spirited rejection of patriarchal norms as well as their financial independence, and accusations of witchcraft risk becoming purely devices used by avaricious men to keep women down, part of a long pattern of oppression and violence, in a way which is at odds with the specific historical insights of the book.

The work of Marianne Hester in the 1990s is more explicitly founded on the assumption that women did indeed represent a particular threat to patriarchal order during this period, and that witchcraft accusations were one of the central mechanisms through which this threat was controlled and eventually dissipated.[21] To summarise very briefly, she argues that women's sexuality was seen as dangerous by the ideologues of the new godly society, while women's earlier active participation in the processes of production was threatening to the craft guilds, increasingly constituting themselves as exclusively male. Consequently, she suggests, the sixteenth and seventeenth centuries saw campaigns against women on both fronts, and strategically placed accusations of witchcraft effectively drove women back into the domestic sphere, where both sexual and economic activity could be policed. Witch hunting ended at the point when patriarchy had successfully reconstructed femininity as passive and helpless, rather than powerful and menacing. Hester's view of gender relations, however, and in particular of sexuality, is open to criticism as both reductive and ahistorical; her interpretation of witch persecution is based entirely on secondary sources, and is essentially a counter-reading of the Essex trial materials analysed by Macfarlane, which is hardly sufficient grounds on which to hypothesise an entire explanation of witch persecution. Hester's topic is not so much witchcraft as misogyny, and she is interested in witchcraft only as the effect of that cause.

This is even more true of Anne Barstow, whose *Witchcraze*, first published in 1994, gives an uncompromising account of witch persecution as wholly the effect of misogyny.[22] She disputes Larner's sex-related/sex-specific distinction on the grounds that men accused of witchcraft were either connected to previously identified female witches, or were caught up in a mass hunt that spiralled out of control. Her book, based entirely on secondary readings, concentrates on the violence, torture and terror inflicted on the accused, and on the structuring opposition between the accused woman and her male jailers, torturers and judges. The focus on witch persecution as a form of systematic male violence against women, however (shared to some degree by Karlsen and Hester as well), is problematic. It locates witch persecution in a generalised narrative of male violence, which obscures the particular conditions and characteristics of what happens in the early modern witch trials. Violence in the form of torture was a central part of the judicial process in most of early modern Europe and New England, directed against men as well as women, and against those tried for other offences such heresy or treason as much as against those tried for witchcraft. Violence was equally a favoured form of judicial punishment, whether in the form of the death penalty, torture, branding, being put in the stocks, whipping, or other local variations. To equate this structured and generalised judicial use of violence against both sexes with for example rape or wife-beating seems unhelpful.

Misogyny's mode of operation in these accounts is violence and terror; but its primary preoccupation is sexual. Witch persecution is equated with hostility to women, which itself is equated with a fear of women's sexuality. The sexual themes of classical demonological theory – fornication with the Devil, orgiastic Sabbats, and the immorality and sexual voraciousness of the female sex in general – become the underlying rationale for witch persecution: it is men's fear of women as sexual beings, in this reading, that is responsible ultimately for practically any form of male oppression of women, but in particular for violence. This is the model that shapes the argument of Hester and Barstow in particular, with the result that their work curiously mirrors that of the demonologists in its willingness to see femininity as defined by sexuality and victimhood.

psychoanalysing witchcraft: ill will, confession, and the maternal body

Given that psychoanalysis as a theory could be said to be founded on human sexuality as dynamic and determining force, it is perhaps surprising that the turn to psychoanalysis on the part of feminist historians of witchcraft involved precisely a displacing of sexuality as central in thinking about gender, and a focus instead on more elementary aspects of psychic life – in particular, on the maternal. 'I had expected', notes Lyndal Roper, in an essay first published in *History Workshop Journal* in 1991,

> to find in witchcraft a culmination of the sexual antagonism which I have discerned in sixteenth- and seventeenth-century German culture. The idea of flight astride a broom or pitchfork, the notions of a pact with the Devil sealed by intercourse, the sexual abandonment of the dance at the witches' sabbath, all seemed to suggest that witchcraft had to do with sexual guilt and attraction between men and women ...[23]

Instead, what she identified at the heart of witchcraft accusation was anxiety about motherhood; and it was to understand this that she made the turn towards psychoanalysis.

The relation between history and psychoanalysis is complex, and much debated; without going into the details of those debates, it is worth noting a few key issues.[24] The question of cultural and historical specificity remains one of the most intractable areas; arguments over the extent to which psychoanalytic interpretation is limited by or to its particular cultural moment are returned to over and over again. If family structures and childrearing practices are different, can the emotional dynamics of early modern people be read in the language of a theory which founds its narrative of human growth and development in a particular triangulation of father/mother/child? In a related area, the argument that concepts and experiences of selfhood are historical rather than universal also raises issues for psychoanalysis; an uncompromising insistence on subjectivity as historically constructed is not easily reconcilable with the transhistorical, or super-historical, psyche of psychoanalysis. And besides such theoretical difficulties, there are substantial problems with the interpretation of sources. Can we read witness statements or confessions as if these haphazardly surviving legal documents were open to interpretation in the same way as the spoken communications of a person in analysis? And yet despite these difficulties psychoanalytic approaches to the study of witchcraft have had a considerable impact; even more strikingly, given the often fraught relationship between feminism and psychoanalysis, this has been most apparent in relation to the question of gender.

Witchcraft has to some extent always been more open than other historical fields to interpretations that engage with the psychological, if not the psychoanalytic. The sheer strangeness of many aspects of demonological theory encouraged some historians to attribute it to the disturbed minds of sexually frustrated clerics and hysterical women, although as the emphasis on the irrationality of witchcraft has come into question, the tendency to explain it as in some sense pathological has receded. But even rational and empirically based interpretations assume the material power of the psyche: the reader is required to accept, for example, that guilt can be projected onto another person, making that person the target of anger and hostility. John Demos' 1982 account of the witch trials of Salem, *Entertaining Satan*, made more direct and explicit use of psychoanalytic models, attempting a holistic picture of witch persecution that would draw together economic, cultural and psychic factors.[25] More recently, Robin Briggs's *Witches and Neighbours*, published in 1996, represents the early modern European village as a seething hotbed of emotion, in which neighbourly relationships generated envy, anger, fear, greed and resentment; his description draws on the terminology not only of projection, but of identification, narcissism, regression, and transference, and his explanations refer back to infantile emotions.[26] This more general deployment of psychoanalytic concepts raises further questions – how far are we in the territory of psychoanalytic history here, and how far in a generalised economy of passions? – but it undoubtedly helps to describe some of the energy and fury invested in the process of accusing a witch.

Neither Demos nor Briggs uses psychoanalytic concepts specifically as a means of engaging with the problem of gender. Their concern is more to find a way into the intense emotionality of the process of identifying and trying a witch; to understand, as Lyndal Roper put the question, how human social relationships could go so drastically and dangerously wrong that people were willing to send their neighbours to torture and probable death.[27] For Roper, however, this question leads on to others: how is it that the victims, when things went wrong, were so overwhelmingly female? As she remarks in a review of Briggs' book:

> Briggs is surely right to locate the genesis of witch accusations not in a conspiracy to victimize women on the part of the elite, or on the part of men, but in the psychic experiences of envy and hatred in small communities. But ...while the variety of forms of witchcraft is certainly striking, even more impressive is the consistency of its targets: women, usually slightly older women, nearly always women who had once been or were still married, and who had borne – or miscarried – children.[28]

And motherhood is the focus of her own work on witchcraft, both in *Oedipus and the Devil* (1994), and in her recent *Witch Craze* (2004).

Oedipus and the Devil, based on Roper's research in German archives, was followed by two studies by literary critics working on English published sources, Deborah Willis' *Malevolent Nurture* (1995) and Diane Purkiss' *The Witch in History* (1996).[29] Roper is the only one of the three to engage seriously in archival research, as well as the most thoughtful and critical in her engagement with psychoanalytic approaches; but it would nonetheless be true to say that the combined weight of these three books, with much in common despite their different perspectives and agendas, represented a significant shift in the analysis of gender in academic work on witchcraft. To understand gender, all three argued, it was necessary to go beyond arguments about sex war, misogyny and the relation between sex-related and sex-specific witch hunting. Rather than asking why women were persecuted, and finding the answer in some larger external cause, we should be attentive to what is being said in the trial process, and to what it tells us about how early modern people understood the character of the witch; 'unless we attend to the imaginative themes of the interrogations themselves,' asserts Roper, 'we shall not understand witchcraft'.[30] What specific fears emerge? What are the points of tension, the moments when suspicion crystallises into certainty? And how are we to understand not only the fears and anxieties of those making the accusation, but also the extraordinary details of the witch's confession? All of these bring us closer to an understanding of the ways in which gender is important in thinking about witches.

Motherhood is a key theme for all three. 'Witches were women, I believe,' asserts Willis, 'because women were mothers.'[31] Of the three, Willis is the most explicitly located in a particular field of psychoanalysis, and her discussion ties witchcraft closely to Kleinian models and concepts. Noting the frequency with which milk, suckling, birth and infancy recur in witchcraft sources, she interprets hostility towards witches in relation to infantile aggression towards the mother. The witch is a bad mother: instead of nourishing babies she suckles devils on her witch's mark; instead of nurturing and supporting children, she kills them, often by poison, and eats them in cannibalistic rituals at the witches' Sabbath. In the figure of the witch are fantasised the characteristics of the anti-mother, destructive and full of hate, and the violent fantasies of infancy, involving dismemberment and destruction of the body, underpin the desire to be revenged on the evil witch. If the language and imagery of Kleinian psychoanalysis, however, offer in many ways a convincing fit with witch persecution, the problem of the relation between the psychic and the social, and the question of how appropriately this language may be applied to early modern culture, remain unresolved. Willis offers as the 'social' side of the equation an account of the ideological construction of femininity in the seventeenth century, with its insistence that women should be chaste and obedient, and its anxieties about women's changing position; but the move between these levels remains unclear.[32]

Roper similarly finds tensions around motherhood at the heart of many cases of witchcraft, but her account is much more historically grounded and specific. Her early work focused on the figure of the lying-in maid, who came to look after the new mother and baby, and who was disproportionately likely to find herself the target of accusations of witchcraft. Like Willis, she cites Klein's work on the projection onto an evil other of unacknowledged feelings of hatred and malice, and suggests that the lying-in maid was a readily available receptacle for such emotions. And while her description of the new mother's responses to her own maternity assumes a certain indelible continuity in the experiences of infancy and maternity alike, it is qualified by reference to the specifics of child-rearing in the period: it is 'the social organization of mothering practices' that made this particular response to maternal ambivalence possible, 'so that a certain kind of psychic dramatic script was available should things go wrong'.[33]

In her most recent work Roper both extends her argument about motherhood more broadly, and to a certain degree retreats from the prominent position given to psychoanalysis. Focusing not only on gender but on age, she suggests that just as historians have failed to engage seriously with the predominance of women amongst the accused, so they have failed to recognise the significance of age, and the fact that overwhelmingly the women accused were in their forties and beyond at the time of accusation. Ambivalent feelings about maternity certainly remain part of the picture, but she draws attention also to a much broader cultural hostility towards older women, and specifically older women's bodies, which she locates in the context of an overriding concern over fertility. Early modern German society was largely agricultural, and critically dependent on the productiveness of the land and people. Witches strike not only at children and babies, but at fertility itself, and the possibility of reproduction not just in the human sphere but in crops and animals, in domestic and agricultural labour. Crops fail, milk dries up, butter will not come, beer goes bad, pigs go wild, horses sicken and die. The withered and ageing body of the no longer reproductive witch stands as a symbol for all failures of the fertile world.

The witch's capacity to disrupt the ordering of the domestic space has been noted in passing by a number of historians;[34] it is also an important part of Diane Purkiss' account of the witch as not just anti-mother but anti-housewife, attacking the processes through which the life of the body is carried on – food, cleanliness, domestic order in general. The accounts of witchcraft given by women involved in the process, whether as witnesses, accusers or accused, she suggests, 'constituted a powerful *fantasy* which enabled women to negotiate the fears and anxieties of housekeeping and motherhood', defining a fantasy in a 'quasi-psychoanalytic sense' as 'a story in which people both express and relieve their unconscious (and sometimes conscious) fears, conflicts and anxieties'.[35] And at the unconscious level, like Roper and Willis, she identifies some of the most fundamental of these conflicts as centred on the mother, whose smothering love is both desired and feared.

Purkiss is eclectic in her methodologies; psychoanalytic readings are not her main priority, and she draws on literary criticism and on anthropological and cultural theory as well as on psychoanalysis. Witchcraft, in her reading, threatens the boundaries of both house (women's territory) and body, both heavily invested with symbolic meaning: the woman's body was identified with the house as something to be ordered, bounded and contained, whose boundaries were constantly at risk of being transgressed – and witchcraft cases often focused on moments of transgression, when bodies or houses were invaded by strange afflictions or unwanted visitors. Such readings can be maintained without necessarily relying on a concept of the unconscious to reinforce them; liminality and pollution figure as prominently as projection and identification. But fantasy remains a central concept in her analysis of witchcraft, above all through its relation with storytelling, and the idea that we must look at the stories people tell to make sense of their lives in order to reach a fuller understanding.

This attentiveness to the imaginary world of witchcraft is important not only in analysing accusations, but also in interpreting the witch's confession. Both Roper and Purkiss stress that confessions cannot be explained simply as the stereotypical products of compulsion, shaped by the rigid structure of demonic pacts, Sabbats and *maleficium*. Each confessing witch, Roper points out, was required to add personal details to her narrative to prove that she was indeed telling her own story, not merely saying whatever would get her out of the torture room. The stories they tell, as Roper comments, are 'peppered with detail drawn from the witch's own experience and coloured by her own emotions'; additionally 'they emphasize parts of the witch fantasy which were not necessarily those which demonology stressed'.[36] Following Freud, Roper notes a resemblance between the emotional dynamics of confession and of therapy, and suggests that 'the theory of transference and counter-transference ... remains a helpful tool for understanding how an accused individual and interrogator might together provide a witch "fantasy"'.[37] To produce a coherent and judicially acceptable narrative requires some element of negotiation between the witch and her interrogator, although a negotiation that would all too often end in death.

There are of course many different reasons for witchcraft confessions. If at one extreme there is pure compulsion – the witch resists torture until she can do so no longer, and then she confesses – at the other the identity of witch was one some people might be ready to adopt, and which indeed might give them some power in their own community; a woman might come to believe she had been a witch, or she might gain some satisfaction from telling a credible and vengeful story of her own power as a witch. Purkiss, working with English cases where judicial torture is not officially permitted, reads confession as a means through which women might access some kind of agency or self-fashioning. Thus of one she notes that confessing to witchcraft offered her 'the opportunity ... to be centre-stage for a while, to speak about desire, to speak about maternity, to speak about her dreams, and to be listened to attentively by her neighbours, and by the better sort'.[38] Her focus is less on the psychic dynamics of confession than on its status as a kind of negotiated discourse, in which the preoccupations of accused women reshape and are reshaped by those of the interrogator, to produce a story which is in a sense a shared fantasy. Like Roper's account, this may seem to overemphasise the freedom of the accused witch to express herself in her confession. But for both writers, the story that is told in confession bears investigation for what it can tell us about the psychic dynamics of witchcraft.

Psychoanalytic concepts and methods, then, have been used to deepen our understanding of the emotional dynamics of the early modern village, the relationship between accused and accusers, and the confessions made by the accused witch; they have also been important in rethinking the significance of gender in witch persecution, particularly through an exploration of how early modern culture imagines the feminine and the maternal. What remains a problem, as noted earlier, is the relation between the historically and culturally specific, and the transhistorical assumptions of psychoanalysis. Historians making use of psychoanalytic approaches attempt to find ways around this gap, because, fundamentally, they feel that psychoanalysis in some way illuminates witch persecution; that is to say, the ends are required to justify the means. Where psychoanalysis looks at the past only in order to find reflections of its own claim to truth or to redescribe its own founding concepts, it will not add a great deal to our understanding of what it is trying to interpret. All of the writers I have been discussing attempt to build into their analyses an awareness of the cultural specificity of early modern witchcraft, rather than reducing it to an illustration of eternal truths, but the articulation of history and psyche remains an issue.

For an example of this we might consider the body. The body in psychoanalysis is in a sense always a fantasy; what is significant is not the elementary physical structure, but the ways in which it is imagined and the significances attributed to it. This on one level might imply a discourse of historical likeness rather than difference: Freud's polymorphously perverse infant, Klein's dismembered maternal body, Lacan's *corps morcelé* (body in bits and pieces), all invoke constant imaginative structures that supersede the historical.[39] The increasing sophistication of historical and theoretical work on the body, however, has made it possible to revisit this with an awareness that bodies are always known and experienced culturally. Rather than being a site of transhistorical truth, then, the body is identified as inescapably historical.[40]

Thus in the work of Roper and Purkiss especially, early modern concepts of the feminine body are at the heart of their interpretations. For Purkiss the fantasised body of the mother is drawn directly from Lacan's theorisations of the pre-oedipal infantile Imaginary, and its characteristics in this sense are given; 'the huge, controlling, scattered, polluted, leaky fantasy of the maternal body of the Imaginary' describes something not subject to historical change.[41] But she juxtaposes this psychoanalytic maternal body with a discussion of early modern conceptions of the female body as leaky, polluting and disturbingly unbounded, and makes this unboundedness central to her interpretation of the magical powers of the witch. Similarly Roper's argument about maternity is specifically tied to early modern cultural hostility to the body of the ageing woman and all it represents, and more generally to the cultural meaning of women's bodies. 'Women's bodies', she argues, 'dramatized the struggle between good and evil. Milk that did not flow, blasted crops, poisoned children – all were the signs that the Devil was at work in the world'; and the removal of cosmic significance from the feminine body is part of the reason why witchcraft became incredible to later generations.[42]

If historicity seems to be located in the body rather than the psyche, the emphasis on the importance of the maternal body as a site of desire, fear and loathing does allow us rather easily to suppose that the body is the guarantor of psychoanalytic truth; this is a tension that perhaps has to remain unresolved. Insofar as the question is whether the recourse to psychoanalytic modes of thought enriches our understanding of witchcraft, the answer would surely be that it does. And while psychoanalysis undoubtedly runs the risk of ahistoricism, the careful and critical engagement with it, exemplified especially in the work of Lyndal Roper, can also help us to go deeper into history. Rather than mapping onto historical subjects modes of thought and language alien to them, much of this work is explicitly concerned to understand what is being said by those involved in witchcraft persecution, and to practise a historical equivalent of the focused and concentrated listening that characterises good analysis.

identifying difference: the past as other

While these approaches have been important in reshaping our understanding of witch persecution, other historians have also attempted to explore questions of gender from other perspectives. Given the problems that psychoanalysis poses for history, this is hardly surprising. Psychoanalytic readings assume a similarity between the past and the present; the passions and tensions they identify in witchcraft cases are familiar, and governed by recognisable psychic processes, even if the articulation of those passions and processes now is different (hostility to the maternal is no longer expressed through the figure of the witch, but that does not mean it no longer exists). Other approaches to the problem of witchcraft, however, insist on its radical unfamiliarity, and remind us that the psychic meanings of important life events are culturally specific. Thus Alison Rowlands, querying the stress laid by Roper among others on witchcraft's imaginative link to fertility, and the psychic importance of menopause as a divider, comments that 'we know little about how menopause was experienced and understood by early modern people'; instead, she suggests, in a culture that generally insisted on deference to older people, 'Witchcraft accusations might ... enable younger members of society legitimately to express hostility rather than deference towards the old.'[43] Her aim here is to reinstate the cultural particularity which may be missing in psychoanalytic history.

Stuart Clark, discussed at length elsewhere in this book, is also concerned to explore witch beliefs in their specific context, linguistic and philosophical as much as specifically historical. In the writings of the educated elite of Europe, he suggests, belief in witchcraft is entirely rational; it offers a route into the thought patterns of the entire intellectual culture of Europe rather than a way to trace deep anxieties, conflicts and desires.[44] Thus he argues that the question of gender has been wrongly approached. Historians have looked for reasons for the persecution of women; but a culture which conventionally thought in polarities, and which saw witches as the inversion of everything good, would naturally associate femininity with witchcraft. That witches were predominantly women was a side-effect of these modes of thought, rather than being central to the aim of persecution. To early modern demonologists in particular, it almost went without saying that femininity was correlated with evil, weakness, impurity and inconstancy. Far from harping obsessively on the vices of women, most demonologists took them so much for granted as barely to mention them; the *Malleus Maleficarum* is an exception in the prominence and explicitness of its misogynist presuppositions. 'This general commitment to dualism', he argues, 'provided a context in which polarized conceptions of witchcraft and polarized conceptions of gender could thrive and, more important, reinforce each other by analogical association'; thus, he concludes, 'it was literally unthinkable that witches should typically be male'.[45]

For historians the tension between our own time and that of the other is both inescapable and generative. If historians such as Roper remind us of the proximity of witch beliefs in the language of psychoanalysis, historians such as Clark alert us to their equal and bewildering otherness in the language of early modern theology; both are necessary parts of understanding the past. In this sense the argument between psychoanalysis and history is not one that will ever be resolved. In the immediate context, however, it has undoubtedly had an invigorating effect. The switch from a generalised notion of witchcraft accusations fuelled by misogyny and fear of women's sexuality to a more precise and focused attention to the interaction between historical and psychic constructions of femininity has transformed the study of gender in witchcraft. Moreover, this reopening of the question of gender has carried along with it an increasing interest in its opposite. In addition to asking why the typical witch is an elderly female, historians are turning their attention to the untypical witch.

beyond the maternal: men, children and witchcraft

If the question 'Why are witches women?' has been a recurring problem for witchcraft studies, one of the reasons it has proved so difficult to answer is of course because a significant number of them are not. And if the discussion of gender is in certain respects also a political question, then to bring men into the equation has implications beyond the merely proportional. Male witches have often functioned primarily as pawns in the argument about the significance of female ones. Historians disputing the significance of gender put forward the number of men executed in order to demonstrate that there was no sexual dimension to the persecution of witches; feminist historians countered by arguing that too few men were involved to alter the fundamental gender dynamic of persecution, or that accused men were generally related to accused women, or were practitioners of sorcery rather than witchcraft proper. Over the last few years, however, the male witch has become more prominent. Both Briggs and Sharpe in their surveys of Europe and England respectively discuss the male witch; other recent work includes William Monter's essay on male witches in Normandy, Robert Walinski-Kiehl's research on accused men in Germany (especially Bamberg), and most recently Lara Apps' and Andrew Gow's study, *Male Witches in Early Modern Europe*.[46]

Apps and Gow are concerned to refute the notion that male witches do not count because they are accused only under exceptional circumstances. Pointing out that in many areas men made up a quarter or more of those accused, they insist that despite the arguments of historians such as Clark and Roper, who (for very different reasons) see witchcraft as inextricably entangled with ideas about the feminine, early modern culture had no difficulty conceptualising witches as male. Their argument is that the male witch was feminised, not in the sense of being seen as effeminate, but as weak-minded: 'he was connected with female witches, and femaleness, via the medieval and early modern sense that it was primarily the weak-minded (especially women) who could be duped by the Devil into becoming his servants'. Thus 'biological sex was not, at the conceptual level, the primary characteristic of the witch; gender was'.[47] In this decoupling of gender from biology they echo Willem de Blécourt, who argues that there were different 'stereotypes' of male and female witches, and that 'men could be classified under the female stereotype and women under the male one'.[48] This is a helpful way of thinking about the complexities of gendering in witchcraft. However, Apps and Gow (unlike de Blécourt) insist that male and female witches are imagined identically, rejecting the possibility that while a witch may be either sex there are generally differences in the characteristics of the male and the female witch. While it is important that the concept of witchcraft should be sufficiently broad to include the many witches who do not fit the dominant model, it also should be possible to specify the characteristics of different groups within that broad concept. To insist that there are no differences seems unhelpful in achieving a fuller understanding of the male witch.

Moreover, while investigations of the male witch are undoubtedly important, there is something odd about the way in which the topic has come suddenly into prominence. There is in principle nothing incompatible in asking simultaneously why most witches were women, and why some were men; gender history by definition ought to be interested in both. And certainly historians studying the male witch have been influenced by theories of gender identity. Reference to early modern constructions of masculinity and masculine identity parallel many of the points made by feminist historians about female witches. But while Willem de Blécourt's claim that current interest in the male witch is overemphasised, and 'merely serves as an excuse to ignore gender issues', seems at first sight unduly harsh, the index entry under 'Gender' in the relevant volume of the *Athlone History of Witchcraft and Magic in Europe* suggests that de Blécourt is onto something.[49] In this very recent overview of early modern witchcraft, eight out of the thirteen index references for gender-related topics are concerned with male witches, which in view of the actual proportions of men and women is hard to justify.[50] Clearly the transformative effect of new approaches to the gendering of witchcraft has not gone very deep into the mainstream (despite the fact that the editors of the volume have all in the past taken an interest in the question of gender); women as witches remain a less attractive topic than men.

The anomalous male witch represents one challenge to rethink the classic figure of the witch as female and elderly; the child witch represents another, even less well explored than the male. There is no space here to do more than note the emergence of this as a new field of interest, perhaps most prominently in German historiography. Like men, children were more likely to be accused of witchcraft if they were related to adult female witches, and more likely to be accused in the course of mass trials. There are also some quite distinctive patterns that emerge around children in orphanages or other institutions, where groups of children (or adolescents) might be accused together, or accuse one another, and apparently spontaneous confessions were not uncommon; this trend is most apparent in cases such as Salem, where adolescent girls were at the centre of the outbreak. Child witches, too, are particularly likely to be involved in cases of possession. Clearly there are many questions here about the nature of criminal responsibility, the concept of childhood, and the idea of the family, as well as about fantasy, which historians are beginning to explore.

Responses to the involvement of children also change over time. In Wurtzburg in the mass trials of 1627–29, according to Robert Walinski-Kiehl, children represented about a quarter of the 160 executions.[51] A century later, however, in Augsburg in 1723, around 20 children were taken into custody for having been seduced by the Devil and committing acts of *maleficium*; all were eventually released.[52] Walinski-Kiehl, focusing on the earlier period, sees child witch-trials in the context of the state's attempts 'to enforce social and moral discipline', as 'one of the consequences of the general movement of reform and repression'.[53] Roper, on the other hand, looks at child witches in the context of the decline in witch belief: 'in the final stages of the witch panic, the death of the old woman as a credible witch led to a brief moment in which the fears and fantasies of children themselves ... emerged in pretty much unmediated form'.[54] By the early eighteenth century, this becomes incorporated into a general reconceptualisation of childhood itself; children are increasingly seen as innocents given to fantasy, rather than as creatures of original sin, and execution is no longer appropriate. 'Before witchcraft was finally consigned to the nursery', comments Roper, 'it paradoxically helped to give birth to an ambivalent fascination with children, their games and their fantasies.'[55] Here, once again, the various manifestations of witchcraft can tell us a great deal about broader cultural shifts, as well as about the early modern psyche.

The question of gender has been at the heart of many of the most significant shifts in witchcraft historiography of the last ten years in particular. It has illuminated our understandings of the fears and anxieties of both accusers and accused, of women's place and of concepts of femininity in early modern society, of power relations and their contradictions. And yet the question itself is by no means answered. As is often the case with accounts of witchcraft more generally, the interpretative models used to address the question of gender seem individually both persuasive and insufficient; offering important insights, yet not fully answering the problem. Moreover there are particular issues that might be raised in relation to the recent focus on mind and body, and in particular the turn to psychoanalysis, apart from the more general conceptual problems discussed earlier. Historians using these approaches have been criticised for a failure to engage with the violence and inequality of gendered power relations, implicitly locating witchcraft in a purely female space, or for not giving enough weight to torture and other forms of compulsion in interpreting the witch's confession. The debate, then, seems likely to continue. But undoubtedly, compared to the position 20 years ago, there is not only much more awareness of gender as an issue, but also a much clearer understanding of how it is articulated in witchcraft beliefs.

notes

1. Elspeth Whitney, 'The Witch "She", the Historian "He": Gender and the Historiography of the European Witch-Hunts', *Journal of Women's History* 7 (1995) 77–101; Diane Purkiss, *The Witch in History: Early Modern and Twentieth-Century Representations* (London: Routledge, 1996), esp. chapter 3.
2. For example, Robin Briggs, 'Many Reasons Why: Witchcraft and the Problem of Multiple Explanation', in Jonathan Barry, Gareth Roberts and Marianne Hester (eds), *Witchcraft in Early Modern Europe: Studies in Culture and Belief* (Cambridge: Cambridge University Press, 1996); Stuart Clark (ed.), *Languages of Witchcraft: Narrative, Ideology and Meaning in Early Modern Culture* (Basingstoke: Macmillan, 2001), introduction.
3. On this see Purkiss, *Witch in History*; Katharine Hodgkin, 'Historians and Witches' (review essay), *History Workshop Journal* 45 (1998) 271–7.
4. Keith Thomas, *Religion and the Decline of Magic: Studies in Popular Beliefs in Sixteenth and Seventeenth-Century England* (London: Weidenfeld & Nicolson, 1971); Alan Macfarlane, *Witchcraft in Tudor and Stuart England: A Regional and Comparative Study* (London: Harper & Row, 1970).
5. Macfarlane, *Witchcraft*, p. 161; and see Thomas, *Religion*, pp. 678–80.
6. Thomas, *Religion*, p. 679. Macfarlane similarly remarks: 'There is no evidence that hostility between the sexes lay behind the prosecutions'; *Witchcraft*, p. 160.

7. William Monter, *Witchcraft in France and Switzerland: The Borderlands During the Reformation* (Ithaca: Cornell University Press, 1976); Monter, 'The Pedestal and the Stake: Courtly Love and Witchcraft', in Renate Bridenthal and Claudia Koonz (eds), *Becoming Visible: Women in European History* (Boston: Houghton Mifflin, 1977).

8. *Shrew* was founded by the Women's Liberation Workshop in London in 1969; Virago Books in 1973.

9. Mary Daly, *Gyn/Ecology* (Boston, Mass.: Beacon Press, 1978).

10. Barbara Ehrenreich and Deirdre English, *Witches, Midwives and Nurses: A History of Women Healers* (Old Westbury: The Feminist Press, 1973). For further discussion see Purkiss, *Witch in History*.

11. Purkiss, *Witch in History*; T. H. Luhrmann, *Persuasions of the Witch's Craft: Ritual Magic in Contemporary England* (Oxford: Blackwell, 1989); Ronald Hutton, *The Triumph of the Moon: A History of Modern Pagan Witchcraft* (Oxford: Oxford University Press, 1999).

12. For example, David Harley, 'Historians as Demonologists: The Myth of the Midwife-Witch', *Social History of Medicine* 3 (1990) 1–26; Clive Holmes, 'Women, Witnesses and Witches', *Past & Present* 140 (1993) 45–78; J. A. Sharpe, 'Witchcraft and Women in Seventeenth-Century England: Some Northern Evidence', *Continuity and Change* 6 (1991) 179–99; Merry E. Wiesner, *Women and Gender in Early Modern Europe* (Cambridge: Cambridge University Press, 1993).

13. For this term, see Purkiss, *Witch in History*, chapter 1, 'A Holocaust of One's Own: The Myth of Burning Times'.

14. Christina Larner, *Enemies of God: The Witch Hunt in Scotland* (London: Chatto & Windus, 1981); *Witchcraft and Religion: The Politics of Popular Belief* (Oxford: Blackwell, 1984). The short section on women and witchcraft in this second volume had a considerable impact; at the time it was practically the only concise and scholarly discussion of the question of gender, and it was very widely circulated.

15. Larner, *Witchcraft and Religion*, pp. 84, 87.

16. See Willem de Blécourt, 'The Making of the Female Witch: Reflections on Witchcraft and Gender in the Early Modern Period', *Gender and History* 12, 2 (2000) 287–309, for a critique of this distinction.

17. Larner, *Witchcraft and Religion*, p. 87.

18. Larner, *Witchcraft and Religion*, p. 87.

19. Carol F. Karlsen, *The Devil in the Shape of a Woman: Witchcraft in Colonial New England* (New York and London: Norton, [1987] 1998).

20. 'Most witches in New England were middle-aged or old women eligible for inheritances because they had no brothers or sons'; Karlsen, *Devil in the Shape of a Woman*, p. 117.

21. Marianne Hester, *Lewd Women and Wicked Witches: A Study in the Dynamics of Male Domination* (London and New York: Routledge, 1992); Hester, 'Patriarchal Reconstruction and Witch Hunting', in Barry, Hester and Roberts, *Witchcraft in Early Modern Europe*.

22. Anne Llewellyn Barstow, *Witchcraze: A New History of the European Witch Hunts* (San Francisco and London: Pandora, 1994).

23. Lyndal Roper, *Oedipus and the Devil: Witchcraft, Sexuality and Religion in Early Modern Europe* (London and New York: Routledge, 1994), p. 203. Chapter 9, 'Witchcraft and Fantasy in Early Modern Germany', was initially published in *History Workshop Journal* 32 (1991).

24. This is a very large topic. For debates on history and psychoanalysis in this context, see Roper, *Oedipus and the Devil*, chapter 10, as well as Roper's comments in her

review essay 'Witchcraft and Fantasy', *History Workshop Journal* 45 (1998), and in the preface to her latest book, *Witch Craze: Terror and Fantasy in Baroque Germany* (New Haven and London: Yale University Press, 2004). For a criticism of historians' use of psychoanalysis, see Garthine Walker, review essay, 'Witchcraft and History', *Women's History Review* 7, 3 (1998) 425–32.

25. John Demos, *Entertaining Satan: Witchcraft and the Culture of Early New England* (Oxford: Oxford University Press, 1982).

26. Robin Briggs, *Witches and Neighbours: The Social and Cultural Context of European Witchcraft* (London: HarperCollins, 1996).

27. Lyndal Roper, *Witch Craze*, pp. 3–4.

28. Roper, 'Witchcraft and Fantasy', 269.

29. Deborah Willis, *Malevolent Nurture: Witch-Hunting and Maternal Power in Early Modern England* (Ithaca: Cornell University Press, 1995); Purkiss, *Witch in History*.

30. Roper, *Oedipus and the Devil*, p. 202.

31. Willis, *Malevolent Nurture*.

32. See Roper's comments in 'Witchcraft and Fantasy', 270.

33. Roper, *Oedipus and the Devil*, p. 34.

34. For example, Karlsen, *Devil in the Shape of a Woman*, pp. 144–7.

35. Purkiss, *Witch in History*, p. 93.

36. Roper, *Witch Craze*, p. 52.

37. Roper, *Witch Craze*, p. 58.

38. Purkiss, *Witch in History*, p. 238.

39. See Sigmund Freud, 'Three Essays on the Theory of Sexuality' (1905), *On Sexuality*, Pelican Freud Library Vol. 7 (Harmondsworth: Penguin, 1977); Melanie Klein, 'Love, Guilt and Reparation' (1937), in Klein, *Love, Guilt and Reparation and Other Works 1921–1945* (London: Virago, 1988); Jacques Lacan, 'The Mirror Stage as Formative of the Function of the I in Psychoanalytic Experience' (1949), in *Ecrits: A Selection*, trans. Alan Sheridan (London: Tavistock, 1977).

40. On early modern women's bodies in particular, see Natalie Zemon Davies, 'Women in Top: Symbolic Sexual Inversion and Political Disorder in Early Modern Europe', in her *Society and Culture in Early Modern France* (Stanford: Stanford University Press, 1975); Laura Gowing, *Common Bodies: Women, Touch and Power in Seventeenth-Century England* (New Haven and London: Yale University Press, 2003). See also David Hillman and Carla Mazzio (eds), *The Body in Parts: Fantasies of Corporeality in Early Modern Europe* (New York and London: Routledge, 1997).

41. Purkiss, *Witch in History*, p. 119.

42. Roper, *Witch Craze*, pp. 255–6.

43. Alison Rowlands, 'Witchcraft and Old Women in Early Modern Germany', *Past & Present* 173 (2001) 50–89, 54. Like several other historians she argues that the witch's age was indicative chiefly of the length of time it took to bring a community to the point of making an accusation; women might carry a reputation for witchcraft for decades before it came to crisis.

44. Stuart Clark, *Thinking with Demons: The Idea of Witchcraft in Early Modern Europe* (Oxford: Oxford University Press, 1997); Clark, 'The "Gendering" of Witchcraft in French Demonology: Misogyny or Polarity?', *French History* 5, 4 (1991) 426–37.

45. Clark, '"Gendering" of Witchcraft', 434, 437.

46. William Monter, 'Toads and Eucharists: The Male Witches of Normandy, 1564–1660', *French Historical Studies* 20, 4 (1997) 563–95; Malcolm Gaskill, 'The Devil in the Shape of a Man: Witchcraft, Conflict and Belief in Jacobean England', *Historical Research* 71 (1998) 142–71; Robert Walinski-Kiehl, 'Males, "Masculine Honour"

and Witch-Hunting in Seventeenth-Century Germany', *Men and Masculinities* 6, 3 (2004) 254–71; Lara Apps and Andrew Gow, *Male Witches in Early Modern Europe* (Manchester: Manchester University Press, 2003). See also Briggs, *Witches and Neighbours*, and James Sharpe, *Instruments of Darkness: Witchcraft in England 1550–1750* (London: Hamish Hamilton, 1996).

47. Apps and Gow, *Male Witches*, p. 12.
48. De Blécourt, 'Making of the Female Witch', 298.
49. De Blécourt, 'Making of the Female Witch', 293.
50. Bengt Ankarloo, Stuart Clark and William Monter (eds), *The Athlone History of Witchcraft and Magic in Europe: The Period of the Witch Trials* (London: Athlone Press, 2002).
51. Robert Walinski-Kiehl, 'The Devil's Children: Child Witch-Trials in Early Modern Germany', *Continuity and Change* 11, 2 (1996) 171–89.
52. Roper, *Witch Craze*, pp. 207–8.
53. Walinski-Kiehl, 'The Devil's Children', 184
54. Roper, *Witch Craze*, p. 203.
55. Roper, *Witch Craze*, p. 203.

12

continuity and change:
social science perspectives on european witchcraft

richard jenkins

This chapter explores the social science literature dealing with European witchcraft beliefs and practices. Although I will refer to the literature on non-European witchcraft when appropriate, there is no space here for a systematic survey.[1] For my purposes Europe is the area west of the Urals and north of the Mediterranean, bounded by the Atlantic and Arctic seaboards. During a millennium of conquest and acculturation between the late Roman and early medieval periods, and unevenly in places at the northern and eastern margins, these lands became 'Christendom'. Due to subsequent colonial expansion over several further centuries, the adjective 'European' can also be applied to peoples and ways of life in the Russian Empire and its successors east of the Urals, and the settler states of the Americas, Africa and Australasia. Although Christianity is emphasised here as the defining criterion of Europe and the European sphere, it has always co-existed and interacted with a diversity of non-Christian beliefs and practices.

The meaning of 'witchcraft' also needs clarification. I propose a cross-culturally applicable definition of witchcraft as malicious supernatural aggression, whether employing spells and rituals or innate individual powers, outside the framework of legitimate religion and ritual.[2] The formulation and use of generic concepts has been controversial in some quarters because of the historical and local specificity of meanings and experiences.[3] While this is a genuine problem, it is not incapacitating; it is, in fact, difficult to know how any cumulative insights might be within our grasp if we do not allow ourselves to use broadly applicable general concepts.

The literature on the early modern period draws a useful distinction between *maleficium*, local supernatural aggression aimed at people, animals and things, and diabolic worship and ritual magic. Although this distinction was often imprecise and confused – the two meanings regularly contaminated each other during accusation and trial – it accurately represents two different sets of beliefs. In modern Europe this becomes a distinction between 'traditional witchcraft' and satanism.[4] Both of these, whether as actual practices or as narratives and beliefs about the powers and wickedness of others, are included in the category of witchcraft here. In this discussion, traditional witchcraft includes beliefs about the 'evil eye', for example,[5] but generally excludes ghosts, fairies and their ilk, vampires, werewolves and other such inhabitants of neighbouring 'other worlds' whose mischief has occasionally troubled the human world.[6] My focus is on the supernatural aggression of humans and the preventive and protective practices for keeping it at bay.

A third kind of witchcraft, which does not concern students of early modern Europe, necessarily comes into view occasionally in this chapter. Modern pagan witchcraft comes in rainbow hues, sometimes allied ideologically to feminism and environmentalism, with many different beliefs and practices.[7] Supernatural aggression is generally not the point of these beliefs and practices. On the one hand, modern pagan witchcrafts are distinct from modern satanisms; on the other, they lack much, if any, continuity with earlier beliefs associated with the word 'witchcraft', although they may draw upon historical inspirations.[8] They are best understood as new religious or spiritual movements and/or modern occultism and ritual magic.[9] Their study thus belongs within the mainstream anthropology and sociology of religion, rather than with witchcraft as defined here.

Finally, 'social science' needs to be defined, too. Social anthropology and sociology have been the main social science contributors to our understanding of witchcraft. Most recently under the varied signs of feminism, postmodernism and post-structuralism, and with some support from that loosely specified enterprise, cultural studies, they are the primary source of the social theoretical concepts and frameworks that are useful to 'witchcraft studies'. This doesn't, however, mean that other social sciences can be ignored. Nor is the boundary around 'social science' clear or consensual. Where to place ethnology and folklore, for example, depends on the ethnologist or the folklorist concerned. The same is true for social history and social historians. Bearing in mind Hutton's argument that recently history and anthropology appear to have converged on a consensual generic or 'supranational' understanding of what a 'witch' is, the time may now be ripe for new interdisciplinary cross-fertilisations.[10] Social anthropology was, however, the first discipline to put witchcraft on the social science agenda. Studying exotic 'primitive' peoples whose cosmologies and ways of life were imbued with magical sensibilities, anthropologists were almost bound to return from the field concerned about how to interpret witchcraft beliefs and practices; in particular, how to analyse witchcraft on its own terms, without dismissing it as irrational, atavistic superstition? With hindsight it was probably no less inevitable that this would eventually bring their work to the attention of historians who were facing a similar interpretive challenge, in attempting to understand the European witch hunts.

anthropological and sociological history

The study that, more than any other, established the value of applying social scientific concepts to early modern witchcraft was Alan Macfarlane's detailed account of accusations and prosecutions in sixteenth- and seventeenth-century Essex, England.[11] The framework within which Macfarlane analysed his data derived ultimately from Evans-Pritchard's seminal study of *Witchcraft, Oracles and Magic Among the Azande* of southern Sudan (1937), and more recently from the African research of Gluckman (1956), Marwick (1965) and others.[12] The common ground shared by these anthropologists was the presumption that witchcraft beliefs and practices could be understood as rational within their local context. They helped people to make sense of their world and to explain misfortune, they provided practical responses to problems, they contributed to the resolution of interpersonal tensions by either repairing problematic social relationships or sundering them, and they were part of local processes of social control.

In his account of the Essex witch-finding movement of 1645, involving Matthew Hopkins and John Stearne, Macfarlane drew upon anthropological analyses of millenarian and witch-finding movements to propose that, allowing for the role of the witch-finders in encouraging and making accusations, this particularly intensive local witch-hunting campaign was a consequence of, first, the dislocation during the Civil War of the local institutions of civil society and, second, the economic, religious and other tensions produced by the war.[13] Witchcraft beliefs provided a medium or vocabulary through which these 'social strains' – to use Marwick's term – found expression: under circumstances of tension, conflict and uncertainty, witch hunting publicly dramatised moral boundaries and ambiguities.[14] From this perspective, witchcraft can be understood as an idiom for the working through of communal, as well as interpersonal, problems. Local witchcraft beliefs and practices may thus shed light on wider social patterns and phenomena. In different guises and contexts, these basic ideas appear and reappear throughout this chapter.

Macfarlane's work emerged from a new social history, influenced by Keith Thomas, that recognised the significance of popular culture and the value of social science, particularly anthropology, in its interpretation.[15] With the benefit of hindsight, the greatest significance of the Essex study lies not in the comparisons that Macfarlane drew between African and European witch beliefs, persuasive though some of them are. Rather, tracing out a detailed landscape of interpersonal conflicts and accusations of *maleficium*, Macfarlane convincingly made the case for approaching European witchcraft beliefs as meaningful and understandable in their own local terms and settings – rather than, at best charming and at worst benighted, superstitions – and analysing patterns of accusation and prosecution as fault lines and expressions of social change (in the case of Macfarlane, and indeed Thomas, changing attitudes to the poor and neighbourly charity).

Anthropologists didn't flock to explore the new worlds that Macfarlane opened up for them, in part because the insistence on participant observation fieldwork as a disciplinary *rite de passage* precluded anything other than occasional detailed flirtations with history.[16] Arguments within the discipline about the legitimacy of the comparative method may also have discouraged the faint-hearted. With particular respect to witchcraft, these concerns were exemplified by a debate between anthropologist Hildred Geertz and historian Keith Thomas about whether one could formulate and use a generic analytical category of 'magic'.[17] Geertz's scepticism, and her concern about the gulf between local or 'native' points of view and the concepts used by anthropologists and historians, resonated loudly for at least the next two decades. That more was going on than an anthropological retreat into 'thick description', rooted in and shaped by local folk models and concepts,[18] is, however, suggested by a similar failure of sociologists to follow in the footsteps of Larner's detailed study of the Scottish hunts (1981, 1984).[19] Although a historian by original vocation, her work was heavily influenced by sociological accounts of gender, stratification, the state and deviance, and was carried out within a sociology department. Focusing on the relationship between state formation, ideology and social control, Larner's work is as striking an achievement as Macfarlane's.

I will return below to the issue of why there are so few detailed anthropological and sociological studies of the early modern witch hunts. For the moment, suffice it to say that social historians of early modern witchcraft have continued to heed the key anthropological and sociological lessons: behaviour and beliefs make sense in context, witch-finding or witch hunting dramatises moral ambiguities and boundaries, small-scale patterns can illuminate wider, long-term processes, and the Devil is always in the exploration of the local detail.[20]

cultural studies of witchcraft

Anthropological and sociological history aside, a more diffuse array of social science influences are at work within a broad-brush historical approach to culture in which pointillist micro-level studies of local interaction are replaced by a wider vision and perspective, exemplified by Peter Burke's *Popular Culture in Early Modern Europe*.[21] Drawing inspiration along the way from a multitude of sources – ethnology, Marxist cultural history, the history of ideas, the *mentalités* approach of the *Annales* school, interpretive cultural anthropologists such as Clifford Geertz, and critical theory deriving from structuralism, feminism and postmodernism – the result has been a heterogeneous array of work across several decades and many national traditions. We can describe these works – emphasising meaning and the interrogation and interpretation of text and discourse – as 'cultural studies of witchcraft'.[22]

Anthropological and sociological micro-studies of local relationships and interaction and cultural studies of worlds of meaning are not mutually exclusive, as is clear from the common ground present in a collection such as Ankarloo and Henningsen's *Early Modern European Witchcraft: Centres and Peripheries*, and the work of scholars such as Lyndal Roper, who creatively combine elements of both approaches.[23] It seems likely that the character of the available records and sources shapes the approach adopted as much as the researcher's intellectual convictions. That many historians of witchcraft now take for granted the value of social science perspectives is clear, for example, in studies of witch hunting in early modern New England,[24] or in the body of work created by the network of researchers who emerged in Holland during the 1980s.[25] Whether or not Hutton (2004) is correct to suggest that anthropologists, over-impressed and daunted by the problems of comparison, actually *abandoned* their engagement with historians, social historians have never ceased to engage with anthropology, or with social science more widely.[26]

historical sociology

Of course it has not always been a matter of social science being brought to bear on the interpretation of historical evidence and questions. Early modern European witch hunts and beliefs have regularly been imported, as old evidential grist for contemporary theoretical mills, into social science debates – largely sociological – that are not actually about witchcraft. Painting with a very broad brush indeed, Marvin Harris (1974) used the witch hunts to ornament a crude materialist argument about class oppression. Within social studies of science and technology Easlea (1980) drew upon early modern witchcraft to illuminate the scientific revolution of the Enlightenment, in order to criticise a crude opposition between 'superstition' and 'reason'.[27] Hirst and Woolley (1982), somewhat less subtly, revisited Winch's arguments about relativism on the basis, inter alia, of a not wholly unsympathetic reading of the *Malleus Maleficarum*.[28]

The sociology of deviance has, perhaps unsurprisingly, made a particularly noteworthy contribution to the literature on the early modern witch hunts. Erikson (1966) explored New England witchcraft to make a Durkheimian case that 'crime waves' contribute to the maintenance of social order. Currie (1973) compared witch hunting in England and continental Europe in order to construct a model of variation in social control systems. Ben-Yahuda (1980, 1985) used the early modern European witch hunts to bolster another

Durkheimian argument about the defence and change of moral boundaries during social change. Finally, Oplinger (1990) drew on the European and New England persecutions to argue that the mass production and suppression of deviance is part of the core business of the modern state. These are all variations on the theme that witchcraft beliefs and witch hunting dramatise moral boundaries and ambiguities.[29]

The expression 'witch hunt' has also become an anachronistic generic term, wrenched out of context and shorn of its original meaning.[30] Viewed in this light, Harris and Oplinger belong to a heterogeneous literature that attempts to compare and connect persecutions from different historical periods, with different victims: early modern witch hunting, the Holocaust, the Stalinist purges, the McCarthy trials, and so on.[31] Perhaps the oddest sociological contribution to this literature comes from Andreski (1989): whether theoretically or empirically, his argument that the early modern witch hunts were caused by a syphilis epidemic – and the use that he puts it to in forecasting the social impact of HIV/AIDS – looks shaky.[32] Unsurprisingly, in the same broad vein feminist scholars have found in the witch hunts a congenial case study of patriarchal persecution in action.[33] No less surprisingly, and perhaps in the belief that bad history leads to bad politics, there has been a spirited feminist defence of the notion that there was more to the hunts than misogyny.[34]

In the main, the authors who are summarised immediately above are historical sociologists, reliant on secondary sources, rather than sociologically aware historians. A little historical knowledge may, however, be dangerous, and critics have often dismissed work of this kind as superficial or unsophisticated in the use of its material and poorly grounded in history (in both of the word's senses: what happened in the past, and the discipline that studies what happened in the past).[35] Lacking specialist training, sociologists and anthropologists appear to make bad historians. Two issues here may shed further light on why anthropologists and sociologists have not undertaken rigorously detailed studies of early modern witchcraft. In the first place, disciplinary boundaries are robust and transfer costs high. Professional historians are often impatient with amateurs (and it is worth remembering that both Macfarlane and Larner originally trained as historians). Nor is disciplinary propriety confined to history: it is difficult to imagine Lucy Mair, for example, tolerating a historian who indulged in readings of anthropological studies as superficial as her own musings on early modern European witch hunting.[36] Perhaps more interesting, however, social theory is a meta-discourse that transcends disciplinary boundaries. It travels better and more easily than historical craft and critical historiography. It may, therefore, be easier for historians to absorb social science perspectives than for sociologists and anthropologists to do good history.

In addition to the general points made earlier – that witchcraft only makes sense in context, witch hunting dramatises moral boundaries and ambiguities, micro-patterns illuminate macro-processes, and the local detail is crucial – social science offers three substantive insights into the early modern European witch hunts. First, the significance to much early modern state-formation of religious ideology is fundamental to any explanation. Second, witchcraft accusations were often related to social change (which doesn't mean that social change always produced witch hunts). Third, witch hunting emerged from the interaction between top-down social control and moral entrepreneurship, and the local significance of close, and therefore difficult to resolve, interpersonal conflicts, for example between kin and neighbours.

One last issue concerns gender. That an understanding of gender roles, and indeed misogyny, is a necessary – if not sufficient – part of any explanation of the witch hunts is a point that hardly needs making. This is an insight owed at least in part to two generations of feminist scholarship – not to mention simple attention to the facts – which cannot be chalked up to the credit of sociology and anthropology. In their time, these disciplines, and their social science neighbours, were no less resistant to feminist revision than was history.

witchcraft in modern europe

Recent scholarship has provided detailed overviews of European witchcraft beliefs and practices during the eighteenth, nineteenth and twentieth centuries.[37] There are also some useful, though ageing, bibliographies available.[38] I do not intend to compete with these surveys here. For my more limited purposes, modernity proper begins with the nineteenth century. This focuses our attention on an era during which rural depopulation, urbanisation, mass migration, industrialisation, democratisation, the emergence of the nation-state, mass education, total warfare, science and technology, public health programmes, evidence-based scientific medicine and mass communications transformed everyday life in the European world. Although stepping back occasionally to the nineteenth century, I will concentrate here on the period after 1900, reflecting the shorter history of social science field research. Contemporary field studies offer opportunities that historians generally do not have: to step outside the archive, record beliefs, traditions and oral history, watch witchcraft at work, interview the participants, and experience the local context.

traditional witchcraft

The literature on traditional witchcraft in modern Europe is not extensive, but it is rich and diverse and has recently expanded rapidly. Where there were few specialist studies 20 years ago, national and regional studies are now available for England, Finland, France, Germany, Italy and the Netherlands, among others.[39] Social science perspectives on European traditional witchcraft can be found in a few in-depth studies of witchcraft and, more often, as by-products of studies or research focusing on other topics. I will draw on both sources to explore a set of loosely connected substantive themes: the continuity (or not) of early modern and modern witchcraft beliefs; the conflict between informal folk healing and organised modern scientific medicine; migration and cultural creativity; and the experience and meaning of witchcraft.

The witchcraft of sixteenth- and seventeenth-century European villages often looks similar to much twentieth-century traditional witchcraft. In fact, the continuity of belief and practice appears to be authentic. To take as an example informal local practitioners to whom the afflicted could turn for protection or relief from bewitchment – or, indeed, other troubles and ailments – the 'cunning folk' of early modern England resemble their nineteenth-century equivalents and the surviving practitioners of early twentieth-century Essex and elsewhere in England.[40] The same point can be made about the *duivelbanners* of Dutch Friesland across four centuries and the *hexenbanners* of Franconia, Germany, and doubtless applies more widely.[41] The many twentieth-century field reports of folk practitioners – not all of whom, it must be admitted, professed to deal with supernatural aggression – show that a wide range of rituals, cures and occupational identities bear close comparison with their equivalents of earlier periods.[42] The same point can be made about beliefs about supernatural aggression, and once again there is a broad range of twentieth-century field reports to choose among.[43] Fieldworkers have come into their own as contributors to the historical record of the future.

While there may be some continuity, it is misleading to see these beliefs and practices as quaint survivals, withering quietly on the vine in the face of progress and modernity as portrayed in Sebald's work on mid-twentieth-century Franconia. Modern traditional beliefs in supernatural aggression and preventative and remedial practices are just that, modern *and* traditional. How witchcraft is talked about – 'the way things are said' – has changed over the centuries.[44] What's more, the contexts of modern traditional witchcraft are in significant respects strikingly different from early modern Europe. Local interpersonal jealousies, feuds and conflicts of interest, and the psychodynamics of blame, guilt, resentment and projection that nurture them, have doubtless altered little over the centuries. Thus witchcraft beliefs did not necessarily become suddenly obsolete: they may still *work* (whatever that means in intimate and neighbourly contexts). However, the legal frameworks, political cultures and institutions, and communications media within which such matters might be pursued have changed almost out of all recognition. Elite sponsorship of traditional beliefs in witches or the evil eye *has* largely withered away, although perhaps not totally or everywhere.[45]

Thus it seems that in modern Europe witchcraft beliefs and practices have changed as well as continued. At the same time, they have also declined in significance: although it is impossible to be precise, today many fewer people believe in traditional witchcraft than did a hundred years ago, partly because we live in a more secular world and partly because of changes in the conditions of life. At the risk of being accused of vulgar modernist rationalism,[46] public health initiatives, scientific medicine and new farming techniques have rendered life less uncertain and less routinely dangerous, and most people no longer have to face the anxieties of producing their own food. Although the world may not have become less mysterious, or even less enchanted, problems such as infant mortality and the uncertainties of butter making are no longer widespread everyday concerns.

The conflict between modernity and traditional witchcraft beliefs, bound up as they are with death, illness and somatic or psychological misfortunes, is perhaps most visible with respect to health and illness. De Blécourt argues forcefully against a simple-minded reductionism which sees traditional beliefs as outmoded by more effective modern scientific healthcare: witchcraft beliefs and practices have their 'own logic', providing 'another kind of rationality' to set alongside the expert knowledge of doctors, veterinarians, and so on.[47] In some senses he is clearly right: the practices and knowledge of modern scientific medicine are poorly understood by lay people – perhaps no better than those of cunning folk were in their day – and lay knowledge is itself imperfectly understood;[48] the hegemony of medical science was in the first place probably achieved more by the exercise of institutional and legal muscle than its greater effectiveness; and the decline of folk medicine is easily overstated in an age of flourishing alternatives to scientific medicine.[49] In another sense, however, de Blécourt is simply wrong: it is, for example, difficult to imagine many modern European farmers seeking out a cunning man to determine who was bewitching their cows in preference to seeking diagnosis and treatment from a vet. Modern medicine offers effective and dependable treatments for a wide spectrum of ailments, and this has at least some bearing on the decline of witchcraft beliefs in the twentieth century.

To return to continuity and change, the post-Roman history of Europe is a complex picture of extensive migration and cultural and religious change, with Christianity replacing, colonising and being influenced by existing beliefs and clashing with Islam. The wider European world is the result of a similar history played out on a wider stage, this time also including forced population movements from Africa and elsewhere to various colonial 'new worlds', where syncretic beliefs in supernatural aggression and magic were created, much as, one imagines, new witchcrafts emerged from the encounter between European paganism and conquering Christianity.[50] A range of modern studies of these beliefs is available, covering Africa, the Caribbean, Latin America and the United States.[51] Since 1945, Europe has experienced significant immigration of non-Christians – from Africa, the Near and Middle East, and south and east Asia – and newly naturalised non-Christian or syncretic traditions of witchcraft and magic have appeared, to be medicalised and/or incorporated into European imaginings of the exotic 'Other'.[52]

That all traditions are, in the end, revealed as invented and reinvented, inspires us to think about continuities and changes in the meaning and experience of traditional witchcraft. As Mitchell has argued in his analysis of a nineteenth-century English case of witch mobbing, even though witchcraft beliefs may lose their communal vigour and immediacy, to become more distant 'folk traditions', they nonetheless continue to provide templates that may, in appropriate circumstances, be acted out (this is what folklorists call 'ostensive action').[53] Beliefs and practices may also acquire and generate new meanings, as in van Dongen's empathetic account of the madness of Rosa and her mother, who practised magic to keep at bay an imaginary hostile magician in late twentieth-century Belgium.[54] Favret-Saada's well-known research in rural France (1980, Favret-Saada and Contreras 1981) focuses on the meanings of traditional witchcraft and is particularly concerned with the experiences of victims.[55] However, to revisit the earlier discussion of local or 'native' points of view, in her medicalising reconceptualisation of 'unbewitching' as 'therapy',[56] Favret-Saada's language and concepts are anything but 'experience-near', to use Geertz's expresion.[57] Even so, hers is a rare enterprise. Sociologists and anthropologists have yet to exploit fully the possibilities that fieldwork offers for exploring the experiential worlds of traditional witchcraft in modern Europe.

satanism

Social science coverage of modern European satanism begins with the emergence, from the mid 1960s onwards and mostly in the United States, of a small number of public satanist groups whose philosophies and practices owed much to earlier occultists such as Crowley, though Crowley was not a satanist himself.[58] Organisations such as the Church of Satan or the Temple of Set often centre on charismatic leaders and combine ritual magic, self-conscious flirtation with deviance, an occasional keen eye for publicity, and notions of 'personal growth'.[59] Viewed sociologically as minority new religious movements, modern satanists have more in common with modern pagan witches than either would like to admit, despite their media depiction as poles apart.[60] Three things may be said with some certainty about modern satanists. First, their activities and rituals, even at their most dramatic, appear to have little in common with sinister popular images of blood sacrifice and the Black Mass.[61] Second, they appear to be few in number.[62] Third, they appear to be organised in small independent groups, not extensive or well-integrated networks.

From this perspective, there appears to be little, if any, continuity with early modern ideas about diabolism and little to interest us here. However, the genealogy of the mythology of Satan on which these groups draw extends through the witch trials and demonological texts of early modern Europe to antiquity.[63] Perhaps even more interesting, the mythology of satanism as a secret underground conspiracy of wicked anti-societies, as Norman Cohn showed, also has a long history.[64] The public appearance of modern satanism during the 1960s provided confirming evidence – of the 'tip of the iceberg' kind – for those who believe in that conspiracy and the contemporary vigour of the threat it poses in twentieth-century Europe. This was a conspiracy theory whose hour came again during the 1980s and 1990s, in the form of high-profile accusations of organised satanic or ritual abuse of children in North America, Europe and Australia.

However, this 'satanic abuse mythology',[65] while drawing on long-standing conventional representations of satanism – from Kramer and Sprenger to Dennis Wheatley – had its own character, context and roots. Revivalist Christian fundamentalism responded to the alternative new religions that appeared and flourished during the 1960s with the rebirth of a personal and active Satan who took his place alongside a personal and active God. In the 1970s it was a small step from this to perceiving 'cults' and the occult as threats to society, particularly to children and young people, and defining satanism as an organised, and by definition largely invisible, evil force, to be confronted and destroyed.[66] In the hands of entrepreneurial campaigners within churches, US law enforcement and elsewhere, the satanist threat became a powerful conventional wisdom.[67] Alongside this, although at some distance, the sexual abuse of children increasingly became a focus for the attention of those feminists, child welfare campaigners and social workers who argued that such abuse was much more common than generally supposed, and that children who claimed they had been abused must be believed.

Fuelled by the sensational testimony of 'survivors'[68] and the enthusiasm of prominent fundamentalists, all widely publicised in the media, these separate concerns had coalesced by the early 1980s into a rounded image of satanic ritual abuse (SRA), in which an organised and extensive network of thousands of satanists – the mirror image of Christian fundamentalism – abducted and abused children, impregnated young women and sacrificed the resultant infants as they were born.[69] Throughout the 1980s the number of cases of supposed SRA grew, with attendant prosecutions, and the scare spread to Canada, Europe and Australia.[70] Perhaps reflecting concerns and guilt about childcare outside the family and other moral ambiguities and anxieties focusing on children, a particular feature of the scare as it spread was accusations against childcare institutions and professionals, the long-running McMartin Preschool case in the United States being paradigmatic.[71]

As the 1990s progressed, the collapse of prosecutions and successful appeals against convictions, public concern at the expanding vigour and violence of accusations and prosecutions, and the stubborn absence of the kind of material evidence that extensive rituals, murders and abductions might be expected to generate, eventually caused the campaign's impetus to falter. Even so, accusations continue, as in Lewis, Scotland, during 2003, and on 21 December 2004, BBC Radio 4's *Today* news programme carried an item about controversial SRA Awareness training for London's Metropolitan Police. The SRA scare has not gone away.

Some consistent themes have emerged within the social science literature on modern satanism and SRA. The first is the importance of 'moral entrepreneurs': key people and groups who formulate and promote public discourse and agendas about deviance,[72] generating 'moral panics' or 'crusades' which dramatise particular societal issues and problems, in the course of which standards of proof or evidence may drop and opposition become muted or suppressed.[73] In this respect, the SRA scare has seen four major, sometimes overlapping, sets of 'moral crusaders'[74] influencing public opinion and attempting to steer events: fundamentalist or evangelical Christians,[75] law enforcement agencies and professionals,[76] social work and psychotherapy professionals,[77] and the media.[78] Without suggesting that the pursuit of material interests caused the SRA scare, SRA has undoubtedly been a money-spinner for some and a potent argument for extra resources for others.

A related theme is the importance of the social dynamics of rumour, or 'improvised news', in the transformation of public moral campaigns into local realities that affect everyday life.[79] If moral crusades or panics are consciously organised by moral entrepreneurs, often on a national or regional scale, rumour is typically disorganised and mainly local. Where one is 'top-down', the other is 'bottom-up'. Both can be usefully understood as dramatisations of moral anxiety, ambiguity, crisis and conflict. If the moral and political orders are experienced as out of kilter, 'what's going on' becomes opaque and uncertain and suspicion and rumour may fill the vacuum.[80] In addition to general discussions, there are detailed local and regional studies of late twentieth-century panics and rumours about satanism and SRA.[81] When rumours become codified, standardised folk narratives, current over a wide area, folklorists talk about 'contemporary legends'.[82] Such narratives, as discussed earlier with respect to traditional witchcraft, may provide the basis for 'ostension' or acting out, in this case encouraging teenage 'satanist' pranks and pseudo-rituals that have provided moral crusaders with ammunition and parents cause for anxiety.[83]

To set against these shared analytical themes, there is also some uncertainty. Not least, there is the question of whether the SRA scare has a basis in fact. Many psychotherapists and social workers insist on the authenticity of survivors' accounts and clients' disclosures, and the cunning of satanists: for them the reality of SRA is axiomatic.[84] Most sociologists, anthropologists and folklorists, while not denying the reality of child sexual abuse, or even *ritualised* abuse, regard satanism and SRA as largely, if not wholly, imaginary.[85] A vociferous proponent of this latter view is La Fontaine, who, having undertaken a government-funded study of cases known to the police and child welfare agencies in England and Wales, concluded that there was evidence of a very few cases of organised and/or ritualised abuse, but none at all of SRA.[86] Against this view, Scott (2001), a sociologist, argues – on the basis of the reasonable sociological precept that people are knowledgeable about what they do and therefore their testimony should be taken seriously

– that we should not dismiss SRA as mere mythology, false memory and ideology.[87] Unresolved issues remain in the epistemological gulf between Scott and La Fontaine: on the one hand, there is the obstinate absence of the kind of evidence that one might expect the extensive organised murder and abuse of children to produce – habeus corpus applies; on the other, there is the complexity and uncertain status of confessions and disclosures. The latter may be further illuminated by studies of confession, itself perhaps a species of ostension,[88] and by consideration of the long-standing issue of the role of child accusers in witch hunts.[89]

There are also questions about direction and dynamics: how the SRA scare spread. Phillip Jenkins represents a widely held view that the ideas which transformed concerns about child abuse in Britain into the SRA scare came from America, imported and propagated by fundamentalists and child welfare specialists.[90] Ellis, however, argues that all the elements of the satanic mythology existed in British folk belief long before the SRA scare and that in fact it provided the basic material that was later elaborated into the American model of SRA.[91] This debate misses an important point, however: a significant folk mythology about satanism, involving child abductions and sacrifice, was in place on both sides of the Atlantic before the SRA scare took off.[92] The key question is why these beliefs did not at that time, the early 1970s, produce the kind of satanic abuse hunts that characterised the 1980s and 1990s. To answer that we have to look at the dynamics of moral crusades and panics, and the powerful coalition of interested parties – Christian fundamentalists, child protection professionals, feminists, law enforcement agents, and journalists – which had by then combined to promote the SRA agenda.

Finally, there is another question: who believes what? Different people who believe in SRA actually believe in strikingly different things. Some live in a fully enchanted world in which the powers of supernatural good and evil are real and in mortal conflict, a battle in which they are foot soldiers. For others, however, it is a more prosaic matter of bad people – driven by their beliefs perhaps – hurting and abusing children. This is one of the reasons why the issue of evidence has become fraught, and so important: the latter is something that can, in principle, be proved according to legal or scientific criteria. The former is a matter of faith, not proof.

Looking at the social science of modern European witchcraft beliefs and practices, whether 'traditional' or satanist, the basic themes are familiar:

- we can only make sociological and anthropological sense of witchcraft if we look at it in context
- witch-finding and witch hunting dramatise moral ambiguities and boundaries
- local, regional and national patterns and processes interact with and illuminate each other

- and the detail is still vital if we are to understand properly what is going on.

Continuity and change, in different combinations and degrees, characterise the picture wherever we look: in particular, while the content and context of belief and practice may change, local and interpersonal themes and processes seem to be more consistent (if not actually universal). Focusing specifically on satanism and SRA, some of the substantive themes are surprisingly comparable with the early modern period, too. Vigorous religious ideology has been crucial to the promotion of satanist abuse hunts. Satanism scares and abuse accusations during the 1980s and 1990s occurred in the context of social change with respect to religion and gender roles. Finally, the satanism and SRA scares, at least in large, depended on moral entrepreneurship exercised by those in public positions, and the dramatisation of moral issues and fears, for their spread and force. *Plus ça change, plus c'est la même chose?*

notes

1. See, for example, B. Levack (ed.), *Anthropological Studies of Witchcraft, Magic and Religion* (New York: Garland, 1992); A. Sanders, *A Deed Without a Name: The Witch in Society and History* (Oxford: Berg, 1995); P. J. Stewart and A. Strathern, *Witchcraft, Sorcery, Rumours and Gossip* (Cambridge: Cambridge University Press, 2004).
2. For other definitions of this kind see C. Larner (1984), *Witchcraft and Religion: The Politics of Popular Belief* (Oxford: Blackwell, 1984); R. Hutton, 'Anthropological and Historical Approaches to Witchcraft: Potential for New Collaboration?', *Historical Journal* 47 (2004) 413–34.
3. See H. Geertz, 'An Anthropology of Religion and Magic, 1', *Journal of Interdisciplinary History* 6 (1975) 71–89.
4. W. de Blécourt, R. Hutton and J. La Fontaine, *Witchcraft and Magic in Europe: The Twentieth Century* (London: Athlone Press, 1999).
5. See A. Dundes (ed.), *The Evil Eye: A Folklore Casebook* (New York: Garland, 1981); C. Maloney (ed.), *The Evil Eye* (New York: Columbia University Press, 1976).
6. See, for example, G. Bennett, *Alas Poor Ghost! Traditions of Belief in Story and Discourse* (Logan: Utah State University Press, 1999); P. Narváez (ed.), *The Good People: New Fairylore Essays* (Lexington: University of Kentucky Press, [1991] 1997); P. Barber, *Vampires, Burial and Death: Folklore and Reality* (New Haven: Yale University Press, 1988); C.-U. Schierup, 'Why are Vampires Still Alive? Wallachian Immigrants in Scandinavia', *Ethnos* 51 (1986) 173–98; J. B. Twitchell, *The Living Dead: A Study of the Vampire in Romantic Literature* (Durham: Duke University Press, 1981); W. de Blécourt, *Werewolves* (London: Hambledon and London, 2005); H. A. Senn, *Were-Wolf and Vampire in Romania* (New York: Columbia University Press, 1982).
7. R. Hutton, *The Triumph of the Moon: A History of Modern Pagan Witchcraft* (Oxford: Oxford University Press, 1999); H. A. Berger, *A Community of Witches: Contemporary Neo-Paganism and Witchcraft in the United States* (Columbia: University of South Carolina Press, 1999); Berger (ed.), *Witchcraft and Magic: Contemporary North America* (Philadelphia: University of Pennsylvania Press, 2005); S. Greenwood, 'Feminist Witchcraft: A Transformatory Politics', in N. Charles and F. Hughes-Freeland (eds), *Practising Feminisms: Identity, Difference and Power* (London: Routledge, 1996);

Greenwood, *Magic, Witchcraft and the Otherworld: An Anthropology* (Oxford: Berg, 2000); T. Luhrmann, 'Persuasive Ritual: The Role of the Imagination in Occult Witchcraft', *Archives des Sciences Sociales des Religions* 60 (1985) 151–70; Luhrmann, *Persuasions of the Witch's Craft: Ritual Magic in Contemporary England* (Oxford: Blackwell, 1989); K. Noonan, 'May You Never Hunger: Religious Foodways in Dianic Witchcraft', *Ethnologies* 20 (1998) 151–74; A. Scarboro, N. Campbell and S. Stave, *Living Witchcraft: A Contemporary American Coven* (Westport, Conn.: Greenwood, 1994).

8. As well as the works of R. Hutton already cited, see also Hutton, 'Paganism and Polemic: The Debate over the Origins of Modern Pagan Witchcraft', *Folklore* 111 (2000) 103–16; Hutton, *Witches, Druids and King Arthur* (London: Hambledon and London, 2003).

9. R. C. Fuller, *Spiritual but not Religious: Understanding Unchurched America* (Oxford: Oxford University Press, 2001); W. W. Zellner and M. Petrowsky (eds), *Sects, Cults and Spiritual Communities* (London: Praeger, 1998); D. Cheal and J. Leverick (1999), 'Working Magic in Neo-Paganism', *Journal of Ritual Studies* 13 (1999) 7–20; A. Owen, *The Place of Enchantment: British Occultism and the Culture of the Modern* (Chicago: University of Chicago Press, 2004).

10. Hutton, 'Anthropological and Historical Approaches'.

11. A. Macfarlane, *Witchcraft in Tudor and Stuart England: A Regional and Comparative Study* (London: Routledge and Kegan Paul, 1970); Macfarlane, 'Witchcraft in Tudor and Stuart Essex', in M. Douglas (ed.), *Witchcraft Confessions and Accusations* (London: Tavistock, 1970).

12. E. E. Evans-Pritchard, *Witchcraft, Oracles and Magic among the Azande* (Oxford: Clarendon Press, 1937); M. Gluckman, *Custom and Conflict in Africa* (Oxford: Blackwell, 1956); M. G. Marwick, *Sorcery in its Social Setting* (Manchester: Manchester University Press, 1965).

13. Macfarlane, *Witchcraft*, pp. 135–44; M. Douglas, 'Techniques of Sorcery Control in Central Africa', in J. Middleton and E. H. Winter (eds), *Witchcraft and Sorcery in East Africa* (London: Routledge and Kegan Paul, 1963); M. G. Marwick, 'Another Modern Anti-Witchcraft Movement in East Central Africa', *Africa* 20 (1950) 100–12. See also A. Richards, 'A Modern Movement of Witch-Finders', *Africa* 8 (1935) 448–61; R. G. Willis, 'Kamcape: An Anti-Sorcery in South-West Tanzania', *Africa* 39 (1968) 1–15.

14. M. G. Marwick, 'Witchcraft as a Social Strain-Gauge', *Australian Journal of Science* 26 (1964) 263–8.

15. K. Thomas, 'The Relevance of Social Anthropology to the Historical Study of English Witchcraft', in M. Douglas (ed.), *Witchcraft Confessions and Accusations* (London: Tavistock, 1970); Thomas, *Religion and the Decline of Magic: Studies in Popular Beliefs in Sixteenth and Seventeenth-Century England* (London: Weidenfeld and Nicolson, 1971).

16. For example, K. Hastrup, 'Iceland: Sorcerers and Paganism', in B. Ankarloo and G. Henningsen (eds), *Early Modern European Witchcraft: Centres and Peripheries* (Oxford: Oxford University Press, 1990).

17. Geertz, 'An Anthropology of Religion and Magic'; K. Thomas, 'An Anthropology of Religion and Magic. II', *Journal of Interdisciplinary History* 6 (1975) 91–109.

18. C. Geertz, *The Interpretation of Cultures* (New York: Basic Books, 1973), pp. 3–30.

19. C. Larner, *Enemies of God: The Witch-Hunt in Scotland* (Oxford: Blackwell, 1981); Larner, *Witchcraft and Religion: The Politics of Popular Belief* (Oxford: Blackwell, 1984).

20. See, for example, W. Behringer, *Witchcraft Persecutions in Bavaria: Popular Magic, Religious Zealotry and Reason of State in Early Modern Europe* (Cambridge: Cambridge University Press, 1997), translation of *Hexenverfolgung in Bayern* (Munich: Oldenbourg, 1987); P. Boyer and S. Nissenbaum, *Salem Possessed: The Social Origins of Witchcraft* (Cambridge, Mass.: Harvard University Press, 1978); R. Briggs, *Witches and Neighbours: The Social and Cultural Context of European Witchcraft*, 2nd edn (Oxford: Blackwell, 2002); H. de Waardt, *Toverij and Samenleving: Holland 1500–1800* (The Hague: Stichting Hollands Historische Reeks, 1991); G. Henningsen, *The Witches' Advocate: Basque Witchcraft and the Spanish Inquisition* (Reno: University of Nevada Press, 1980).

21. P. Burke, *Popular Culture in Early Modern Europe* (London: Temple Smith, 1978).

22. J. Barry, M. Hester and G. Roberts (eds), *Witchcraft in Early Modern Europe: Studies in Culture and Belief* (Cambridge: Cambridge University Press, 1996); C. Ginzburg, *The Night Battles: Witchcraft and Agrarian Cults in the Sixteenth and Seventeenth Centuries* (London: Routledge and Kegan Paul, [1966] 1983); Ginzburg, *Ecstasies: Deciphering the Witches' Sabbath* (London: Hutchinson, 1990); G. Klaniczay, *The Uses of Supernatural Power: The Transformation of Popular Religion in Medieval and Early Modern Europe* (Cambridge: Polity Press, 1990); W. Stephens, *Demon Lovers: Witchcraft, Sex and the Crisis of Belief* (Chicago: University of Chicago Press, 2002).

23. L. Roper, *Oedipus and the Devil: Witchcraft, Sexuality and Religion in Early Modern Europe* (London: Routledge, 1994); Roper, '"Evil Imaginings and Fantasies": Child-Witches and the End of the Witch Craze', *Past & Present* 167 (2000) 107–39; Roper, *Witch Craze: Terror and Fantasy in Baroque Germany* (New Haven: Yale University Press, 2004).

24. Boyer and Nissenbaum, *Salem Possessed*; E. G. Breslaw, *Tituba, Reluctant Witch of Salem: Devilish Indians and Puritan Fantasies* (New York: New York University Press, 1996); J. P. Demos, *Entertaining Satan: Witchcraft and the Culture of Early New England* (New York: Oxford University Press, 1982); R. Godbeer, *The Devil's Dominion: Magic and Religion in Early New England* (Cambridge: Cambridge University Press, 1992); C. F. Karlsen, *The Devil in the Shape of a Woman: Witchcraft in Colonial New England* (New York: Norton, 1987); E. Reis, *Damned Women: Sinners and Witches in Puritan New England* (Ithaca: Cornell University Press, 1997); R. Weisman, *Witchcraft, Magic and Religion in 17th-Century Massachusetts* (Amherst: University of Massachusetts Press, 1984).

25. W. de Blécourt, *Termen van Toverij: De veranderende betekenis van toverij in Nordoost Nederland tussen de 16de en 20ste eeuw* (Nijmegen: SUN, 1990); de Blécourt and M. Gijswijt-Hofstra (eds), *Kwade Mensen: Toverij in Nederland*, special issue of *Volkskundig Bulletin* 12, 1 (1986); de Waardt, *Toverij and Samenleving*; M. Gijswijt-Hofstra and W. Frijhoff (eds), *Witchcraft in the Netherlands: From the Fourteenth to the Twentieth Centuries* (Rotterdam: Universitaire Pers, [1987] 1991).

26. Hutton, 'Anthropological and Historical Approaches to Witchcraft'.

27. M. Harris, *Cows, Pigs, Wars and Witches: The Riddles of Culture* (New York: Random House, 1974); B. Easlea, *Witch-Hunting, Magic and the New Philosophy: An Introduction to Debates of the Scientific Revolution 1450–1750* (Brighton: Harvester, 1980).

28. P. Hirst and P. Woolley, *Social Relations and Human Attributes* (London: Tavistock, 1982).

29. K. T. Erikson, *Wayward Puritans: A Study in the Sociology of Deviance* (New York: John Wiley, 1966); E. P. Currie, 'The Control of Witchcraft in Renaissance Europe', in D. Black and M. Mileski (eds), *The Social Organization of Law* (New York: Seminar Press,

1973); N. Ben-Yahuda, 'The European Witch Craze of the 14th to 17th Centuries: A Sociologist's Perspective', *American Journal of Sociology* 86 (1980) 1–31; Ben-Yahuda, *Deviance and Moral Boundaries: Witchcraft, the Occult, Science Fiction, Deviant Science and Scientists* (Chicago: University of Chicago Press, 1985), pp. 23–73; J. Oplinger, *The Politics of Demonology: The European Witchcraze and the Mass Production of Deviance* (Selinsgrove: Susquehanna University Press, 1990), pp. 34–125.

30. Larner, *Witchcraft and Religion*, pp. 88–90.

31. A. R. Cardozo, 'A Modern American Witch-craze', in M. Marwick (ed.), *Witchcraft and Sorcery: Selected Readings* (Harmondsworth: Penguin, 1968); E. Terray, 'Witchcraft Trials and Stalinist Trials: Reflection on a Parallel', in C. W. Gailey (ed.), *The Politics of Culture and Creativity: A Critique of Civilization*, Vol. 2 (Gainesville: University of Florida Press, 1992).

32. S. Andreski, *Syphilis, Puritanism and Witch Hunts: Historical Explanations in the Light of Medicine and Psychoanalysis with a Forecast about Aids* (Basingstoke: Macmillan, 1989).

33. B. Ehrenreich and D. English, *Witches, Midwives, and Nurses: A History of Women Healers* (New York: The Feminist Press, 1973); M. Hester, *Lewd Women and Wicked Witches: A Study of the Dynamics of Male Domination* (London: Routledge, 1992).

34. R. Hasted, 'The New Myth of the Witch', *Trouble and Strife* 2 (1984) 10–17; Hasted, 'Mothers of Invention', *Trouble and Strife* 7 (1985) 17–25; Larner, *Witchcraft and Religion*, pp. 84–8; L. Mitchell, 'Enemies of God or Victims of Patriarchy?', *Trouble and Strife* 2 (1984) 18–20; Roper, *Oedipus and the Devil*.

35. Ben-Yahuda, for example, found himself involved in a spat with Hoak, a historian, about the factual accuracy of his account of the hunts (*American Journal of Sociology* 86: 1270–9). Hester's 'revolutionary feminist' analysis of the witch hunts was criticised in a review by Mack (*American Historical Review* 98: 1246) for not linking the English hunts to events elsewhere, downplaying the role of the state, ignoring masculinity and not examining the views of the accused women themselves, while another reviewer, Woolley (*Sociology* 27: 560), calls Hester's work to task for 'pressing the evidence into the service of a theory which is at best a sweeping generalisation and at worst partisan'. In a similar vein, Stephen Mitchell has recently described Harris' account of the witch hunts as 'one-dimensional', a judgement which is probably charitable: Mitchell, 'A Case of Witchcraft Assault in Early Nineteenth-Century England as Ostensive Action', in W. de Blécourt and O. Davies (eds), *Witchcraft Continued: Popular Magic in Modern Europe* (Manchester: Manchester University Press, 2004), p. 14.

36. L. Mair, *Witchcraft* (London: Weidenfeld and Nicolson, 1969), pp. 222–37.

37. M. Gijswijt-Hofstra, B. P. Levack and R. Porter, *Witchcraft and Magic in Europe: The Eighteenth and Nineteenth Centuries* (London: Athlone Press, 1999); W. de Blécourt, 'The Witch, her Victim, the Unwitcher and the Researcher: The Continued Presence of Traditional Witchcraft', in de Blécourt, Hutton and La Fontaine, *Witchcraft and Magic in Europe*, pp. 141–219.

38. C. N. Kies, *The Occult in the Western World: An Annotated Bibliography* (London: Mansell, 1986); J. G. Melton, *Magic, Witchcraft and Paganism in America: A Bibliography* (New York: Garland, 1982).

39. See, for example, the essays in de Blécourt and Davies, *Witchcraft Continued*; O. Davies, *Witchcraft, Magic and Culture 1736–1951* (Manchester: Manchester University Press, 1999); Davies, *A People Bewitched: Witchcraft and Magic in Nineteenth Century Somerset* (Bruton: O. Davies, 1999); de Blécourt, *Termen van Toverij*; J. Devlin, *The Superstitious Mind: French Peasants and the Supernatural in the Nineteenth Century*

(New Haven, Conn.: Yale University Press, 1987); H. Sebald, *Witchcraft: The Heritage of a Heresy* (New York: Elsevier, 1978); Sebald, 'Franconian Witchcraft: The Demise of a Folk Magic', *Anthropological Quarterly* 153 (1980) 173–87.

40. Macfarlane, *Witchcraft*, pp. 115–34; Thomas, *Religion*, pp. 252–300; Davies, *Witchcraft, Magic and Culture*, pp. 214–29; Davies, *A People Bewitched*, pp. 27–91; Davies, *Cunning-Folk: Popular Magic in English History* (London: London & Hambledon, 2003); Davies, *Murder, Magic, Madness: The Victorian Trials of Dove and the Wizard* (Harlow: Pearson, 2005); Davies and L. Tallis, *Cunning Folk: An Introductory Bibliography* (London: FLS Books, 2005); E. Maple, *The Dark World of Witches* (London: Hale, 1962), pp. 171–88.

41. W. de Blécourt, 'Vier eeuwen Friese duivelbanners', in M. Gijswijt-Hofstra and W. Frijhoff (eds), *Nederland betoverd: Toverij and Hekserij van de veertiende tot in de twintigste eeuw* (Amsterdam: De Bataafsche Leeuw, 1987); de Blécourt, 'Duivelbanners in de noordelijke Friese Wouden, 1860–1930', *Volkskundig Bulletin* 14 (1988) 159–87; Sebald, *Witchcraft*, pp. 81–99; Sebald, 'Franconian Witchcraft'.

42. On Denmark, see H. P. Hansen, *Kloge Folk: Folkemedicin og overtro i Vestjylland*, 2 vols (Copenhagen: Rosenkilde og Bagger, [1942] 1960); B. Rørbye 'Allmän etnomedicinsk översikt', in B. G. Alver et al. (eds), *Botare: En bok om etnomediicin i Norden* (Stockholm: Nordic Institute of Folklore, 1980). On France see J. Favret-Saada, *Deadly Words: Witchcraft in the Bocage* (Cambridge: Cambridge University Press, 1980), translation of *Les mots, la mort, les sorts: la sorcellerie dans le Bocage* (Paris: Éditions Galimard, 1975); Favret-Saada, 'Unbewitching as Therapy', *American Ethnologist* 16 (1989) 40–56. On Ireland, see T. C. Correll, 'Believers, Sceptics and Charlatans: Evidential Rhetoric, the Fairies, and Fairy Healers in Irish Oral Narrative and Legend', *Folklore* 116 (2005) 1–18; E. Lenihan, *In Search of Biddy Early* (Cork: Mercier, 1987); G. Ó Crualaoich, 'Reading the *Bean Feasa*', *Folklore* 116 (2005) 37–50; M. Ryan, *Biddy Early: The Wise Woman of Clare* (Cork: Mercier, 1978). On Norway, see O. Nordland, 'The Street of "The Wise Women": A Contribution to the Sociology of Folk-Medicine', *ARV: Journal of Scandinavian Folklore* 18–19 (1962–63) 263–74; I. Torstenson, 'Signekjerringer i storbyen', unpublished Master's thesis, Institute of Folklore Research, Oslo University, 1979. On Sweden, see B. G. Alver, 'Andreas', in B. G. Alver et al., *Botare*; C.-H. Tillhagen, *Folklig läkekonst* (Stockholm: Nordiska Museet, 1958). On the US, see V. Randolph, *Ozark Superstitions* (New York: Columbia University Press, 1947), pp. 92–161, 287–92. On Wales, see S. Philpin, 'Wool Measurement: Community and Healing in Rural Wales', in C. A. Davies and S. Jones (eds), *Welsh Communities: New Ethnographic Perspectives* (Cardiff: University of Wales Press, 2003).

43. On France, see Favret-Saada, *Deadly Words*; J. Favret-Saada and J. Contreras, *Corps pour corps: Enquête sur la sorcellerie dans le Bocage* (Paris: Gallimard, 1981); Davies, 'Witchcraft Accusations in France'. On Germany, see Sebald, *Heritage*. On Greece, see R. Dionisopoulos-Mass, 'Greece: The Evil Eye and Bewitchment in a Peasant Village', in C. Maloney (ed.), *The Evil Eye* (New York: Columbia University Press, 1976); M. Herzfeld, 'Meaning and Morality: A Semiotic Approach to Evil Eye Accusations in a Greek Village', *American Ethnologist* 8 (1981) 560–74; C. Stewart, *Demons and the Devil: Moral Imagination in Modern Greek Culture* (Princeton: Princeton University Press, 1991). On Hungary, see É. Pócs, 'Evil Eye in Hungary: Belief, Ritual, Incantation', in J. Roper (ed.), *Charms and Charming in Europe* (Basingstoke: Palgrave Macmillan, 2004). On Ireland, see C. M. Arensberg, *The Irish Countryman: An Anthropological Study* (New York: Macmillan, 1937), pp. 28–33; R. Harris, *Prejudice and Tolerance in Ulster: A Study of Neighbours and 'Strangers' in*

a Border Community (Manchester: Manchester University Press, 1972), pp. 101–2. On Italy, see W. Appel, 'Italy: The Myth of the *Jettatura*', in Maloney, *The Evil Eye*; S. Magliocco, 'Witchcraft, Healing and Vernacular Magic in Italy', in de Blécourt and Davies, *Witchcraft Continued*. On Newfoundland, see J. C. Faris, *Cat Harbour: A Newfoundland Fishing Settlement* (St Johns: Memorial University of Newfoundland, 1971), pp. 135–42. On Portugal, see J. Cutileiro, *A Portuguese Rural Society* (Oxford: Clarendon Press, 1971), pp. 273–8. On Romania, see É. Pócs, 'Curse, *Maleficium*, Divination: Witchcraft on the Borderline of Religion and Magic', in de Blécourt and Davies, *Witchcraft Continued*. On Sicily, see C. G. Chapman, *Milocca: A Sicilian Village* (London: George Allen and Unwin, 1973), pp. 196–209. On Sicilian Canadians, see S. Migliore, *Mal'uocchiu: Ambiguity, Evil Eye and the Language of Distress* (Toronoto: University of Toronto Press, 2000). On Spain, see J. C. Baroja, *The World of Witches* (London: Weidenfeld and Nicolson, 1964), pp. 227–38, translation of *Las brujas y su mundo* (Madrid: Revista de Ociodente, 1961); W. A. Christian, *Person and God in a Spanish Valley*, 2nd edn (Princeton: Princeton University Press, 1989), pp. 195–8; J. A. Pitt-Rivers, *The People of the Sierra*, 2nd edn (Chicago: University of Chicago Press, 1971), pp. 189–201.

44.   See de Blécourt, *Termen van Toverij* [*Words of Witchcraft*]. However, for a discussion of the historical genealogy of contemporary exorcism within a mainstream European Protestant Christian church, which stresses continuities and changes, see M. J. Maagaard, *En Satans Uvæsen: Eksorcisme i Folkekirken* (Copenhagen: Forlaget ANIS, 2003).

45.   For a journalistic account of a recently notorious Italian case see C. Compton and G. Cole, *Superstition: The True Story of the Nanny They Called a Witch* (London: Ebury, 1990).

46.   De Blécourt, 'The Witch, her Victim, the Unwitcher and the Researcher', p. 212.

47.   De Blécourt, 'Boiling Chickens and Burning Cats: Witchcraft in the Western Netherlands, 1850–1925', in de Blécourt and Davies, *Witchcraft Continued*, p. 89; de Blécourt, 'The Witch, her Victim, the Unwitcher and the Researcher', p. 213.

48.   On these points see L. Prior, 'Belief, Knowledge and Expertise: The Emergence of the Lay Expert in Medical Sociology', *Sociology of Health and Illness* 25 (2003) 41–57; I. Shaw, 'How Lay are Lay Beliefs?', *Health* 6 (2002) 287–99.

49.   K. Bakx, 'The "Eclipse" of Folk Medicine in Western Society', *Sociology of Health and Illness* 13 (1991) 20–38.

50.   See, for example, E. G. Breslaw, *Tituba, Reluctant Witch of Salem: Devilish Indians and Puritan Fantasies* (New York: New York University Press, 1996); L. A. Lewis, *Hall of Mirrors: Power, Witchcraft and Caste in Colonial Mexico* (Durham: Duke University Press, 2003); S. MacCormack, *Religion in the Andes: Vision and Imagination in Early Colonial Peru* (Princeton: Princeton University Press, 1991); I. Silverblatt, *Moon, Sun, and Witches: Gender Ideologies and Class in Inca and Colonial Peru* (Princeton: Princeton University Press, 1987).

51.   See, for example, G. C. Bond, 'Ancestors and Witches: Explanations and the Ideology of Individual Power in Northern Zambia', in G. C. Bond and D. M. Ciekawy (eds), *Witchcraft Dialogues: Anthropological and Philosophical Exchanges* (Athens: Ohio University Press, 2001); M. Brendbekken, 'Beyond Vodou and Anthroposophy in the Dominican-Haitian Borderlands', *Social Analysis* 46 (2002) 31–74; M. S. Laguerre, *Voodoo and Politics in Haiti* (London: Macmillan, 1989); A. Métraux, *Voodoo* (London: André Deutsch, 1959); R. Romberg, *Witchcraft and Welfare: Spiritual Capital and the Business of Magic in Modern Puerto Rico* (Austin: University of Texas Press, 2003); M. T. Taussig, *The Devil and Commodity Fetishism*

in South America (Chapel Hill: University of North Carolina Press, 1980); K. McC. Brown, *Mama Lola: A Vodou Priestess in Brooklyn* (Berkeley: University of California Press, 1991); St C. Drake and H. R. Cayton, *Black Metropolis: A Study of Negro Life in a Northern City*, second edition (New York: Harper and Row, 1962), pp. 474–8; C. J. Jacobsen, '¿Espiritus? No. Pero la maldad existe: Supernaturalism, Religious Change, and the Problem of Evil in Puerto Rican Folk Religion', *Ethos* 31 (2003) 434–67; M. Simmons, *Witchcraft in the Southwest: Spanish and Indian Supernaturalism on the Rio Grande* (Flagstaff: Northland, 1974).

52. D. Pierre, 'Dieu qu'elle devait être seule! Histoire d'une brève recontre entre deux mondes', *L'Autre* 4 (2003) 53–69; K. Reed, *Worlds of Health: Exploring the Health Choices of British Asian Mothers* (Westport, Conn.: Praeger, 2003), pp. 38–43, 101–4; L. Sachs, 'Evil Eye or Bacteria: Turkish Migrant Women and Swedish Health Care', *Stockholm Studies in Social Anthropology* 12 (Stockholm: University of Stockholm, 1983); T. Sanders (2003), 'Imagining the Dark Continent: The Met, the Media and the Thames Torso', *Cambridge Anthropology* 23 (2003) 53–66; U. Streit, J. Leblanc and A. MekkiBerrada, 'A Moroccan Woman Suffering from Depression: Migration as an Attempt to Escape *Sorcellerie*', *Culture, Medicine and Psychiatry* 22 (1998) 445–64.

53. S. Mitchell, 'A Case of Witchcraft Assault in Early Nineteenth-Century England as Ostensive Action', in de Blécourt and Davies, *Witchcraft Continued*.

54. E. van Dongen, *Walking Stories: An Oddnography of Mad People's Work with Culture* (Amsterdam: Rozenberg, 2002), pp. 109–204.

55. Favret-Saada, *Deadly Words*; Favret-Saada and Contreras, *Corps pour corps*.

56. Favret-Saada, 'Unbewitching as Therapy'.

57. C. Geertz, *Local Knowledge* (New York: Basic Books, 1983), p. 57.

58. J. La Fontaine, 'Satanism and Satanic Mythology', in de Blécourt, Hutton and La Fontaine, *Witchcraft and Magic in Europe*, pp. 94–109.

59. R. H. Alfred, 'The Church of Satan', in C. Glock and R. Bellah (eds), *The New Religious Consciousness* (Berkeley: University of California Press, 1976); W. S. Bainbridge, *Satan's Power: A Deviant Psychotherapy Cult* (Berkeley: University of California Press, 1978); J. Gunn, *Modern Occult Rhetoric: Mass Media and the Drama of Secrecy in the Twentieth Century* (Tuscaloosa: University of Alabama Press, 2005), pp. 172–203; E. J. Moody, 'Magical Therapy: Contemporary Satanism', in I. Zaretsky and M. Leone (eds), *Religious Movements in Contemporary America* (Princeton: Princeton University Press, 1974).

60. L. Rowe and G. Cavender, 'Cauldrons Bubble, Satan's Trouble, But Witches are Okay: Media Constructions of Satanism and Witchcraft', in J. T. Richardson, J. Best and D. G. Bromley (eds), *The Satanism Scare* (New York: Aldine de Gruyter, 1991).

61. See, for example, B. Ellis, 'The Highgate Cemetery Vampire Hunt: The Anglo-American Connection in Satanic Cult Lore', *Folklore* 104 (1993) 13–39.

62. G. Harvey, 'Satanism in Britain Today', *Journal of Contemporary Religion* 10 (1995) 353–66.

63. La Fontaine, 'Satanism and Satanic Mythology', 83–93; J. B. Russell, 'The Historical Satan', in Richardson, Best and Bromley, *The Satanism Scare*; P. Stevens, 'The Demonology of Satan: An Anthropological View', in Richardson, Best and Bromley, *The Satanism Scare*.

64. N. Cohn, *Europe's Inner Demons* (London: Heinemann, 1975). See also J. M. Roberts, *The Mythology of the Secret Societies* (St Albans: Paladin, 1974).

65. La Fontaine, 'Satanism and Satanic Mythology', 115–38.

66. J. Beckford, *Cult Controversies: The Societal Response to New Religious Movements* (London: Tavistock, 1985).

67. D. G. Bromley, 'Satanism: The New Cult Scare', in Richardson, Best and Bromley, *The Satanism Scare*.

68. P. Jenkins and D. Maier-Katkin, 'Occult Survivors: The Making of a Myth', in Richardson, Best and Bromley, *The Satanism Scare*.

69. B. Ellis, *Raising the Devil: Satanism, New Religions and the Media* (Lexington: University Press of Kentucky, 2000); B. Ellis, *Lucifer Ascending: The Occult in Folklore and Popular Culture* (Lexington: University Press of Kentucky, 2004); D. Nathan and M. Snedeker, *Satan's Silence: Ritual Abuse and the Making of a Modern American Witch Hunt* (New York: Basic Books, 1995); Richardson, Best and Bromley, *The Satanism Scare*; J. S. Victor, *Satanic Panic: The Creation of a Contemporary Legend* (Chicago: Open Court, 1993); S. A. Wright, 'Satanic Cults, Ritual Abuse and Moral Panic: Deconstructing a Modern Witch-Hunt', in Berger, *Witchcraft and Magic*.

70. La Fontaine, 'Satanism and Satanic Mythology', 129–32; J. La Fontaine, *Speak of the Devil: Tales of Satanic Abuse in Contemporary England* (Cambridge: Cambridge University Press, 1998), pp. 56–75; Victor, *Satanic Panic*, pp. 355–61.

71. J. Best, 'Endangered Children and Antisatanist Rhetoric', in Richardson, Best and Bromley, *The Satanism Scare*; P. Jenkins, *Intimate Enemies: Moral Panics in Contemporary Great Britain* (New York: Aldine de Gruyter, 1992); Nathan, 'Satanism and Child Molestation'; Victor, *Satanic Panic*, pp. 103–94; Nathan and Snedeker, *Satan's Silence*, pp. 67–92.

72. H. S. Becker, *Outsiders: Studies in the Sociology of Deviance* (New York: Free Press, 1963), pp. 147–63.

73. E. Goode and N. Ben-Yahuda, *Moral Panics: The Social Construction of Deviance* (Cambridge, Mass.: Blackwell, 1998); Jenkins, *Intimate Enemies*; K. Thompson, *Moral Panics* (London: Routledge, 1998).

74. Victor, *Satanic Panic*, pp. 207–71.

75. Ellis, *Lucifer Ascending*.

76. B. Crouch and K. Damphousse, 'Law Enforcement and the Satanic Crime Connection: A Survey of "Cult Cops"', in Richardson, Best and Bromley, *The Satanism Scare*; R. D. Hicks, 'The Police Model of Satanism Crime', in Richardson, Best and Bromley, *The Satanism Scare*; J. T. Richardson, 'Satanism in the Courts: From Murder to Heavy Metal', in Richardson, Best and Bromley, *The Satanism Scare*.

77. G. Clapton, *The Satanic Abuse Controversy: Social Workers and the Social Work Press* (London: University of North London Press, 1993); S. Mulhern, 'Satanism and Psychotherapy: A Rumor in Search of an Inquisition', in Richardson, Best and Bromley, *The Satanism Scare*; R. Ofshe and E. Watters, *Making Monsters: False Memories, Psychotherapy, and Sexual Hysteria* (London: Andre Deutsch, 1995).

78. Clapton, *The Satanic Abuse Controversy*; Ellis, *Raising the Devil*; Jenkins, *Intimate Enemies*; Rowe and Cavender, 'Cauldrons Bubble, Satan's Trouble'.

79. G. W. Allport and L. Postman, *The Psychology of Rumour* (New York: Holt, Rhinehart and Winston, 1947); A. Corbin, *The Village of Cannibals: Rage and Murder in France, 1870* (Cambridge, Mass.: Harvard University Press, 1992); A. Farge and J. Revel, *The Rules of Rebellion: Child Abductions in Paris in 1750* (Cambridge: Polity Press, 1991); G. A. Fine and P. A. Turner, *Whispers on the Color Line: Rumor and Race in America* (Berkeley: University of California Press, 2001); E. Morin, *Rumour in Orléans. Jews Accused of White Slaving: A Modern Myth Examined* (London: Anthony Blond, 1971); P. A. Turner, *I Heard it Through the Grapevine: Rumor in Afro-American Culture* (Berkeley: University of California Press, 1993); T. Shibutani, *Improvised News: A Sociological Study of Rumor* (Indianapolis: Bobbs-Merrill, 1966).

80. H. G. West and T. Sanders (eds), *Transparency and Conspiracy: Ethnographies of Suspicion in the New World Order* (Durham: Duke University Press, 2003).
81. For example, R. W. Balch and M. Gilliam, 'Devil Worship in Western Montana: A Case Study in Rumor Construction', in Richardson, Best and Bromley, *The Satanism Scare*; G. Cavaglion and R. Sela-Shayovitz, 'The Cultural Construction of Contemporary Satanic Legends in Israel', *Folklore* 116 (2005) 255–71; T. A. Green, 'Accusations of Satanism and Racial Tensions in the Matamoros Cult Murders', in Richardson, Best and Bromley, *The Satanism Scare*; R. Jenkins, 'Spooks and Spooks: Black Magic and Bogeymen in Northern Ireland, 1973–74', in de Blécourt and Davies, *Witchcraft Continued*; Victor, *Satanic Panic*, pp. 27–56.
82. B. Ellis, *Aliens, Ghosts and Cults: Legends We Live* (Jackson: University Press of Mississippi, 2003).
83. B. Ellis, 'Legend-Trips and Satanism: Ostensive Traditions as "Cult" Activity', in Richardson, Best and Bromley, *The Satanism Scare*; Ellis, *Lucifer Ascending*, pp. 112–41; Victor, *Satanic Panic*, pp. 149–52.
84. V. Sinason (ed.), *Treating Survivors of Satanist Abuse* (London: Routledge, 1994).
85. For example, P. Jenkins and D. Maier-Katkin, 'Satanism: Myth and Reality in a Contemporary Moral Panic', *Crime, Law and Social Change* 17 (1992) 53–75.
86. J. La Fontaine, *The Extent and Nature of Organised and Ritual Abuse* (London: HMSO, 1994); La Fontaine, *Speak of the Devil*.
87. S. Scott, *The Politics and Experience of Ritual Abuse: Beyond Disbelief* (Buckingham: Open University Press, 2001).
88. Ellis, *Raising the Devil*, pp. 143–201; G. Gudjonsson, *The Psychology of Interrogations, Confession and Testimony* (New York: Wiley, 1992); M. Hepworth and B. S. Turner, *Confessions: Studies in Deviance and Religion* (London: Routledge and Kegan Paul, 1982).
89. La Fontaine, *Speak of the Devil*, pp. 112–33; L. Roper, '"Evil Imaginings and Fantasies"'; H. Sebald, *Witch-Children: From Salem Witch-Hunts to Modern Courtrooms* (Amherst: Prometheus, 1995).
90. Jenkins, *Intimate Enemies*.
91. Ellis, 'Legend-Trips and Satanism'; Ellis, *Raising the Devil*, pp. 202–39.
92. Balch and Gilliam, 'Devil Worship in Western Montana'; Jenkins, 'Spooks and Spooks'.

13
writing witchcraft:
the historians' history, the practitioners' past

jo pearson

This chapter is concerned with the relationships that exist between stories of the past written by academic historians and by practitioners. It focuses on three issues. The first is the writing of recent history, or of history that impinges upon people still living. In the case of witchcraft, this includes not only the recent history of Wicca, but also the notion of 'the witch' further back in the past. The second issue, that of practitioner understanding and use of history, concerns Wiccan responses to the work of academic historians, particularly Ronald Hutton. It is also necessary to examine practitioner accounts of Wiccan history, and the use of more remote pasts by Wiccan authors such as Janet and Stewart Farrar. The third considers questions raised by these relationships, the ownership of history in the postmodern era and the ethics involved in writing about it.

the historians' history

It was not until 1980 that a historian of witchcraft considered the growth and presence of Wicca, witchcraft and paganism in the modern world. In *A History of Witchcraft: Sorcerers, Heretics and Pagans*, Jeffrey B. Russell included a concluding chapter on modern witchcraft. This remained the only instance of history engaging with the modern religious phenomenon of witchcraft for a further decade, until Ronald Hutton published his *The Pagan Religions of the Ancient British Isles: Their Nature and Legacy* in 1991. As with Russell's work, the concluding chapter dealt with modern Pagan religions. A few years later, feminist literary critic Diane Purkiss, though not a historian per se, then contributed to the academic discourses on histories of witchcraft with *The Witch in History: Early Modern and Twentieth Century Representations* (1996), followed shortly by Hutton's book-length study of modern witchcraft, *The Triumph of the Moon: A History of Modern Pagan Witchcraft* (1999).[1] It is these latter two contributions that this chapter initially considers, as examples of important recent work that reflect and respond to the revival of witchcraft in contemporary society, rather than relegating its practice to a now-disappeared premodern era. Beginning with Purkiss, for no better reason than chronology, this chapter engages with the texts themselves and with their reception in both academic and practitioner circles.

Purkiss' *The Witch in History* plays simultaneously on three areas. The first is those accused of witchcraft *in* history (that is, the Great Witch Hunt). The second is current beliefs concerning the historical figure of the witch among feminist witches and other contemporary witchcrafts. The third is the way in which historians engaged in the academic discipline of history have studied and represented the witches of the past. In a review for the Institute of Historical Research's electronic journal *Reviews in History*, Hutton described the book as, 'a firework display of destructive polemic, in three bursts'.[2] The first part of the book attacks the misuse of the history of the Great Witch Hunt by feminist witches, and by other modern witches, and critiques historians' studies of early modern witchcraft. This critique of the misuse of the history of the Great Witch Hunt is really aimed at radical feminist witches in North America, where the discourse of the witch as fore-sister in the struggle against a patriarchy is strongest, and shows no sign of loosening its grip.[3] The excesses of radical feminist witchcraft which cause Purkiss to entitle the chapter concerning them 'A Holocaust of One's Own' are not so influential on witchcraft in the UK, and the chapter may therefore seem to be less relevant to them.

This said, among Alexandrian and Gardnerian coven witches in Britain, there is a great fondness for Charlie Murphy's song *Catch the Fire* (1980),[4] a gloriously ahistorical piece of romanticism which commemorates the 'nine million European women [who] died' in 'a war against the women who's power they [the Christians] feared'.[5] It is sung, not for any particular 'historical' content or claims to 'what really happened', but because, 'it's one of our protest songs'.[6] Knowledge of historical fact does not preclude the use of 'false' history in mythopoetic stances; indeed, one would be hard-pressed to find a religious tradition in which such 'false' history does not play an important part. The presence of such motifs as the witch of the past joining those of the present in some kind of protest at various aspects of contemporary life in UK witchcraft cannot, then, be denied. The lack of influence of radical feminist witchcraft from the US does not mean that British witches are untouched by feminism, or that they do not see in 'the witch' a figure that can be manipulated to fit various agendas. This is despite the fact that the Murrayite lineage of witchcraft was collapsing among British Wiccans from the 1970s on.

Nevertheless, it is undoubtedly partly because of Purkiss' apparent focus on US radical feminist witchcraft and her tendency to regard this as the norm, collapsing all contemporary witchcrafts into it, that her impact on the Wiccan community in the UK appears to have been slight. This may well have been exacerbated by the fact that, though she had contact with three – rightly anonymised, but as a consequence decontextualised – British witches, she lacked any prior connection with the Wiccan community. Her book received a relatively long review of two columns in the Imbolc 1997 issue of the journal of the Pagan Federation, *Pagan Dawn*, where it was variously described in a review by Pagan practitioner David Cook as

> a book for academics ... it is useful if the reader has some knowledge of such concepts as post-structuralism, relativism, object relations theory, Freud and psychoanalysis ... She assumes a knowledge of Shakespeare and other playwrights; these chapters are ones for English students.

Whilst accepting her criticisms of feminists and Pagans for their disregard for historical accuracy, and of 'stuffy' historians who 'depose folk myth and magic', the review concludes that her theories 'will have little interest for Pagans' and that Keith Thomas and Carlo Ginzburg, for all their faults as historians of witchcraft, 'are a much better read'. The opinion of this reviewer suggests a need for entertainment, a 'good read' rather than the jargon of critical theory in academic writing, if one is to be an influence on those whose history one is writing.[7] A previous track record also helps.

In her response to the review of *The Witch in History* by Hutton in *Reviews in History*, Purkiss suggests that 'Ronald Hutton probably underestimates the extent to which the pagans and witches he encounters are aware of his own publications on the topic of Pagan history, and the way this inflects what they are willing to say to him.' He cannot fail, however, to be aware of the influence of his work on Wicca's reassessment of its own history, despite his altogether too humble insistence that he wasn't telling witches anything they didn't know already, at least when it comes to the previously debunked 'creation myths'.[8] Yet reviews in the Beltane 2000 issue of *Pagan Dawn* reveal the extent to which witches are assumed to have given up their once-cherished myths of origin, and a taste for a 'ripping yarn'.

Three years later, Hutton's *The Triumph of the Moon* was considered by practitioners to be 'a very controversial work', and *Pagan Dawn* commissioned two writers to give opposite views of this 'extraordinary powerful book'.[9] Hutton is described by the first reviewer, Tony Geraghty, as 'the gentle iconoclast[10] ... [who has] banished our illusions; exorcised our icons' whilst retaining a 'love affair with the great, instinctive "What if ..." of magic, sorcery, witchcraft and an all-embracing, nature-based, fertility-dancing religion'. After this book, claims Geraghty,

> the Craft will never be quite the same again ... That we can believe in the Craft in future without Gardner's bluff or Murray's blindfold is no bad thing. For that, we should thank Hutton, while rubbing some soothing oil ... onto the affected parts of our collective ego.[11]

'Gardner's bluff' and 'Murray's blindfold' are, however, precisely those parts of the myth which Hutton claims had been debunked in Wiccan circles well before the publication of his book, and indeed, this point is picked up by the second reviewer, John Macintyre, who criticises Geraghty's 'apparent belief that the historical claims advanced by Gardner are not only still generally accepted amongst Wiccans but that they constitute a cornerstone of the Craft'. *The Triumph of the Moon* has not, argues Macintyre, 'burst upon us like a bolt from the blue, to shatter the tower of our "cherished concepts" into ruin'.[12] He asserts that early historical claims have been disintegrating for the past 20 years,[13] as they had in academic circles, and praises Hutton for producing a book that affords 'a powerful, scholarly affirmation of the intrinsic value of modern Pagan Witchcraft'.

Reviews of Hutton's *Triumph of the Moon*, by both academics and practitioners, return again and again to his soundbite 'the only religion England has ever given the world',[14] and acknowledge the book's surprising readability (given that it is written by an academic). It is a fact that historical narrative is far closer to traditions of storytelling than critical theory can ever be.[15] The first creates an atmosphere of wonder and excitement at finding out something new; the second uses a jargonistic language few can (or want) to understand.[16] For all her skill with language, this puts Purkiss at a disadvantage when it comes to widespread influence amongst witches – she is not so easy to read. More importantly, she challenges in a way that demolishing myths once held dear but, by many practitioners, long-since set aside cannot.[17] Hutton offers witches more information about their own history, facts he's discovered, and writes it all down in the manner of a novelist, drawing his readers in to his adventure of discovery. Purkiss, on the other hand, asks witches to think about their own history, to explore the implications of those beliefs supposedly now debunked, and to critically analyse their use of the figure of the witch and of witchcraft historiography. This is not so easily swallowed. Hutton is preaching to the converted, providing answers; Purkiss is asking questions – and uncomfortable ones at that – to historians (like Hutton) as well as to practitioners. This is not to dismiss either approach – both have their strengths. One provides the facts, in a fantastically good read; the other does something with them. Indeed, it would have been interesting to see just what Purkiss would have done with them had Hutton's book preceded hers by three years instead of vice versa.

the practitioners' past

Early in 2000, the first book-length study of Wiccan history written by a practitioner was published: Philip Heselton's *Wiccan Roots: Gerald Gardner and the Modern Witchcraft Revival*.[18] In his foreword to the book, Hutton praises it as 'a major landmark in the recovery of [Wicca's] origins ... It gives a quantum leap to knowledge.'[19] This is praise indeed for an author hailing from outside the academic discipline of history. However, although Heselton's book marked the first attempt by a practitioner to write a volume on the history of Wicca, it was by no means the only scholarly foray into witchcraft's past. Doreen Valiente had tried to find facts about a mysterious figure from the early days of Wicca, Dorothy Clutterbuck, which was published as 'The Quest for Old Dorothy' as an appendix to *The Witches' Way* by Janet and Stewart Farrar (1984).[20] Prior to the 1990s, however, introductions to and chapters in books written by practitioners presented an accepted (and acceptable) mythic history, contrary to the claims that such histories had been disintegrating since the 1970s.

This 'history' consisted of two threads, which were either combined or used separately according to the agenda of the author/s. Both can be classed as myths of origin and continuity, which operate as a discourse of descent, creating links to the past as desired. The first arose within the vacuum of historical knowledge about the witch trials prior to 1970. The second derives from the broader feminist consciousness and Goddess spirituality movements in the US, and provides a feminist alternative to Wicca's heritage from founding fathers and secret, fraternal, magical societies.[21] The myths are not mutually exclusive, however – 'the burning times' and the constructed figure of the early modern witch form a vital part of the myth of the golden age of matriarchy (by showing what occurs when it is lost), and so the two threads are often combined in feminist witchcraft.

In 1995, Loretta Orion used the Holocaust remembrance phrase 'Never Again' in the title of her book *Never Again the Burning Times*, an examination of Gardnerian witchcraft in North America.[22] A year later, as already noted, Purkiss included a chapter entitled 'A Holocaust of One's Own' in *The Witch in History*. Both authors reflect the tendency in some branches of modern witchcraft, specifically feminist witchcraft and often in North America and North American-derived witchcrafts, to see the persecutions of the Great Witch Hunt in early modern Europe as a holocaust against women. Such a view was once prevalent amongst Wiccans and Pagans more generally. It had become part of their 'history' via a plaque commemorating the 'nine million women' executed as witches erected at the Museum of Witchcraft and Magic on the Isle of Man (where Gardner was 'resident witch') and via the inclusion of this figure in Mary Daly's *Gyn/Ecology* (1978).[23] Both gleaned the figure from the feminist writer Matilda Jocelyn Gage, who had stated the number in her 1893 work *Women, Church and State* to emphasise the crimes of the Church against women.[24] Although it was not until a decade after Daly's book, in the late 1980s, that more reliable estimates of the death toll became available, the current scholarly estimates of 40,000–60,000 deaths have not been universally accepted by modern practitioners of witchcraft.

This refusal is based largely on the reclamation of female power through elective affinity with those accused of witchcraft in the early modern period; it is a psychological link rather than a claim for direct descent, yet it undoubtedly contributes to the misrepresentation of history. The repackaging of the Great Witch Hunt as a holocaust against women – as 'gendercide rather than genocide'[25] – constitutes a political demand by feminist witches that the same level of respect be given to the death toll of those executed as witches as to the 6 million Jews murdered in the Nazi Holocaust of the Second World War. As such, it embodies a political recuperation of the past framed as an ethic of remembering.

The myths of continuity found within witchcraft and feminist witchcraft act as genealogies, as legitimising lineages that aid the formation and continuation of communities and individual identities geared towards a particular, anti-patriarchal political stance. The myth of the golden age of matriarchy, never strong in Britain,[26] retains its hold on the imagination of feminist witches for whom the point of the myth is to behave *as if* it were true. Such readings do not tend to require historical accuracy any more than the use of 'ancient' calendar festivals within paganism more generally, or the former UK focus on the idea of the witches and witchcraft as the remnants of indigenous, pre-Christian paganism which survived the persecution of the Great Witch Hunt, and was revived in the twentieth century.

Whether or not the 'witches' of the Great Witch Hunt are central, ideas of a matriarchal age are not so prevalent in non-feminist witchcrafts in the UK. The matriarchal thread, however, is present in the work of two prolific practitioner authors, Janet Farrar and her late husband Stewart. The suggestion that 'what works now' is as or more important than historical accuracy was used by the Farrars to justify their use of a matriarchal emphasis. Three pages of the introduction to *Eight Sabbats for Witches* ([1981], 1992) is given over to a recounting of the matriarchal structure of past society, its being overthrown by patriarchy, the cost of this to the world/humankind, and the importance of its re-establishment.[27] This is the standard form of the story recounted in popular works of the feminist consciousness movement, based on the reinterpretation of the archaeological work of Marija Gimbutas,[28] by other feminist writers such as Mary Daly, Merlin Stone, Charlene Spretnak, and Adrienne Rich in the 1970s and 1980s.[29] Indeed, it is Merlin Stone's *The Paradise Papers* that the Farrars use as the basis of their myth. However, buried as it is in the introduction to a book which is essentially a 'how-to' ritual manual, it seems to remain a somewhat marginal influence, especially when compared to the centrality of the myth in feminist witchcraft such as that practised by the extremely influential witch-author, Starhawk.[30] The Farrars seem to undermine the stance they have taken with their admission that this 'is over-simplification enough to make a historian's hair stand on end, but food for thought'. Likewise, the rituals outlined in the volume are ahistoricised by the inclusion of a retainer: 'this is a book of suggested rituals, not a work of detailed historical analysis; so it is not the place to explain in depth how we extracted the … pattern [of the seasonal cycle they expound]'.[31]

Eight Sabbats provides an example not only of the myth of the golden age of matriarchy in a major publication by English witches, but also represents another ahistorical thread important to contemporary witchcraft and paganism – that of a systematised, eightfold calendar of festivals combining four ancient Celtic agricultural feasts and four solar festivals imported by invading Saxons as the Roman Empire decayed. These 'Greater' and 'Lesser' Sabbats, respectively, have been celebrated continuously for thousands of years,[32] with witchcraft's use of them being simply the latest revival. According to the Farrars, the 'modern witches' calendar … is rooted, like that of their predecessors through untold centuries, in Sabbats, seasonal festivals which mark key points in the natural year …', the Celts providing 'the actual ritual shape in which it has survived in the West'.[33]

Margaret Murray, Charles Leland, Sir James Frazer and Robert Graves all appear in the bibliography, with Graves acknowledged, along with Doreen Valiente, as the most helpful resource for their research. Regarding Graves, it is worth noting that he made clear in a subtitle that *The White Goddess* was a 'historical grammar of poetic myth', written under inspiration, rather than a book of historical research. There is also a reliance upon Alexander Carmichael's collection of Scottish folklore and charms, the *Carmina Gadelica*, first published in the early 1900s.[34] What was not available to them, even at the time of the paperback publication in 1992, was Hutton's *The Stations of the Sun* (1996). Here, the idea of 'fire festivals' (an alternative name for the Greater Sabbats) and an ancient Celtic year is uncovered as an eighteenth-century construction which was popularised with the 1970s attraction of all things Celtic, continues to be employed in popular works today, but was never accepted by academia. In the end, Hutton suggests that

> the notion of a distinctive 'Celtic' ritual year, with four festivals at the quarter-days and an opening at Samhain, is a scholastic construction of the eighteenth and nineteenth centuries which should now be considerably revised or even abandoned altogether.[35]

Unlike *Pagan Religions of the Ancient British Isles* and *The Triumph of the Moon*, Hutton's uncovering of the origins of the Celtic calendar has made little difference to its use as a ritual cycle within modern Pagan witchcraft and other forms of contemporary paganism. Doubtless many now realise that it never existed as an all-embracing structure adhered to by all Celtic peoples but, as the Farrars argue, 'attention should be paid to the role of the Old Religion in today's conditions; in short, to what works best *now*',[36] and the festivals making up what has become known as the Wheel of the Year are generally deemed to 'work', regardless of their provenance or lack of it.[37]

This way of reading and writing the imagined past was repeated throughout the 1970s and 1980s,[38] and continues in some more recent practitioner publications.[39] But, as I have argued elsewhere,

> despite the historical inaccuracy of the early Wiccan view of its history ... it would be unfair to summarily dismiss a community's use of history as devious simply because that history was incorrect, and as wilfully ignorant when such ignorance was, until the 1970s, entirely in line with the current knowledge of the academy itself in an area of research which has only recently started to come of age.[40]

One could argue, however, that as the academic history of witchcraft has become more accessible to Pagans through the writings of Hutton and, to a lesser extent, Purkiss, the unwitting 'bad history' of the 1970s and 1980s has given way to a disregard for history. Its use as a mythical story is, in effect, a deliberate misrepresentation of the history of witchcraft, which has become endemic since the mid 1990s, as the information age really began to take off. As Edith Wyschogrod has argued, '[o]ne of the ambiguities that attaches to the historian's role stems from a widening of the field of history in the culture of information and image so that persons with a variety of concerns ... construct and disseminate interpretations of the past'.[41] Such a view has certainly been exacerbated by the rise of the internet, alternative publishing and education, and the popularity of television broadcasts on history and prehistory. All have eroded the traditional boundaries of knowledge, repackaging a commodified past. Sociologists of religion have been engaged in debates concerning the religious consumer in the spiritual supermarket, and the related commodification of religion, since the 1980s.[42] Their studies have, on the whole, celebrated the ramifications of postmodern consumer culture in the religious arena – the choice and freedom available once traditional religious institutions lose their authority.[43] Similarly, history is marketed and commodified; the consumer culture consumes history, as the past is packaged in the manufacturing of a history that never existed.[44]

What, then, lies behind the consumption of 'false' history in an era when so much information is readily available at the click of a mouse button? Is it simply a recalcitrant stance, an ongoing need to legitimise religious traditions still only 50 or so years old? Is it the sheer amount of information available that overwhelms Wiccan authors and leads to the adoption of older, more comfortable histories? Or does it represent a stand-off between academic history and practitioner history in which ownership is at issue? The writing of historical narratives that are concerned with still-living or remembered characters, where historians have direct contact with contemporary people and groups who both practise witchcraft and, increasingly, research and write its history, seems to require the historian to acknowledge their entry into a community and its sense of self.[45] It necessitates, as Purkiss indicates, an awareness of the effect of historical research in the area of witchcraft on those practising witchcraft today, and the use that is made of such research.

the ownership of history

There is little doubt that the development of historical research regarding witchcraft has had various effects on modern practitioners and their view of their own history. The responses, very generally, can be summed up as two extremes: embrace or reject. Hutton claimed in the introduction to the 1993 edition of *Pagan Religions* that Wicca 'has proved capable of re-evaluating its own [historical] claims with a genuine scholarly rigour'.[46] Nevertheless, there remain some practitioners of witchcraft who still assume an unbroken line of descent from prehistory to the present day. Why such recalcitrance in the face of overwhelming historical facts to the contrary? In a sense, such a stance may be related to the pre-Christian definition of religion put forward by Cicero, in which *religio*, derived from *relegere* (to retrace, or reread) is almost synonymous with *traditio*; based on an injunction to maintain the rituals of one's ancestors,[47] it requires a recognisable tradition and a clearly established body of traditional practices. Thus, it could be argued that those modern witches who subscribe to the older view of their own history are simply being guided, albeit unwittingly, by the older, Pagan idea of what religion actually is.[48] Beyond such conjecture, however, is a further issue touched on in a previous publication – that of ownership of history and, related to this, the ethics of remembering.[49]

If history, as Walter Benjamin argues, is 'a time of the now, a moment drawn from the past into the present because of some urgent claim it has upon us',[50] then we may begin to see why modern practitioners of witchcraft lay claim to the history not just of their own contemporary traditions but also of the beliefs and practices of those accused of witchcraft before the emergence of modern Wicca and witchcraft. It is important that a distinction is made between the history of witchcraft in early modern Europe, the history of modern witchcraft, and perceived connections between the two. The identificatory connections and the continuing conflations beg the following questions: whose truth is being told, to whom, by whom, and to what end?

Such questions are of importance to modern practitioners of witchcraft, who are concerned with issues of representation and wish to protect, and to project, a certain image of 'the witch'.[51] The tendency to link psychologically or emotionally with those accused of witchcraft in the early modern period leads some to impose restrictions on who is allowed to do so. Laurie Cabot, for example, who considers herself to be the 'official witch of Salem',[52] takes great delight in dressing as a stereotypical fairy-tale witch, complete with long black cloak, pointy hat, broomstick and black cat. At the same time, she denounces others – particularly Hollywood – for using and promoting stereotypes of 'the witch'. One can only presume that her use of the stereotype is legitimate because she considers herself to *be* a witch. In addition, since there was no such thing as 'the Great Witch Hunt' until it was constructed after the events of the early modern period, there exists an underlying imperative for practitioners to claim it, to ensure its reification as 'an' event, to prevent it from being swallowed up by institutional, academic history which is deemed to have no feeling for the individuals persecuted and executed.

The witch hunt thus becomes 'an imaginative creation, like all historical events',[53] but one that is 'owned' by modern witches with an imaginative link to those persecuted and executed as witches in the past in an attempt to truly experience, or even recreate that past. Thus, whilst no modern witch wants to relive or re-create a time of persecution (though there is great awareness of persecutions which pertain to this day), there is an effort to construct the witch as a figure of alterity that can then be mapped on to the self-marginalisation of contemporaries seeking the kudos of radicalism and rebellion. Participating in this construction of identity and taking some control of representations with little concern for 'reality' aids in the reclamation of the power of the witch of the past in the present,[54] aided by the ambiguities of postmodernism.

The supposed end of the grand narrative brought by postmodernism has led many to question the narratives once accepted without question from those in positions of (academic) power, writing history from above. Postmodern theory is often seen as antithetical to history, arguing as it does that there exists no 'objective', 'pure', 'neutral' history and no fixed past. Postmodernity, complains Deborah Lipstadt, 'fosters deconstructionist[55] history at its worst'. In the postmodern era, '[n]o fact, no event, and no aspect of history has any fixed meaning or content. Any truth can be retold. Any fact can be recast. There is no ultimate historical reality ...',[56] the very foundations of academic history seem to be undermined.[57] The 'very notion of the past as a recuperable object of study',[58] seems to disappear, legitimising the collapse of boundaries in the 'creation' of historical pasts drawn into the present by, for example, witches writing as 'we'.[59] Characteristics of postmodernity such as 'indeterminacy and immanence; ubiquitous simulacra, pseudo-events; ... a polychronic sense of history; a ... deconstructive approach to knowledge and authority'[60] are easily spotted in the use of history by contemporary practitioners of witchcraft, both in their rituals and in their perception of the history of witchcraft. Writing the story of the past in answer to the needs of the present necessitates the adoption of a 'polychronic sense of history' and the deconstruction of knowledge and the authorities that lie behind it and produce it. The work of the historian, it seems, must now be negotiated with those professing a personal interest in, or identifying with, the subjects of academic enquiry. Otherwise, their work will be ignored as practitioners produce their own, acceptable literature on their history. As Gabrielle Spiegel has argued:

> what is the past but a once material existence, now silenced, extant only as a sign and as a sign drawing to itself chains of conflicting interpretations that hover over its absent presence and compete for possession of the relics, seeking to inscribe traces of significance upon the bodies of the dead.[61]

Historians and practitioners want to inscribe different traces, tell different stories to different people, according to their own agendas and using appropriate (or, sometimes, inappropriate) 'evidence'. In such a climate, competing representations are multiplied and the idea of some 'authentic' retelling of the past becomes more and more distant,[62] as *ficciones* are adopted instead. In an article on 'The Myth of the Burning Times and the Politics of Resistance in Contemporary American Wicca', Glenn William Shuck draws on Wyschogrod's use of the *ficción*, an account written in a heterological sense, defying traditional conceptions of 'what happened', at least from the perspective of mainstream historiography. Wyschogrod writes:

> [b]y ficciones I mean what fictions become when they take themselves up into the story of their own ontological errancy. Both fact and fiction are transformed when shards of metaphysical history through which they have passed, in a return of the repressed, percolate at the surface of the narrative.[63]

Wiccan ficciones, according to Shuck, do not 're-present that which happened in a fact or fiction model, but rather present the kernel of contemporary concerns read through a historical lens',[64] in an attempt to draw out those things deemed to be hidden in academic historiography. Rather than operating as 'false history', the *ficción* represents an attempt to explore possibilities, avoiding such essentialist statements as 'it could only have been thus'. According to Wyschogrod, the 'historian's responsibility is mandated by another who is absent, cannot speak for herself, one whose actual face the historian may never see, yet to whom "giving countenance" becomes a task.' There must be not only a passion for the past, but also 'an ardor for the others in whose name there is felt an urgency to speak', a 'promise to the dead to tell the truth about the past'.[65] Wyschogrod questions whether such a promise can be fulfilled, and it might be argued that broken promises and a lack of the necessary ardour might be sensed by some witchcraft practitioners, specifically feminist witches, within many accounts of academic history concerned with witchcraft.[66] Instead, the required passion and determination to 'give countenance' to the dead other might be deemed the preserve of those who claim to continue their practices and beliefs, and who see themselves as representing the survival of witchcraft[67] – those who, 'before saying "Thus it was" or "It could not have been thus", [are] willing to pronounce the words "It *ought* not to have been thus."'[68]

As far as academic historians are concerned, such 'ought not' claims constitute a false historical narrative. To Diane Purkiss, for example, feminist witchcraft is simply crippled by a false historical narrative that supports a permanent status of victimhood. According to Shuck, however, the identification with the suffering and torture of the Burning Times, is more a 'manifesto for growth' than 'a crippling apologia for victimization', and Purkiss, he claims, misunderstands the role of the *ficciones*.[69] In its *ficciones*, witches are not remembering – recalling or retelling – 'the past'. The *ficciones* were not ever 'past'. But if the fictions or mythologisations do not 're-member', then it could be argued that they fall outside Wyschogrod's ethic of remembering. As stories for being, dwelling and thinking, the *ficciones* can still be beneficial, though they can also, of course, be dangerous.[70] But they are not history, at least in the sense in which it is traditionally understood; they are a different genre altogether.

In their adoption, there is a refusal to accept the discipline of the historian who is fact-based and empirical, who perhaps sees myth as nothing more than a poetic lie in relation to a history based on reason. Instead, practitioners of modern witchcraft, by virtue of their elective emotional link to the silenced witches of the early modern period, might well claim that they are the ones – perhaps the only ones – who can effectively engage with the lacunae of the past, in 'a game of more historical than thou'.[71] In doing so, they challenge the idea that certain modes of representation are more reasonable than others. They subvert the idea of history – dominant since the seventeenth century[72] – as the most authoritative means of representing the past, and at the same time undermine the notion of myth as contrary to the noble pursuit of history. In a return to the pre-Enlightenment mindset, the boundaries between myth and history are collapsed in a subversion of the Enlightenment dissolution of myths and substitution of knowledge for imagination.[73]

For historians, however, the crucial point about the 'ethics of remembering' is whether the people claiming to 'speak for' or about the dead have knowledge in addition to imagination. Do they have any regard for source material and the testability of evidence, two of the main conventions of the genre of history? It is telling that those who claim to empathise most with the victims of the witch trials seem to have little interest in recent scholarly literature on the subject, or in conducting research of their own. There is thus no real attempt to get close to the dead, to speak for them, but instead to employ the latter solely as ingredients in a political myth.[74] In preferring to remain disengaged from the horrors of the past, they are free to fantasise, to imagine without recourse to knowledge. Like Borges, it may be that, in their *ficciones*, witches 'deliberately confuse recognised genres in order to unsettle', to force a re-examination of the means by which knowledge is conventionally imparted.[75] But in the practitioner desire to deviate from what are seen as the externally imposed norms of the genre of history, the ethics of subversion trump the ethics of remembering.

notes

1. Jeffrey B. Russell, *A History of Witchcraft: Sorcerers, Heretics and Pagans* (London: Thames & Hudson, [1980] 1991); Ronald Hutton, *The Pagan Religions of the Ancient British Isles: Their Nature and Legacy* (Oxford: Blackwell, [1991] 1993); Diane Purkiss, *The Witch in History: Early Modern and Twentieth Century Representations* (London: Routledge, 1996); Ronald Hutton, *The Triumph of the Moon: A History of Modern Pagan Witchcraft* (Oxford: Oxford University Press, 1999).
2. Institute of Historical Research (IHR), <www.history.ac.uk/reviews/paper/hutton.html>.
3. As Purkiss herself acknowledges in her response to Hutton's review, 'the will to believe in the hoariest and oldest myths of origins is stronger in the US than here, perhaps strongest among US feminist witches' (IHR).
4. Lyrics on <www.thegreenfuse.org>.
5. See, for example, the popular Pagan goth-rock band Incubus Succubus (renamed Inkubus Sukkubus in 1995) song 'The Burning Times' from the album *Wytches* (1994): 'They came to bring the good news / To burn Witches, Pagans Jews / They said they were the shepherd's sheep / They whipped old women through the street / Then the turning of the tide / From the truth they could not hide / Now the darkest age is past / The Goddess has returned at last!' Their official website argues as follows: 'While the figures quoted may vary, all are agreed that the pan-European witch hunts of the middle ages, driven by the Church and enacted by the evil and the greedy, claimed many victims. Here is a song that both tells the story, and reminds us to resist any return to the horrors of the past – as certain elements of the 'Religious Right' would love us to do ...

> Children resist, a return to the burning times
> People be wise, to the power of their lies
> Be not fooled, as those who were fooled before
> Children, oh children, be free, be wild!'

6. This was repeated to me many times during the course of fieldwork conducted for my PhD in the period 1995–99.
7. This was the only review I could find in the Pagan press, though I have no doubt others are extant.
8. For Hutton's extensive analysis of his own work on witchcraft, see Ronald Hutton, 'Living with Witchcraft', in Hutton, *Witches, Druids and King Arthur* (London: Hambledon & London, 2003), pp. 259–94.
9. *Pagan Dawn* 135 (2000) 37.
10. As opposed to Purkiss' 'battering-ram' (Hutton, review in IHR).
11. Tony Geraghty, 'The Bull (and a Few Sacred Cows) Sacrificed: Mithras, Hutton and Others Invite You to the Wake', review of *The Triumph of the Moon*, in *Pagan Dawn* 135 (2000) 37–8.
12. John Macintyre, 'Rebutting Bull! An Alternative View of the Triumph of the Moon', review of *The Triumph of the Moon*, in *Pagan Dawn* 135 (2000) 39–40, 39.
13. Those who still adhere to the idea of Wicca as the ancient religion of the British Isles simply ignored Hutton's research. See Hutton, 'Living with Witchcraft', p. 281:

'they have tended to ignore my work completely and to carry on writing as though it had never appeared'.

14. Hutton, *Triumph*, p. vii.
15. See Lionel Gossman, *Between History and Literature* (Cambridge, Mass.: Harvard University Press, 1990), who argues that historical narratives 'have much more in common with fictional knowledge that historians are normally willing to allow' (p. 286).
16. Perhaps the very reason why historians have for so long hidden themselves from the impact of critical theory in a way that other humanities disciplines – English, Religious Studies, Philosophy – have not.
17. Of course, Hutton's work is challenging to those who still hold to those myths, as well as to readers outside paganism and witchcraft who may still consider witchcraft to be satanic or in some way evil.
18. Followed in 2003 by Heselton's *Gerald Gardner and the Cauldron of Inspiration: An Investigation into the Sources of Gardnerian Witchcraft* (Milverton: Capall Bann Publishing).
19. Philip Heselton, *Wiccan Roots: Gerald Gardner and the Modern Witchcraft Revival* (Milverton: Capall Bann Publishing, 2001), p. 1.
20. Doreen Valiente, 'The Search for Old Dorothy', in Janet Farrar and Stewart Farrar, *The Witches' Way* (London: Hale, 1984), pp. 282–93.
21. See Jone Salomonsen, *Enchanted Feminism: Ritual, Gender and Divinity Among The Reclaiming Witches of San Francisco* (London: Routledge, 2002), pp. 6–7.
22. Loretta L. Orion, *Never Again the Burning Times: Paganism Revived* (Prospect Heights: Waveland Press, 1995).
23. The plaque at the Museum influenced practitioners of Wicca, whilst Mary Daly's use of the figure was influential on the early development of feminist witchcraft.
24. Gage in turn had obtained the figure from German historians, following a computation based on false extrapolation of local records by a Quedlinburg antiquarian in the late eighteenth century. See Hutton, 'How Myths are Made', in Hutton, *Witches, Druids and King Arthur*, pp. 1–38, 30–1. Matilda Joscelyn Gage, *Women, Church and State* (New York: Arno Press, [1893] 1972).
25. Robin Briggs, *Witches and Neighbours: The Social and Cultural Context of European Witchcraft* (London: HarperCollins, 1996), p. 8.
26. Although the myth was influential on Margaret Murray.
27. Formulated as a cultural theory of Paradise, Fall, Persecution and Regeneration by Starhawk; see Starhawk, *The Spiral Dance: A Rebirth of the Ancient Religion of the Great Goddess* (London and New York: Harper & Row, [1979] 1989).
28. Gimbutas' own personal leanings towards the myth are expressed in the comment she provided for the back cover of Monica Sjöö and Barbara Mor's *The Great Cosmic Mother: Rediscovering the Religion of the Earth* (London: Harper & Row, 1987), in which she describes the book as 'a vivid picture of the tragic consequences caused by the savage priests of savage patriarchy'.
29. Mary Daly, *Gyn/Ecology: The Metaethics of Radical Feminism* (London: Women's Press, [1978] 1991); Merlin Stone, *The Paradise Papers: The Suppression of Women's Rites* (London: Virago, 1976); Charlene Spretnak, *Lost Goddesses of Early Greece* (Boston, Mass.: Beacon, [1978] 1981); Adrienne Rich, *Of Woman Born: Motherhood as Experience and Institution* (London: Virago, 1979).
30. Starhawk, *The Spiral Dance*. For a detailed examination of Starhawk's use of the myth, see Salomonsen, *Enchanted Feminism*, pp. 67–96.
31. Janet Farrar and Stewart Farrar, *Eight Sabbats for Witches* (London: Hale, [1981] 1992), pp. 20, 24.

32. The Farrars claim them as 'the highly sophisticated preoccupation of the mysterious Megalithic peoples who pre-dated Celt, Roman and Saxon on Europe's Atlantic fringe by thousands of years'; Farrar and Farrar, *Eight Sabbats*, p. 14

33. Farrar and Farrar, *Eight Sabbats*, pp. 13, 14.

34. Alexander Carmichael, *Carmina Gadelica* (Edinburgh, 1900).

35. R. Hutton, *The Stations of the Sun: A History of the Ritual Year in Britain* (Oxford: Oxford University Press, 1996), p. 411.

36. Farrar and Farrar, *Eight Sabbats*, p. 17.

37. The ritual year was not necessarily considered 'Celtic' within Wicca until the 1980s, principally through the mutual influences of the Farrars (who had moved to Ireland in 1976) and North American witchcrafts which emphasised Celtic ancestry. Many older witches to whom I have spoken still call Samhain 'Hallowe'en' or 'Hallows', and Imbolc 'Candlemass', use the Saxon word 'Lammas' instead of 'Lughnasadh', and refer to Beltane as 'May Eve/Day'. Heathens and some witches who feel more of an affinity with the Saxon or Nordic traditions use a different system drawing on other myths, histories and folklore. The idea of a Celtic calendar followed by all witches and Pagans today is no more true than the notion of an ancient Celtic calendar adhered to by all Celtic peoples.

38. See, for example, Vivianne Crowley, *Wicca: The Old Religion in the New Millennium* (London: Thorsons, [1989] 1996); Farrar and Farrar, *Eight Sabbats*. See also Starhawk, *The Spiral Dance*, pp. 16–22.

39. For example, Ann Moura, *The Origins of Modern Witchcraft: The Evolution of a World Religion* (St. Pauls, Minn.: Llewellyn, 2000).

40. Joanne Pearson, 'Wicca, Paganism and History: Contemporary Witchcraft and the Lancashire Witches', in R. Poole (ed.), *The Lancashire Witches: Histories and Stories* (Manchester: Manchester University Press, 2002), p. 191.

41. Edith Wyschogrod, *An Ethics of Remembering: History, Heterology and the Nameless Others* (Chicago: University of Chicago Press, 1998), p. 4.

42. For example, Zygmunt Bauman, *Freedom* (Milton Keynes: Open University Press, 1998); Bauman, 'Postmodern Religion?', in Paul Heelas (ed.), *Religion, Modernity, and Postmodernity* (Oxford: Blackwell, 1999), pp. 55–78; Anthony Giddens, *Modernity and Self-Identity* (Stanford: Stanford University Press, 1991). For a more recent assessment of consumer spirituality, see Jeremy Carrette and Richard King, *Selling Spirituality: The Silent Takeover of Religion* (London: Routledge, 2005).

43. Lyon, for example, regards consumerism as a democratising force, whilst Moore deems greater religious tolerance to be directly related to the idea of religion as a commodity, or commodities, to be bought and sold in the open market: David Lyon, *Jesus in Disneyland: Religion in Postmodern Times* (Cambridge: Polity Press, 2000); R. Moore, *Selling God: American Religion in the Marketplace of Culture* (Oxford: Oxford University Press, 1994).

44. This is particularly pertinent to the discussion of religion and consumerism when considering witchcraft, which regards itself as free from traditional religious institutions, yet at least in part is now constrained by the quasi-religious institutions of the market; Wouter J. Hanegraaff, 'New Age Spiritualities as Secular Religion: A Historian's Perspective', *Social Compass* 46 (1999) 145–60, 148. Recent work in this area includes Ezzy, who draws a distinction between Wicca, on the one hand, and 'commodified witchcraft', on the other; Doug Ezzy, 'The Commodification of Witchcraft', *Australian Religion Studies Review* 14, 1 (2001) 31–44. See also Ezzy, 'White Witchcraft and Black Magic: Ethics and Consumerism in Contemporary Witchcraft', *Journal of Contemporary Religion* 21, 1 (2006) 15–31. My own perusal

of publications and websites relating to the latter form suggests that it is in commodified witchcraft that one finds a related commodified history of witchcraft, one which has largely disappeared from the publications of Wiccan authors.

45. Wyschogrod, *An Ethics of Remembering*, p. xvi.

46. Hutton, *Pagan Religions*, p. xiii.

47. Such ancestors do not have to be related by blood or marriage. As Marion Bowman has argued convincingly, elective affinity is important in the identity construction of Pagans, her best-known example being the 'cardiac Celts'. See Marion Bowman, 'Cardiac Celts: Images of the Celts in Paganism', in Graham Harvey and Charlotte Hardman (eds), *Paganism Today: Wiccans, Druids, the Goddess and Ancient Earth Traditions for the Twenty-First Century* (London: Thorsons, 1996), pp. 242–51; Bowman, 'Contemporary Celtic Spirituality', in Joanne Pearson (ed.), *Belief Beyond Boundaries: Wicca, Celtic Spirituality and the New Age* (Aldershot: Ashgate, 2002), pp. 55–101.

48. In contrast with the definition provided by the Christian Lactantius in the third century which is generally accepted without question in the West – that *religio* derives from *re-ligare*, and thus refers to the 'binding' of oneself in covenant to the one true God.

49. Pearson, 'Wicca, Paganism and history', pp. 190–1.

50. Paraphrased in Wyschogrod, *An Ethics of Remembering*, p. xv.

51. Albeit there is no agreement over what that image should be.

52. The dust jacket of her 1990 book *Power of the Witch* states that she is 'known as the "official witch of Salem"'. Cabot founded the Witches' League for Public Awareness, who protested against the film *The Witches of Eastwick* because of 'the inaccurate image it would present of witches'; Laurie Cabot, with Tom Cowan, *Power of the Witch* (London: Joseph, 1990), p. 79. It is impossible to tell whether she means contemporary or early modern witches.

53. Hans Kellner, '"Never Again" is Now', in Brian Fay, Philip Pomper and Richard T. Vann (eds), *History and Theory: Contemporary Readings* (Oxford: Blackwell, 1998), p. 236.

54. In this context it is interesting to note that, whilst recent works on the history of modern witchcraft are popular amongst witches, there tends to be a distinct lack of interest in academic works concerning the early modern 'witch hunt', knowledge of which is gleaned from whatever is included in the former and from popular works written by practitioners.

55. Though deconstruction is more properly a technique of post-structuralism, which effectively destroyed the idea that texts had 'a' meaning and thus allowed for the multiple interpretations that have so irritated historians as well as encouraging practitioners to take the writing of the past into their own hands. See Jacques Derrida, *Of Grammatology*, trans. Gayatri Spivak (Baltimore: Johns Hopkins University Press, [1967] 1976).

56. Cited in Robert Eaglestone, *Postmodernism and Holocaust Denial* (Cambridge: Icon Books, 2001), p. 7.

57. Lipstadt falls into the trap of thinking that postmodern theorists deny that history can be written. As Shelley Walia points out, 'Postmodernists do not ignore logical arguments, verification and archival research. But neither do they maintain that all interpretations are valid. Postmodernism only asserts that there is never *only* one meaning. Reason is not forsaken, as is argued by many conventional historians; only its dogmatic representation of itself as timeless certainty is cast aside ... They are not of the view that history is only creative fiction, as is commonly assumed,

or that every perspective on the past is as valid as the other.' Shelley Walia, *Edward Said and the Writing of History* (Cambridge: Icon Books, 2001), p. 60, emphasis added.

58. Gabrielle Spiegel, 'History and Postmodernism', in Keith Jenkins (ed.), *The Postmodern History Reader* (London: Routledge, 1997), p. 262.

59. See Eaglestone, *Postmodernism and Holocaust Denial*, p. 49: 'Another convention … is that historians generally write in the third person in a style recognised as "realist". Not writing in this style is frowned upon. Those who don't choose this style are not seen as historians.'

60. I. Hassan, *The Postmodern Turn: Essays in Postmodern Theory and Culture* (Columbus: Ohio State University Press, 1987), p. xvi, cited in Spiegel, 'History and Postmodernism', p. 270, n.3.

61. Spiegel, 'History and Postmodernism', p. 270.

62. The persecution of persons accused of witchcraft in early modern Europe, for example, is variously named as the 'Great Witch Hunt', the 'Witch Craze' or the 'Burning Times', and the 'nine million' dead are remembered. As with the Holocaust/Shoah/Churban/Final Solution, in the list of names 'there are too many possible and competing representations of these events' (Kellner, '"Never Again" is Now', pp. 234–5).

63. Wyschogrod, *Ethics of Remembering*, p. 32.

64. Glenn William Shuck, 'The Myth of the Burning Times and the Politics of Resistance in Contemporary American Wicca', *Journal of Religion and Society* 2 (2000), 2, <http://moses.creighton.edu/JRS/toc/2000–4.pdf>.

65. Wyschogrod, *Ethics of Remembering*, pp. xii, xi. According to Wyschogrod, Jules Michelet, author of *La sorcière* (1862), an important literary source for the formulation of Wicca, 'believed the historian to be "one who enters into a compact with the dead such that life is restored to them"' (p. 99).

66. Though perhaps not Hutton, whose *Triumph of the Moon* is described as 'passionate yet written with calm and clarity', according to the review in the *Journal of Ecclesiastical History* – though one must remember this is not a book about the Great Witch Hunt.

67. Indeed, Hans Kellner argues that the highest ethical/rhetorical position in Holocaust representations belongs to the survivor, which might, when extrapolated, suggest the importance of making witchcraft survive and of representing contemporary witchcraft practitioners as survivors; Kellner, '"Never Again" is Now', p. 227.

68. Wyschogrod, *Ethics of Remembering*, p. xvii.

69. Shuck, 'The Myth of the Burning Times', 7.

70. One only needs to consider the use of such myths made by the Nazis.

71. Shuck, 'The Myth of the Burning Times, 4.

72. See Theodor W. Adorno and Max Horkheimer, *Dialectic of Enlightenment*, trans. John Cumming (London: Verso, 1997). Walia argues that 'the discipline of history is [currently] surrounded by confusion as the traditional analytical and conceptual Enlightenment structures of historical knowledge stand eroded'. Walia, *Edward Said*, p. 12.

73. Adorno and Horkheimer, *Dialectic of Enlightenment*, pp. 3, 25, 31.

74. This is not to suggest that history isn't as susceptible to political interests as myth, though it is only recently that historians have begun questioning their ability to retain political neutrality, and to realise that history is anything but apolitical.

75. Daniel Carpenter, 'The Foundations of Modern Druid Spirituality', unpublished MA dissertation, Cardiff University, 2006, p. 12.

index